NATIONAL ACADEMIES *Sciences Engineering Medicine*

NATIONAL ACADEMIES PRESS
Washington, DC

Progress Toward Restoring the Everglades

The Tenth Biennial Review—2024

Committee on Independent Scientific Review of Everglades Restoration Progress

Water Science and Technology Board

Division on Earth and Life Studies

Consensus Study Report

NATIONAL ACADEMIES PRESS 500 Fifth Street, NW Washington, DC 20001

This activity was supported by contracts between the National Academy of Sciences and the United States Department of the Army under Cooperative Agreement No. W912EP-15-2-0002 and by the South Florida Water Management District and the U.S. Department of the Interior. Any opinions, findings, conclusions, or recommendations expressed in this publication do not necessarily reflect the views of any organization or agency that provided support for the project.

International Standard Book Number-13: 978-0-309-72305-3
International Standard Book Number-10: 0-309-72305-1
Digital Object Identifier: https://doi.org/10.17226/27875

This publication is available from the National Academies Press, 500 Fifth Street, NW, Keck 360, Washington, DC 20001; (800) 624-6242 or (202) 334-3313; http://www.nap.edu.

Copyright 2024 by the National Academy of Sciences. National Academies of Sciences, Engineering, and Medicine and National Academies Press and the graphical logos for each are all trademarks of the National Academy of Sciences. All rights reserved.

Printed in the United States of America.

Suggested citation: National Academies of Sciences, Engineering, and Medicine. 2024. *Progress Toward Restoring the Everglades: The Tenth Biennial Review—2024.* Washington, DC: The National Academies Press. https://doi.org/10.17226/27875.

The **National Academy of Sciences** was established in 1863 by an Act of Congress, signed by President Lincoln, as a private, nongovernmental institution to advise the nation on issues related to science and technology. Members are elected by their peers for outstanding contributions to research. Dr. Marcia McNutt is president.

The **National Academy of Engineering** was established in 1964 under the charter of the National Academy of Sciences to bring the practices of engineering to advising the nation. Members are elected by their peers for extraordinary contributions to engineering. Dr. John L. Anderson is president.

The **National Academy of Medicine** (formerly the Institute of Medicine) was established in 1970 under the charter of the National Academy of Sciences to advise the nation on medical and health issues. Members are elected by their peers for distinguished contributions to medicine and health. Dr. Victor J. Dzau is president.

The three Academies work together as the **National Academies of Sciences, Engineering, and Medicine** to provide independent, objective analysis and advice to the nation and conduct other activities to solve complex problems and inform public policy decisions. The National Academies also encourage education and research, recognize outstanding contributions to knowledge, and increase public understanding in matters of science, engineering, and medicine.

Learn more about the National Academies of Sciences, Engineering, and Medicine at **www.nationalacademies.org**.

Consensus Study Reports published by the National Academies of Sciences, Engineering, and Medicine document the evidence-based consensus on the study's statement of task by an authoring committee of experts. Reports typically include findings, conclusions, and recommendations based on information gathered by the committee and the committee's deliberations. Each report has been subjected to a rigorous and independent peer-review process and it represents the position of the National Academies on the statement of task.

Proceedings published by the National Academies of Sciences, Engineering, and Medicine chronicle the presentations and discussions at a workshop, symposium, or other event convened by the National Academies. The statements and opinions contained in proceedings are those of the participants and are not endorsed by other participants, the planning committee, or the National Academies.

Rapid Expert Consultations published by the National Academies of Sciences, Engineering, and Medicine are authored by subject-matter experts on narrowly focused topics that can be supported by a body of evidence. The discussions contained in rapid expert consultations are considered those of the authors and do not contain policy recommendations. Rapid expert consultations are reviewed by the institution before release.

For information about other products and activities of the National Academies, please visit www.nationalacademies.org/about/whatwedo.

COMMITTEE ON INDEPENDENT SCIENTIFIC REVIEW OF EVERGLADES RESTORATION PROGRESS

JAMES SAIERS (*Chair*), Yale University, New Haven, CT
CASEY BROWN, University of Massachusetts Amherst
JOHN CALLAWAY, University of San Francisco, CA
PHILIP M. DIXON, Iowa State University, Ames
CHARLES T. DRISCOLL, JR. (NAE), Syracuse University, NY
MARLA R. EMERY, Norwegian Institute for Nature Research
MARGARET W. GITAU, Purdue University, West Lafayette, IN
WENDY D. GRAHAM, University of Florida, Gainesville
  (*resigned November 2023*)
MATTHEW C. HARWELL, U.S. Environmental Protection Agency, Newport, OR
WILLIAM A. HOPKINS III, Virginia Tech, Blacksburg
TRACY QUIRK, Louisiana State University, Baton Rouge
K. RAMESH REDDY, University of Florida, Gainesville
HELEN M. REGAN, University of California, Riverside
ALAN D. STEINMAN, Grand Valley State University, Allendale, MI
JEFFREY R. WALTERS, Virginia Tech, Blacksburg
DAVID L. WEGNER, Woolpert Engineering, Tucson, AZ

Staff of the National Academies of Sciences, Engineering, and Medicine

STEPHANIE E. JOHNSON, Study Director
NOEL WALTERS, Associate Program Officer
EMILY BERMUDEZ, Senior Program Assistant *(until September 2024)*
SAMUEL KRAFT, Senior Program Assistant *(as of April 2024)*

WATER SCIENCE AND TECHNOLOGY BOARD

DAVID L. SEDLAK (NAE) (*Chair*), University of California, Berkeley
NEWSHA AJAMI, Lawrence Berkeley National Laboratory, CA
PEDRO J. ALVAREZ (NAE), Rice University, Houston, TX
MARTIN DOYLE, Duke University, Durham, NC
JORDAN R. FISCHBACH, The Water Institute, Baton Rouge, LA
SHEMIN GE, University of Colorado Boulder
ELLEN GILINSKY, Ellen Gilinsky, LLC, Richmond, VA
ROBERT M. HIRSCH, U.S. Geological Survey, Reston, VA
BRANKO KERKEZ, University of Michigan, Ann Arbor
YUSUKE KUWAYAMA, University of Maryland, Baltimore County
VENKATARAMAN LAKSHMI, University of Virginia, Charlottesville
CAMILLE PANNU, Columbia University, New York, NY
AMY PRUDEN, Virginia Tech, Blacksburg
JENNIFER TANK, University of Notre Dame, IN
CRYSTAL L. TULLEY-CORDOVA, Navajo Nation Department of Water Resources, Window Rock, AZ

Staff of the National Academies of Sciences, Engineering, and Medicine

DEBORAH GLICKSON, Director
LAURA J. EHLERS, Senior Program Officer
STEPHANIE E. JOHNSON, Senior Program Officer
M. JEANNE AQUILINO, Financial Business Partner
CHARLES BURGIS, Program Officer
MARGO REGIER, Program Officer
JONATHAN M. TUCKER, Program Officer
NOEL WALTERS, Associate Program Officer
MAYA FREY, Senior Program Assistant
SAMUEL KRAFT, Senior Program Assistant
MILES LANSING, Senior Program Assistant
BRYAN RUFF, Senior Program Assistant

Reviewers

This Consensus Study Report was reviewed in draft form by individuals chosen for their diverse perspectives and technical expertise. The purpose of this independent review is to provide candid and critical comments that will assist the National Academies of Sciences, Engineering, and Medicine in making each published report as sound as possible and to ensure that it meets the institutional standards for quality, objectivity, evidence, and responsiveness to the study charge. The review comments and draft manuscript remain confidential to protect the integrity of the deliberative process.

We thank the following individuals for their review of this report:

MICHELLE BAUMFLEK, U.S. Department of Agriculture Forest Service
BILL BOGGESS, Oregon State University
KATE BUENAU, Pacific Northwest National Laboratory
DAWN D. DAVIS, Idaho National Laboratory
FRANK W. DAVIS, University of California, Santa Barbara
BENJAMIN KIRTMAN, University of Miami
JOHN KOMINOSKI, Florida International University
BETH ROSE MIDDLETON MANNING, University of California, Davis
DENISE J. REED, University of New Orleans
CHADWIN B. SMITH, Headwaters Corporation
DENICE WARDROP, The Pennsylvania State University
JOHN WIENS, Colorado State University

Although the reviewers listed above provided many constructive comments and suggestions, they were not asked to endorse the conclusions or recommendations of this report nor did they see the final draft before its release.

The review of this report was overseen by **CHRIS T. HENDRICKSON (NAE)**, Carnegie Mellon University, and **CATHERINE L. KLING (NAS),** Cornell University. They were responsible for making certain that an independent examination of this report was carried out in accordance with the standards of the National Academies and that all review comments were carefully considered. Responsibility for the final content rests entirely with the authoring committee and the National Academies.

Acknowledgments

The study committee and staff would like to respectfully acknowledge the Miccosukee and the Seminole Peoples, past and present, and the Calusa, Tequesta, Jeaga, Ais, and Mayaimi Peoples before them—the original and current caretakers of Everglades' land, water, and air. The study committee and staff would also like to gratefully acknowledge the Native peoples on whose ancestral homelands they live and work. The National Academies of Sciences, Engineering, and Medicine are physically housed on the traditional land of the Nacotchtank (Anacostan) and Piscataway Peoples, past and present. The committee and staff honor and respect the enduring relationship that exists between Native peoples and nations and this land. The committee and staff thank these peoples for their resilience in protecting this land and aspire to uphold our responsibilities to their example.

Many individuals assisted the committee and staff in their task to create this report. We would like to express our appreciation to Robert Johnson, U.S. Department of the Interior (DOI); Gina Ralph, U.S. Army Corps of Engineers (USACE); and Amanda Kahn, South Florida Water Management District (SFWMD), who served as agency liaisons to the committee. We would also like to thank the following people who gave presentations, participated in panel discussions, provided public comment to the committee, or served as field trip guides.

Andrea Atkinson, National Park Service (NPS)
Nick Aumen, U.S. Geological Survey (USGS)
Steve Baisden, USACE
Bill Baker, MacVicar Consulting
Marcel Bozas, Florida International University (FIU)
Ken Bradshaw, USACE
Laura Brandt, U.S. Fish and Wildlife Service (FWS)
Liz Caneja, SFWMD

Amy Castaneda, Miccosukee Tribe
Eric Cline, SFWMD
Carlos Coronado, SFWMD
Kelly Cox, Everglades Coalition
Dan Crawford, USACE
Kevin Cunniff, Miccosukee Tribe
Reverend Houston Cypress, Miccosukee Tribe
Laura D'Acunto, USGS
Steve Davis, Everglades Foundation
Tylan Dean, NPS
Kevin Donaldson, Miccosukee Tribe
Angie Dunn, USACE
Mikhael Elfenbein, Cypress Chapter Izaak Walton League of America
Tabitha Elkington, USACE
Gretchen Ehlinger, USACE
Morgan Elmer, NPS
Jason Engle, USACE
James Erskine, Florida Fish and Wildlife Conservation Commission (FWC)
Chrissie Figueroa, USACE
Tom Forsyth, NPS
Michael Frank, Miccosukee Tribe Everglades Advisory Committee
Tom Frazer, University of South Florida
Cristina Gauthier-Hernandez, NPS
Adam Gelber, DOI
Nicholas Gonzalez, Miccosukee Tribe
Timothy Gysan, USACE
Officer Holland, Wildlife Police Officer
Bonnie Irving, FWS
Marci Jackson, USACE
Jennifer John, USACE
Jennifer Jurado, Broward Co.
Amanda Kahn, SFWMD
Fahmida Khatun, NPS
Phyllis Klarman, RECOVER
Corporal Rusty Lacy, Wildlife Police Officer
Jacob Larsson, FWC
Jennifer Leeds, SFWMD
Tom MacVicar, MacVicar Consulting
Carolina Maran, SFWMD
Jenna May, USACE

Amanda McKenzie, SFWMD
Miles Meyer, FWS
Brenda Mills, SFWMD
Kristen Mills, Miccosukee Tribe
Melodie Naja, NPS
Sergeant Sergio Najera, Wildlife Police Officer
Melissa Nasuti, USACE
Sue Newman, SFWMD
Nicole Niemeyer, SFWMD
Thomas Oates
Jayantha Obeysekera, FIU
Edward Ornstein, Miccosukee Tribe
Betty Osceola, Miccosukee Tribe Everglades Advisory Committee
Mindy Parrott, SFWMD
Rajendra Paudel, NPS
Mark Perry, Everglades Coalition
Armando Ramirez, SFWMD
Stephanie Romañach, USGS
Eve Samples, Friends of the Everglades
Colin Saunders, SFWMD
Debby Scerno, USACE
Matthew Schwartz, University of West Florida
Fred Sklar, SFWMD
Ed Smith, Florida Department of Environmental Protection (FDEP)
Mailin Sotolongo-Lopez, FDEP
Erik Stabenau, NPS
John Stamm, USGS
Donatto Surratt, NPS
Jordan Tedio, FDEP
Cindy Thomas, USACE
Michael Tompkins, SFWMD
Craig van der Heiden, Miccosukee Tribe
Zulamet Vega-Liriano, USACE
Eva Velez, USACE
Kim Vitek, USACE
Leslye Waugh, SFWMD
Erin White, NPS
Walter Wilcox, SFWMD
Larry Williams, FWS
Chris Wittman, Captains for Clean Water

Contents

PREFACE		xv
ACRONYMS AND ABBREVIATIONS		xix
SUMMARY		1

1 INTRODUCTION 15
  The National Academies and Everglades Restoration, 16
  Report Organization, 21

2 RESTORATION PROGRESS 23
  Programmatic Progress, 23
  Natural System Restoration Progress, 26
  Issues That May Impact Progress: Water Quality, 84
  Conclusions and Recommendations, 103

3 APPLYING INDIGENOUS KNOWLEDGE IN THE
  COMPREHENSIVE EVERGLADES RESTORATION PLAN 107
  Place of Miccosukee and Seminole Tribes in the Everglades and
    Its Significance to Their Livelihoods and Cultures, 108
  Collaboration with Everglades Tribes, 113
  Assessment of CERP Tribal Consultation and Collaboration
    Over Time, 116
  Indigenous Knowledge and the Scientific Process, 126
  Best Practices for Integration of Indigenous Knowledge in
    Florida Everglades Restoration, 134
  CERP-Specific Examples of Ways to Improve Application of
    Indigenous Knowledge, 143
  Conclusions and Recommendations, 149

4 APPLICATION OF TOOLS TO EVALUATE THE EFFECTS OF
  CLIMATE CHANGE   153
  Climate Scenarios for Restoration Planning and Management, 154
  Sediment Accretion Modeling for Restoration Decision Making in
     Light of Sea-Level Rise, 158
  Models for Assessing Ecological Responses to Climate Change, 165
  Integration, 169
  Application of Climate Tools for Adaptation of Operations, 172
  Conclusions and Recommendations, 173

5 ADAPTIVE MANAGEMENT AND USE OF NEW INFORMATION IN
  DECISION MAKING   177
  Adaptive Management Process Guidance from RECOVER, 180
  Evaluation of CERP Adaptive Management and Incorporating
     New Information into CERP Decision Making, 184
  Revisiting the Building Blocks of Adaptive Management, 207
  Conclusions and Recommendations, 214

REFERENCES   219

APPENDIXES

A The Restoration Plan in Context   247
B The National Academies of Sciences, Engineering,
   and Medicine Everglades Reports   263
C Biographical Sketches of Committee Members
   and Staff   273

Preface

The Everglades is a treasure. It supports a remarkable diversity of birds, fish, reptiles, and mammals and encompasses landscapes unlike those anywhere else on the planet. As the homeland of the Seminole and Miccosukee, the Everglades has provided sustenance and shelter essential to their lives and serves as a center of cultural activities and sacred traditions that have been passed down through the generations. The Everglades also benefits the burgeoning population along South Florida's coastline by providing storm-surge protection, recharging drinking-water aquifers, attracting tourism that contributes to local economies, and offering recreational opportunities that enrich the lives of the region's residents.

That this natural wonder was once regarded as a nuisance and impediment to progress now seems unfathomable, but this was the undeserved reputation that the Everglades endured through much of the 20th century as it was replumbed to accommodate agriculture and development. Nearly every part of the Everglades was affected. At the top of the watershed, the once-meandering Kissimmee River was straightened and channelized, cutting off its connection with the floodplain. Lake Okeechobee, the so-called liquid heart of the Everglades, was isolated from the surrounding marsh by an earthen levee, and the lion's share of sheet flow that once sustained the ridge-and-slough landscape and fed the southern estuaries was short-circuited through a maze of canals to the Atlantic Ocean. With drainage enabling the expansion of agriculture, excess phosphorus runoff from poorly managed fertilizer applications caused water quality impairments that propagated throughout much of the system, from Lake Okeechobee and the coastal estuaries to the freshwater marshes of the central and western Everglades. The loss of water storage and hydrologic connectivity, coupled with water quality degradation, led to widespread habitat loss and left ecosystems throughout the Everglades struggling to support their wildlife and sustain ecosystem services upon which South Florida residents rely.

Marjory Stoneman Douglas and other visionaries recognized that the Everglades was in trouble even before large-scale drainage began in 1948 with the launch of the Central and Southern Florida Project for Flood Control. Unfortunately, more than half the original Everglades was gone and much of the remaining fraction impacted by the time a plan to halt Everglades degradation was conceived. In 2000, Congress authorized the Comprehensive Everglades Restoration Plan (CERP), the world's largest ecosystem restoration effort intended to restore, preserve, and protect the South Florida ecosystem by addressing the quantity, quality, timing, and distribution of water while providing for water supply and flood protection.

The CERP is daring and ambitious, originally consisting of 68 projects outlined in the Yellow Book that were projected to take 30 years to complete. Recognizing the scale of the challenge, Congress recommended that an independent scientific review be conducted on progress toward restoration on a regular basis. In response, the National Academies of Sciences, Engineering, and Medicine formed the Committee on Independent Scientific Review of Everglades Restoration Progress (CISRERP) in 2004. This report represents the tenth biennial review of the CERP by this committee.

CISRERP is comprised of natural scientists, social scientists, and engineers specializing in ecosystem restoration; wetland ecology; water resources; climate change; environmental policy; adaptive management; program administration; and ecological, water quality, and hydrological modeling. These experts were selected for their eminence in their fields and for their record of accomplishment in addressing issues relevant to Everglades restoration. Over a 12-month period, the committee met in-person on four occasions and in many additional virtual meetings, when it heard oral presentations on the various dimensions of Everglades restoration and had discussions with federal and state personnel, Tribal representatives, academic scientists, interest groups, members of nongovernmental organizations, and the public. Outside of these meetings, the committee read thousands of pages of reports and peer-reviewed literature, synthesized and drafted its findings, and made revisions based on committee-wide feedback. I am extremely grateful for the energy and thoughtful efforts that this distinguished group dedicated to these important tasks and have been inspired by the respectful way the members engaged with one another and worked collegially to produce this document. The 2024 CISRERP report represents the consensus assessment of the committee on restoration accomplishments and challenges that have emerged primarily over the past 2 years but also over the 24 years since the CERP was authorized.

CISRERP could not have completed its work without Stephanie Johnson, Emily Bermudez, and Noel Walters, talented staff of the National Academies of Sciences, Engineering, and Medicine. This is the tenth iteration of CISRERP, and Stephanie Johnson has served as study director for every one of them. It is a great fortune to this committee and, more generally, to the quality of the review process that Stephanie is able to reprise this important role. Stephanie has an extraordinary command of the science, engineering, and policy that underpin the CERP, and she has built a well-respected reputation that enables her to convene members across the Everglades community to familiarize CISRERP of the latest developments and unaddressed challenges. The information that Stephanie is uniquely able to provide was essential to mapping the course of this report, and her leadership, analytical thinking, and determination were critical in assisting the committee to develop its ideas and insights into a coherent narrative. Emily Bermudez provided key technical and logistical support before, during, and after each meeting, and Noel Walters provided support for production of the report. On behalf of the entire committee, I wish to express our gratitude and admiration for the exceptional abilities and valuable contributions of the National Academies of Sciences, Engineering, and Medicine staff. The task would have been impossible without them.

The committee is indebted to many people for information and resources they provided. The committee's technical liaisons—Gina Ralph, U.S. Army Corps of Engineers; Robert Johnson, U.S. Department of the Interior; and Amanda Kahn, South Florida Water Management District—responded to numerous information requests and facilitated the committee's access to agency resources and expertise. The committee also wishes to thank the numerous individuals who shared their views on Everglades restoration through presentations, field trips, and public comments (see Acknowledgments).

James Saiers, *Chair*
Committee on Independent Scientific
Review of Everglades Restoration Progress

Acronyms and Abbreviations

AF	acre-foot
AFR	Adaptive Foundational Resilience
AMI	Active Marsh Improvement
ASR	aquifer storage and recovery
BBCW	Biscayne Bay Coastal Wetlands
BBSEER	Biscayne Bay and Southeastern Everglades Ecosystem Restoration
BISECT	Biscayne and Southern Everglades Coastal Transport
BMP	best management practice
CEM	conceptual ecological model
CEPP	Central Everglades Planning Project
CEQ	Council on Environmental Quality
CERP	Comprehensive Everglades Restoration Plan
CFR	Code of Federal Regulations
CISRERP	Committee on Independent Scientific Review of Everglades Restoration Progress
CMIP	Coupled Model Intercomparison Project
COP	Combined Operational Plan
CROGEE	Committee on the Restoration of the Greater Everglades Ecosystem
DOI	U.S. Department of the Interior
DPM	Decomp Physical Model
EAA	Everglades Agricultural Area
EAV	emergent aquatic vegetation

ECB	existing conditions baseline
EDEN	Everglades Depth Estimation Network
ENSO	El Niño–Southern Oscillation
EDR	Engineering Documentation Report
EIS	environmental impact statement
EPA	U.S. Environmental Protection Agency
ERTP	Everglades Restoration Transition Plan
ESA	Endangered Species Act
EVA	Everglades Vulnerability Analysis
FCE	Florida Coastal Everglades
FDEP	Florida Department of Environmental Protection
FEB	flow equalization basin
FWC	Florida Fish and Wildlife Conservation Commission
FWM	flow-weighted mean
FWO	future without
FWS	U.S. Fish and Wildlife Service
FY	fiscal year
GCM	General Circulation Model
GRR	General Reevaluation Report
IDS	Integrated Delivery Schedule
IPRL	Invasive Plant Research Laboratory
IQA	Information Quality Act
IRL-South	Indian River Lagoon-South
JEM	Joint Ecosystem Modeling
LILA	Loxahatchee Impoundment Landscape Assessment
LNWR	Arthur R. Marshall Loxahatchee National Wildlife Refuge
LOCAR	Lake Okeechobee Component A Reservoir
LORS	Lake Okeechobee Regulation Schedule
LOSOM	Lake Okeechobee System Operating Manual
LOWRP	Lake Okeechobee Watershed Restoration Project
LRR	Limited Reevaluation Report
LTER	Long-Term Ecological Research
MGD	million gallons per day

NEPA	National Environmental Policy Act
NGVD	National Geodetic Vertical Datum
OSTP	Office of Science and Technology Policy
PACR	Post-Authorization Change Report
PED	Preconstruction Engineering and Design
PIR	Project Implementation Report
POM	Project Operating Manual
ppb	parts per billion
RCW	red-cockaded woodpecker
RECOVER	Restoration, Coordination, and Verification
RSM	Regional Simulation Model
RSM-GL	Regional Simulation Model for the Glades and Lower East Coast Service Areas
SAV	submerged aquatic vegetation
SFWMD	South Florida Water Management District
SOM	System Operating Manual
SOP	Standard Operating Procedure
STA	stormwater treatment area
TBD	to be determined
TMDL	total maximum daily load
TP	total phosphorus
USACE	U.S. Army Corps of Engineers
USDA	U.S. Department of Agriculture
USGS	U.S. Geological Survey
WCA	Water Conservation Area
WERP	Western Everglades Restoration Project
WIIN Act	Water Infrastructure Improvements for the Nation Act
WPA	Water Preserve Area
WQBEL	water quality–based effluent limit
WRDA	Water Resources Development Act
WRRDA	Water Resources Reform and Development Act
WY	water year

Summary

The Everglades, one of the world's treasured ecosystems, once encompassed about 3 million acres of slow-moving water, which supported an array of wetland and estuarine habitats from Lake Okeechobee in the north to the Florida Keys in the south. During the past century, the Everglades has been dramatically altered by drainage and water management infrastructure to improve flood management, urban water supply, and agricultural production. The remnants of the original Everglades now compete for water with urban and agricultural interests and are impaired by contaminated runoff from these two sectors. The Comprehensive Everglades Restoration Plan (CERP), a joint effort launched by the state and federal governments in 2000, seeks to reverse the decline of the ecosystem. This multibillion-dollar project was originally envisioned as a 30- to 40-year effort with 68 individual projects including water storage reservoirs, water quality treatment using constructed wetlands, seepage management, and removal of barriers to sheet flow (e.g., canals, levees). Collectively these CERP projects aim to achieve ecological restoration by reestablishing the natural hydrologic characteristics of the Everglades—quality, quantity, timing, distribution, and flow—where feasible and by creating a water system that serves the needs of both the natural and the human systems of South Florida.

The National Academies of Sciences, Engineering, and Medicine established the Committee on Independent Scientific Review of Everglades Restoration Progress in 2004 in response to a request from the U.S. Army Corps of Engineers (USACE), with support from the South Florida Water Management District (SFWMD) and the U.S. Department of the Interior, based on Congress's mandate in the Water Resources Development Act of 2000. The committee is charged to submit biennial reports that review the CERP's progress in restoring the natural ecosystem. This is the committee's tenth report. Each report provides an update on progress toward natural system restoration during the previous 2 years, describes substantive accomplishments, and reviews developments in research, monitoring, and assessment that inform restoration decision making. The committee also identifies issues for in-depth evaluation stemming from new CERP program developments, policy initiatives, or improvements in scientific knowledge that have implications for restoration progress (see Chapter 1 for the committee's full statement of task). For the 2024 report, the committee evaluated the use of Indigenous Knowledge, application of tools for understanding climate change, and progress in application of adaptive management and incorporation of new information in the CERP.

During 2023–2024, Everglades restoration progress has continued at a remarkable pace, supported by record-high levels of federal and state funding. CERP projects are under construction or have been completed in nearly every region of the Everglades (Figure S-1), and important early benefits are evident. Notably, the recent completion of seepage management, combined with non-CERP projects, has enabled increased flows to Northeast Shark River Slough, a long-time restoration goal. With restoration of flows, however, concerns about new or increasing phosphorus impacts have arisen and warrant further study. With many projects coming online and evidence of both restoration benefits and unanticipated outcomes, there is a need for timely use of the best available information to improve the CERP. In this context, the committee presents opportunities to further boost restoration gains in three specific areas focused on (1) best practices for including Indigenous Knowledge in project planning, monitoring, and management; (2) application of climate change information and tools; and (3) improvements in adaptive management and more nimble incorporation of new information into restoration decision making. Each of these areas offers substantial potential to enhance restoration outcomes but requires new investment of CERP resources to achieve them, including increased staff, a more robust science enterprise, and leadership to build agency cultures that support these objectives.

RESTORATION PROGRESS

In Chapter 2, the committee outlines the major accomplishments of restoration, with an emphasis on natural system restoration progress from the CERP in the past 2 years.

The pace of restoration implementation has reached historic levels, based on record state and federal investments in fiscal year (FY) 2022 and FY 2023. Six CERP projects are under construction, one CERP project and two major project components have been completed, and one additional project is essentially complete. The Central Everglades Planning Project (CEPP) continues to progress rapidly, as befits the project that is the keystone to restoring the central heart of the Everglades. Maintaining this pace of progress requires both

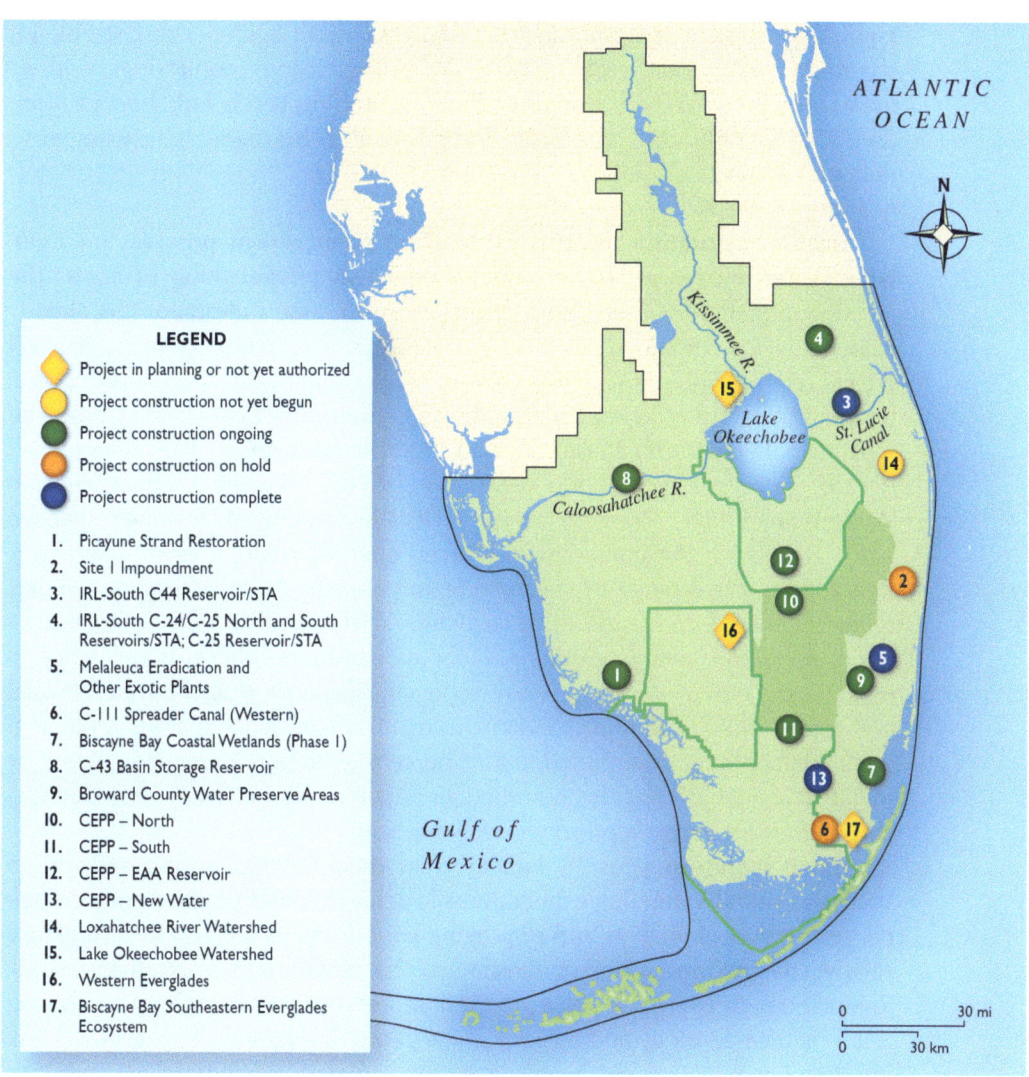

FIGURE S-1 Locations and status of CERP projects.

SOURCE: Map by International Mapping.

continued construction funding and support for other agencies responsible for facilitating restoration implementation (e.g., permitting, monitoring). With so many projects under construction, if future funding levels fall short of those used for planning, difficult decisions will need to be made as to whether to delay all projects equally or, preferably, to expedite those with the greatest near-term benefits.

Sizable restoration benefits are evident from recent progress on CEPP New Water, Picayune Strand, and the Melaleuca Eradication projects. The recent completion of CEPP New Water is a restoration milestone. It is already evident that the combined effect of the recently constructed seepage barriers will greatly reduce, and perhaps even eliminate, flood control constraints imposed by the 8.5 Square Mile Area within the CEPP footprint. Prior to the Combined Operational Plan (COP), these constraints completely stymied every attempt over decades to restore the historic distribution of flow between Northeastern and Western Shark River Slough. However, with the addition of these new seepage barriers, no such constraints have affected operations since water year 2022. Restoration of hydrology in Picayune Strand appears to be generating benefits to the local flora and fauna, with vegetation and macroinvertebrate communities responding favorably. Additional longitudinal monitoring will be needed to continue documentation of recovery, especially given the magnitude of seasonal and inter-annual variation. CERP investments in invasive species biological control efforts have contributed to a 75 percent reduction in area dominated by Melaleuca and have largely controlled air potato reproduction to the extent that air potato is no longer a priority invasive species.

Initial monitoring results indicate that the COP has been a restoration success, generally meeting expectations in achieving hydrological and ecological restoration objectives and improving conditions in the central Everglades relative to previous water management. The rehydration of Northeast Shark River Slough in Everglades National Park represents the largest step yet toward restoring the hydrology and ecology of the central Everglades. However, achieving complex objectives that involve creating fairly precise hydrologic conditions over extended periods of time—for example, optimal recession rates and water levels during the nesting season for threatened and endangered species—has been more challenging. Like any system operations plan, the COP likely lacks the capacity to adjust sufficiently to meet the restoration targets under all rainfall regimes. Thus, the COP is proving to be what it was intended to be, not a complete solution but rather the first major step toward restoring the central heart of the Everglades. The changes wrought by the COP have revealed some issues with water quality and have confirmed anticipated adverse effects on endangered Cape Sable seaside sparrows. This latter issue requires creative solutions, sooner rather than later.

Information on natural system restoration progress relative to expectations and project objectives remains difficult to find and interpret. The CERP lacks a mechanism for centralized multi-agency reporting of project-level restoration outcomes. Data, when available, are often presented in permit reports produced by a single agency or in monitoring reports produced by contractors. Increased attention to multi-agency data synthesis and interpretation is needed to support assessment and learning. The COP Biennial Report is an example of effective multi-agency analysis of extensive monitoring data on hydrologic, water quality, and ecological conditions in support of learning and adaptive management. Simplified and straightforward analyses of key metrics will provide increased transparency of restoration outcomes to the public and Congress.

The Western Everglades Restoration Project (WERP) as proposed offers important benefits to the western Everglades, but implementation progress largely depends on non-CERP source control implemented by private landowners, which could lead to large delays without implementation and performance requirements. In general, WERP features, if implemented as planned, should improve hydration, hydrologic and ecological connectivity, and water quality, which have been longstanding concerns in the WERP study area. Yet, issues regarding flood risk of the Looneyville community still need to be addressed to meet Savings Clause requirements.

Downward trends in total phosphorus (TP) concentrations in stormwater treatment area (STA) outflow reflect extensive recent Restoration Strategies efforts, but timely attainment of the stringent water quality–based effluent limit (WQBEL) will depend on how effectively the STAs respond to these efforts and the extent to which data collection (including cell-by-cell monitoring), data analysis, modeling, and synthesis are rigorously applied to inform adaptive management decisions. As noted in NASEM (2023), CERP progress and the timely delivery of restoration benefits, particularly for CEPP North and the Everglades Agricultural Area Reservoir, depend on meeting the WQBEL in all STAs. High TP concentrations in Lake Okeechobee further the challenge of STA performance given the future plans to move more lake water south. The cumulative performance of STAs during the recent 5-year period is impressive and generally trending in the right direction, but only STA-3/4 currently meets the WQBEL. Maintaining low phosphorus and hydraulic loading rates should improve STA treatment efficiency, but high inflow concentrations in the Western and Eastern Flow Paths could pose particular challenges for WQBEL attainment.

Additional research is recommended to explore the potential biogeochemical effects from the CERP through increased flows, flow velocities, canal-to-marsh interactions, and other factors that may mobilize legacy phosphorus and impact periphyton and plant communities. The S-333 Working Group identified sediment mobilization in the L-29 Canal as a key driver of increasing concentrations and loads into Everglades National Park with recent increased flows during the dry season, and agencies are implementing strategies to reduce sediment mobilization. Research has also illuminated potential connections between increased flows and increased phosphorus loading and/or vegetation impacts, although discrepancies between the SFWMD and Long-Term Ecological Research data need to be resolved. The potential of ecological effects from increased CERP flows also merits further study. Specifically, research should examine the dominant TP sources (e.g., sediment, floc, suspended), determine the mechanisms by which these concentrations and loads may be exacerbated under the CERP through increased flows and/or canal-to-marsh interactions, and, if necessary, identify approaches to mitigate phosphorus impacts.

USE OF INDIGENOUS KNOWLEDGE IN THE CERP

The lands and waters of the Everglades are the geographic and spiritual home of the Miccosukee and Seminole peoples. The health and well-being of the Everglades is synonymous with that of the Miccosukee Tribe of Indians of Florida (hereafter the Miccosukee Tribe) and the Seminole Tribe of Florida (hereafter the Seminole Tribe). Therefore, the Tribes have a wealth of knowledge about the South Florida ecosystem based on their intimate reciprocal relationship with the biophysical environment that has been developed through lived experience and passed down through generations. The following conclusions and recommendations are provided in response to a request from CERP agencies for advice on how Indigenous Knowledge could be better included in CERP planning and management, which was prompted by recent Executive Office requirements to include Indigenous Knowledge in federal scientific and policy decision making.

Indigenous Knowledge, like western science, is a "body of observations, oral and written knowledge, innovations, practices, and beliefs" about the natural world that has much to offer Everglades restoration (EOP, 2022a). For example, the Miccosukee and Seminole Tribes' extensive personal and Tribal knowledge of tree islands, if applied to restoration efforts, would benefit both the ecosystem and the Tribes. Indigenous Knowledge spans much longer time frames than western scientific studies and can therefore enhance understanding of historical ecological conditions and modern deviations from baseline conditions. Indigenous Knowledge should also be considered and applied in efforts to refine the RECOVER monitoring plan and conceptual models to better develop and incorporate performance measures and metrics that are relevant to biocultural restoration.

Consistent and meaningful engagement between CERP agencies and Tribal Nations is necessary to ensure a partnership in which Indigenous Knowledge is recognized, considered, and applied in restoration decision making, and

notable progress has recently been made to improve the quality of Tribal engagement and cooperation in the CERP. The Everglades has offered refuge and sustenance to the Miccosukee and Seminole peoples for generations. Historically, the U.S. and state governments attempted to remove all Tribal people from the southern Florida landscape through coercive and violent means. This painful history is part of the living memory of Miccosukee and Seminole Tribal elders and remains an under-current in all consultation, engagement, and coordination with the Tribes. The lack of meaningful engagement with the Tribes historically heightens the importance today of building trustful relationships, based on integrity and with careful adherence to laws, regulations, and guidance in consultation and cooperation with the Tribes. Over the past decade, consultation has become less proforma and more meaningful, as exemplified by the application of Indigenous Knowledge to inform decision making in WERP and in a 2023 temporary deviation from the seasonal closures of the S-12A and S-12B structures. The work required to shift agency cultures to further elevate meaningful Tribal engagement will be labor- and resource-intensive for both agencies and Tribes but will reap rewards for Everglades restoration.

The recently developed Miccosukee internal peer-review process is an important step toward facilitating consideration of Indigenous Knowledge in Everglades restoration processes and provides a potential model for others throughout the nation. This process results from extensive effort on the part of the Miccosukee Tribe to speak to western scientific norms on data quality and transparency in culturally appropriate ways. The Miccosukee Tribe developed the process considering federal government requirements for data quality to ensure that the Indigenous Knowledge cannot be rejected because of quality assurance/quality control concerns, while protecting Indigenous data sovereignty and governance. The onus is now upon the agencies to meet the Tribes "where they are" and develop protocols that effectively consider and apply Indigenous Knowledge even when it does not conform to western scientific norms and presentation.

RECOVER and other CERP staff should implement best practices in their efforts to engage Tribes and apply Indigenous Knowledge in Everglades restoration planning, operations, monitoring, and adaptive management. Staff should consult with the Miccosukee and Seminole Tribes to determine their desired timing and level of engagement for specific projects. Some best practices include the following:

- Recognize that Tribes are autonomous nations and interact with Indigenous partners as such, not as stakeholder groups.
- Involve Tribal members in planning, research, and monitoring efforts, with funding where necessary, to foster co-stewardship and co-production of knowledge to support restoration and priorities of value to the Tribes.

This may require new contracting mechanisms that should be informed by a review of other federal-state-Tribal efforts to improve capacity building and enhance partnerships.
- Provide opportunities for the Miccosukee and Seminole Tribes to engage in CERP processes and share knowledge in meaningful and culturally sensitive ways, recognizing that independent internal discussions among Tribal members may be necessary to provide input to the decision-making process.
- Include Tribal members in meetings in thoughtful ways that avoid unintentional tokenism and encourage participation by Tribal members in planning meeting agendas.
- Create data-sharing agreements that center Indigenous data sovereignty and governance and support the culturally sensitive integration of Indigenous Knowledge into reports and agency actions while protecting sensitive Tribal cultural information from public disclosure.
- Build, support, and maintain an inclusive agency and CERP culture that establishes high expectations for meaningful engagement and incorporation of Indigenous Knowledge throughout agency hierarchy and operations.

To continue to improve the quality of Tribal engagement and inclusion of Indigenous Knowledge, training should be developed, in consultation with the Miccosukee and Seminole Tribes, and required on a recurring basis for all agency staff who interact with Tribal members. The integration of Indigenous Knowledge in Everglades restoration planning, operations, and adaptive management is in its infancy, and the acceptance of its application in long-established federal and state restoration processes varies within the CERP community. The CERP is not alone—restoration programs throughout the United States are grappling with how to meaningfully engage with Indigenous Knowledge at various scales. Formal training should be provided for CERP agency personnel who interact with the Tribes, including leadership, scientists, engineers, and field staff, to strengthen a culture of meaningful engagement by all restoration practitioners. This training could cover the history and governance of the Miccosukee and Seminole Tribes of Florida; Indigenous Knowledge; laws, regulations, and guidance; case studies that highlight the complementarity of Indigenous Knowledge and western science; and best practices for consultation, cooperation, and meaningful engagement. Also valuable would be education on the appropriate protection of sensitive Indigenous Knowledge to build a relationship of trust and cooperation.

Attention is needed to ensure that Tribal input and opportunities for meaningful collaboration and inclusion of Indigenous Knowledge are not lost because of staffing resource constraints. The pace of Everglades restoration planning, operations, and adaptive management and the associated requirements for effective engagement may exceed the existing staff resources of both the CERP agencies and the Tribal Nations. CERP agencies' dependence on a single staff member to coordinate Tribal relations creates vulnerabilities. CERP agencies could improve their capacity through increased agency training and staffing to ensure a breadth of expertise and uninterrupted relations in the event of personnel turnover. Other restoration programs have provided grants to Tribes, when needed, to increase Tribal staff availability for consultation and engagement.

CLIMATE CHANGE

Attempting to include climate change information in CERP planning could overwhelm even the most intrepid project planner, and CERP planners have been cautious in their attempts to do so. The amount of climate change information, the unknown credibility of the information, the number and complexity of models used in CERP planning, and the complications of incorporating climate information into models not designed for that purpose add up to a very challenging endeavor. Although there is danger in using climate projections without a carefully considered plan that is consistent across the diverse set of analysis and models involved in the CERP, the greatest danger is making no attempt to plan for climate change. The committee offers the following recommendations to advance the use of climate change information in CERP planning and operations.

A strategy to understand the impacts of climate change should be developed with a curated set of scenarios that are used consistently across all components of planning and restoration implementation. This set of climate change scenarios should represent the range of plausible changes based on review of the scientific literature and available climate projections and be used to assess project and system vulnerability to changes in temperature, precipitation, and sea-level rise. Validated methods for the use of climate scenarios, of which there are several in the scientific literature, should be used to stress test restoration plans. These planning scenarios should be applied through existing hydrologic and ecological models to provide insights on the potential vulnerability of flora, fauna, and infrastructure to these plausible changes. This strategy can be implemented now, and the results can lead to better estimation of benefits of planning and ultimately better outcomes in the long run.

A dynamic model that predicts coastal wetland elevations through time informed by empirical data is needed to provide more accurate predictions of coastal restoration outcomes and guide investment decisions. Sediment accretion models are currently being used to compare alternative restoration plans for the Biscayne Bay and Southeastern Everglades Ecosystem Restoration (BBSEER) project in the context of sea-level rise. Because the models are based on approximated relationships between flooding, salinity, and accretion, the predicted and even relative outcomes for each alternative may not reflect reality. A dynamic predictive model of accretion and wetland elevations over time will be more accurate and could provide helpful confirmation of the potential ecological return on the large, expected infrastructure investments in BBSEER while also informing future planning of the Southern Everglades project.

Existing ecological models should be used to a greater extent and further developed to anticipate the effects of climate change, including temperature, on the wildlife indicators of Everglades restoration success. Because wildlife species and habitats are the ecological endpoints of Everglades restoration, the output of ecological models should be considered early in the process when evaluating restoration plans. Ecological models should be developed and applied to evaluate the effects of projected changes in precipitation and temperature on biotic indicators of restoration success. Because confidence in temperature projections is greater than that for precipitation, attention should be paid to the effects of increased temperature on life history, phenology, and physiology of wildlife species using tools such as mechanistic niche models. Furthermore, a more thorough accounting of uncertainty in ecological models with respect to climate change impacts should be undertaken. How changing climate is incorporated into models based on historical data that are unlikely to hold in the future, as well as the reliability of tools under changing conditions, should be evaluated.

A more cohesive integration of ecological and physical modeling and monitoring that draws together existing data, models, and efforts should be pursued to understand and mitigate the effects of climate change on Everglades restoration to better support restoration decisions. In an integrated framework, management decisions can be informed by physical and ecological models, which in turn can be updated with monitoring and other data to further inform management, monitoring, and model refinement, in an ongoing cycle of learning that can reduce uncertainty in projections. Such an integrated framework that enables updating based on new information, such as in the Everglades Vulnerability Analysis (EVA) modeling framework, can better support decision makers as they weigh risks and benefits of alternatives, manage trade-offs, and prepare for the effects of climate change on Everglades restoration. A long-term commitment and careful coordination and communication among teams overseeing models, monitoring, management, and decision making will be necessary to achieve this objective.

Regular revisions to the System Operating Manuals and other operational plans should incorporate the evolving understanding of climate variability and change, including extreme events, to ensure anticipation of and planning for a wide range of conditions. The System Operating Manuals represent the flexibility inherent in infrastructure operations and can be leveraged for that purpose if monitoring and periodic updating are systemized and linked to operations. The evolution of the COP to CEPP 1.0 represents a prime opportunity to apply learning from several years of COP operations and consider a subset of future climate scenarios to test the response of operations to changing climate conditions.

ADAPTIVE MANAGEMENT AND USE OF NEW INFORMATION IN DECISION MAKING

When the CERP was authorized in 2000, Congress directed the use of adaptive management to ensure continued restoration progress amid uncertainties and to improve restoration outcomes through the incorporation of new information. After 24 years of CERP program development and implementation, the committee evaluates the effectiveness of the incorporation of new information into the CERP at four stages: (1) project-level adaptation during design and construction, (2) project-level adaptive management after operations begin, (3) adaptation of operations at regional scales, and (4) program-level adaptive management. Adaptive management itself is not a stationary process. Increasing input from the Tribes, experience with restoration, and the challenges associated with climate change all demand that the adaptive management program evolve to meet the needs of agency decision makers and stakeholders.

The CERP has developed thorough project-level adaptive management guidance, but the process to incorporate new information is often time-consuming and burdensome, which limits its application and effectiveness at the full scale of the CERP. RECOVER's adaptive management guidance is sound conceptually and sufficiently detailed to be effective and has informed the development of well-crafted project-level adaptive management plans. The guidance promotes a collaborative relationship between RECOVER and Project Delivery Teams that spans design and operation to support adaptive management. However, effective adaptation of authorized designs and operational plans for implemented projects has often been a very lengthy process. The USACE approval processes required to modify a congressionally authorized project contribute to the delays. Modifications can be made more quickly when some flexibility is built into the Project Implementation Reports (PIRs) or when uncertainties are anticipated in project adaptive management plans. As many more projects come online, more timely responses and processes are needed that facilitate adaptation to new information.

Adaptation of operations to new information is a particular strength of water management for restoration, and the USACE and the SFWMD should continue their increased use of conditions-based operations to better adapt to changing conditions. Adaptive approaches have been used effectively to provide and improve operational flexibility, in large part because of regular and transparent communication between scientists and decision makers. Many regional operations, such as the Lake Okeechobee System Operating Manual and the COP, rely on conditions-based operations to provide maximum flexibility and responsiveness to changing conditions, but other parts of the system (e.g., S-12A-D along Tamiami Trail) continue to rely on specific calendar-based operations. Conditions-based operations, where feasible, would enable optimization of water management in light of prior hydrologic and ecological conditions. However, such actions would require modification of existing operational plans and likely require additional resources for staff to review data and make management recommendations or to upgrade control structures to enable remote operations. Structured opportunities for learning from operations, such as the COP Adaptive Management Plan, will further improve future operations.

Program-level adaptive management has not been implemented in detail. The focus of adaptive management to date has instead been at the project level where the objective appears to be to modify projects to achieve the expected outcomes of those projects. At the program level, adaptive management has potential to maximize the benefits of the restoration as a whole. RECOVER identified program-level "mission-critical uncertainties" in 2015, including the implications of climate change and available water storage on restoration outcomes, but no coordinated program-level adaptive management strategy has been implemented. Past System Status Reports have lacked the necessary analysis of how and why the system is changing over time relative to goals and expectations to support program-level decision making. The potential for learning through the long-anticipated Second Periodic CERP Update is substantial and should support renewed attention to program-level adaptive management.

Additional efforts are needed to provide the necessary foundation for successful adaptive management in the CERP. The following three areas are essential to achieve this goal:

1. **Prioritize building expertise and a culture of adaptive management.** Guidance alone is insufficient to support effective implementation of adaptive management. Clear endorsement from leadership of the importance, objectives, and value of adaptive management is essential to promoting a culture of adaptive management. In addition, CERP staff at all levels need additional experience in adaptive management. Building capacity and a culture of adaptive management across agencies and stakeholders can take many years but can be jump-started with strategic hiring of experienced adaptive management practitioners from other restoration programs. Regular workshops have been used in other large-scale restoration efforts across the country to build a culture of adaptive management and should be incorporated into the CERP.

2. **Develop a robust, integrated science enterprise to support adaptive management,** consisting of effective monitoring, modeling, synthesis, and research activities with inclusion of Indigenous Knowledge. This requires adequate staffing of appropriately trained scientists, including a critical mass of expertise in statistics, data analysis, visualization, synthesis, and communication. With the shift from planning to implementation, responsibilities for adaptive management in both RECOVER and in the agencies have expanded substantially; however, staffing levels have not increased to match these needs. The creation of an Environmental Assessment Team within the SFWMD charged with collecting relevant knowledge and providing it to Project Delivery Teams is a positive step toward supporting adaptive management. A CERP Science Plan would facilitate systemwide coordination of science, which is necessary for effective adaptive management.
3. **Improve communication of restoration performance (relative to expectations and objectives) and implications for decision making.** Effective adaptive management relies on the timely sharing of data and knowledge both within and across projects and integration of this information into management decision making. Multi-agency reporting on project outcomes, such as Project Delivery Team biennial reports and/or a CERP dashboard built on key metrics that are collectively developed, would improve tracking of project outcomes and communication to management and the public.

All of these efforts will require strong direction from USACE and SFWMD leadership that adaptive management is a CERP priority that adds value and improves efficiency toward restoration success. Such communication should include not only messaging but also provision of the necessary resources for adaptive management implementation, including staffing to support the rapid increase in the number of CERP projects.

USACE headquarters should review required approval processes associated with incorporating new information into design, construction, and project-level adaptive management processes to ensure timely use of new information. Such an effort would benefit not only the CERP but all USACE restoration projects. Under the current framework many decisions related to effective implementation of adaptive management are cumbersome, slow, and often not timely enough to address current or future challenges. Implementation of existing processes could deter staff from making valuable changes because of the administrative burden necessary to affect that change or the implication of delays on other projects. As a result, new information that could substantially improve restoration outcomes may not be used if it is unrelated to legal drivers such as the Endangered Species Act. Therefore, a more timely and effective administrative process for incorporating new information into decision making is needed. The USACE has used the 3×3×3 approach in the project planning process to reduce the time associated with review and decision making. The committee encourages USACE headquarters to identify lessons learned from this approach that could be applied to project-level adaptive management and the incorporation of new information. By reviewing when decisions must be elevated to the district, division, or headquarters for action, the USACE can determine whether the effort for these approvals is appropriate to the risks posed, including impacts to ecosystem integrity and Tribal resources associated with lengthy delays. The USACE headquarters review should consider opportunities to incorporate flexibility into the decision-making process and empower decision making at the lowest reasonable level. USACE headquarters should also explore mechanisms to increase post-authorization flexibility within the constraints of existing processes, such as incorporating more flexibility into PIRs and operation manuals.

1

Introduction

The Florida Everglades, formerly a large and diverse aquatic ecosystem, has been dramatically altered over the past 140 years by an extensive water control infrastructure originally designed to increase regional economic productivity through improved flood management, urban water supply, and agricultural production (Davis and Ogden, 1994). Shaped by the slow flow of water, its vast terrain of sawgrass plains, ridges, sloughs, and tree islands supported a high diversity of plant and animal habitats. This natural landscape also served as a sanctuary for Native Americans. However, large-scale changes to the landscape have diminished the natural resources, and by the mid- to late-20th century many of the area's defining natural characteristics had been lost. The remnants of the original Everglades (Figure 1-1) now compete for vital water with urban and agricultural interests, and contaminated runoff from these two activities impairs the South Florida ecosystem.

Recognition of past declines in environmental quality, combined with continuing threats to the natural character of the remaining Everglades, led to initiation of large-scale restoration planning in the 1990s and the launch of the Comprehensive Everglades Restoration Plan (CERP) in 2000. This unprecedented project envisioned the expenditure of billions of dollars in a multidecadal effort to achieve ecological restoration by reestablishing the hydrologic characteristics of the Everglades, where feasible, and to create a water system that simultaneously serves the needs of both the natural and the human systems of South Florida. Within the social, economic, and political latticework of the 21st century, restoration of the South Florida ecosystem is now under way and represents one of the most ambitious ecosystem renewal projects ever conceived. An overview of the CERP in the context of other ongoing restoration activities and the restoration goals that guide the overall effort is provided in Appendix A as background for readers new to the Everglades. This report represents the tenth independent assessment of the CERP's progress by the Committee on Independent Scientific Review of Everglades Restoration Progress (CISRERP) of the National Academies of Sciences, Engineering, and Medicine.

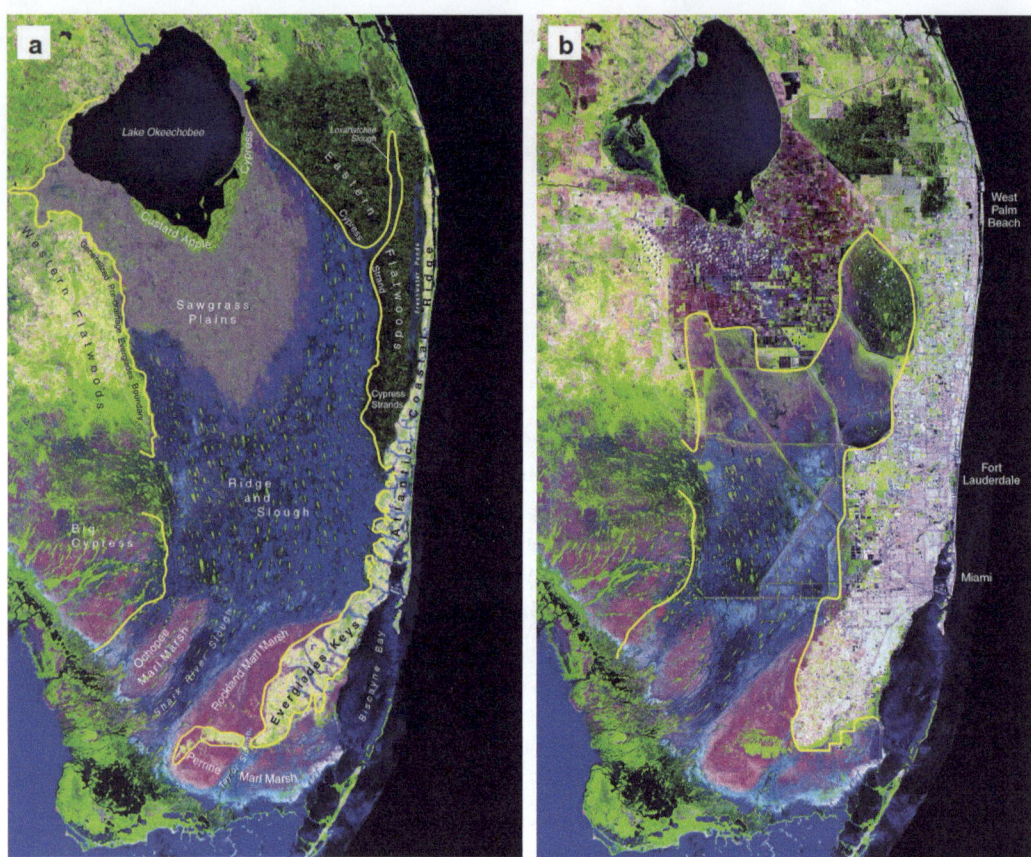

FIGURE 1-1 Reconstructed image of (a) predrainage (circa 1850) conditions compared to (b) a 1994 satellite image of the Everglades ecosystem.

NOTE: The yellow line in (a) outlines the historical Everglades ecosystem, and the yellow line in (b) outlines the remnant Everglades ecosystem as of 1994.

SOURCE: Courtesy of C. McVoy, J. Obeysekera, and W. Said, South Florida Water Management District.

THE NATIONAL ACADEMIES AND EVERGLADES RESTORATION

The National Academies have provided scientific and technical advice related to the Everglades restoration since 1999. The National Academies' Committee on the Restoration of the Greater Everglades Ecosystem (CROGEE), which operated from 1999 to 2004, was formed at the request of the South Florida Ecosystem Restoration Task Force (hereafter, simply the Task Force), an intergovernmental body established to facilitate coordination in the restoration effort, and the committee produced six reports (NRC, 2001, 2002a,b, 2003a,b, 2005). The National Academies' Panel to Review the Critical Ecosystem Studies Initiative produced an additional report in 2003 (NRC, 2003c; see Appendix B). The Water Resources Development Act of 2000 (WRDA 2000) mandated that the United States Department of the Army, the U.S. Department of the Interior, and the State of Florida, in consultation with the Task Force, establish an independent scientific review panel to evaluate progress toward achieving the natural system restoration goals of the CERP. The National Academies' CISRERP was therefore established in 2004 under contract with the U.S. Army Corps of Engineers. After publication of each of the first nine biennial reviews (NASEM, 2016, 2018, 2021, 2023; NRC, 2007, 2008, 2010, 2012, 2014; see Appendix B for the report summaries), some members rotated off the committee and some new members were added. The committee is charged to submit biennial reports that address the following items:

1. an assessment of progress in restoring the natural system, which is defined by section 601(a) of WRDA 2000 as all the land and water managed by the federal government and state within the South Florida ecosystem (Figure 1-2 and Box 1-1);
2. a discussion of significant accomplishments of the restoration;
3. a discussion and evaluation of specific scientific and engineering issues that may impact progress in achieving the natural system restoration goals of the plan; and
4. an independent review of monitoring and assessment protocols to be used for evaluation of CERP progress (e.g., CERP performance measures, annual assessment reports, assessment strategies).

The primary audience for the report is Congress, as well as agency staff who are involved in Everglades restoration and stakeholders who are engaged with or deeply interested in restoration efforts.

Given the broad charge, the complexity of the restoration, and the continually evolving circumstances, the committee did not presume it could cover all issues that affect restoration progress in any single report. This report builds on the past reports by this committee and emphasizes restoration progress since 2022, high-priority scientific and engineering issues that the committee judged to be relevant to this time frame, and other issues that have impacted the pace of progress. The committee focused particularly on issues for which the "timing was right"—that is, the committee's advice could be useful relative to the decision-making time frames—and on topics that had not been fully addressed in past National Academies' Everglades reports. Interested readers should look to past reports by this committee to find detailed discussions of important topics, such

FIGURE 1-2 Land and waters managed by the State of Florida and the federal government as of December 2005 for conservation purposes within the South Florida ecosystem.

SOURCE: Map by International Mapping. Based on data compiled by Florida State University's Florida Natural Areas Inventory (http://www.fnai.org/gisdata.cfm).

> **BOX 1-1**
> **Geographic Terms**
>
> This box defines some key geographic terms used throughout this report.
>
> - The **Everglades,** the **Everglades ecosystem,** or the **remnant Everglades ecosystem** refers to the present areas of sawgrass, marl prairie, and other wetlands and estuaries south of Lake Okeechobee (Figure 1-1b).
> - The **original, historical,** or **predrainage Everglades** refers to the areas of sawgrass, marl prairie, and other wetlands and estuaries south of Lake Okeechobee that existed prior to the construction of drainage canals beginning in the late 1800s (Figure 1-1a).
> - The **Everglades watershed** is the drainage that encompasses the Everglades ecosystem but also includes the Kissimmee River watershed and other smaller watersheds north of Lake Okeechobee that ultimately supply water to the Everglades ecosystem.
> - The **South Florida ecosystem** (also known as the Greater Everglades Ecosystem; see Figure 1-3) extends from the headwaters of the Kissimmee River near Orlando through Lake Okeechobee and the Everglades into Florida Bay and ultimately the Florida Keys. The boundaries of the South Florida ecosystem are determined by the boundaries of the South Florida Water Management District, the southernmost of the state's five water management districts, although they approximately delineate the boundaries of the South Florida watershed. This designation is important and helpful to the restoration effort because, as many publications have made clear, taking a watershed approach to ecosystem restoration is likely to improve the results, especially when the ecosystem under consideration is as water dependent as the Everglades (NRC, 1999, 2004a).
> - The **Water Conservation Areas** (WCAs) include WCA-1 (the Arthur R. Marshall Loxahatchee National Wildlife Refuge), -2A, -2B, -3A, and -3B (Figure 1-3).
>
> The following represent legally defined geographic terms used in this report:
>
> - The **Everglades Protection Area** is defined in the Everglades Forever Act as comprising WCA-1, -2A, -2B, -3A, and -3B and Everglades National Park.
> - The **natural system** is legally defined in WRDA 2000 as "all land and water managed by the Federal Government or the State within the South Florida ecosystem" (Figure 1-2). "The term 'natural system' includes (i) water conservation areas; (ii) sovereign submerged land; (iii) Everglades National Park; (iv) Biscayne National Park; (v) Big Cypress National Preserve; (vi) other Federal or State (including a political subdivision of a State) land that is designated and managed for conservation purposes; and (vii) any tribal land that is designated and managed for conservation purposes, as approved by the tribe."
>
> Many maps in this report include shorthand designations that use letters and numbers for engineered additions to the South Florida ecosystem. For example, canals are labeled as C-#; levees and associated borrow canals as L-#; and structures, such as culverts, locks, pumps, spillways, control gates, and weirs, as S-# or G-#.

FIGURE 1-3 The South Florida ecosystem.

SOURCE: Map by International Mapping.

as Lake Okeechobee (NASEM, 2018; NRC, 2008), estuaries (NASEM, 2021), new information impacting the CERP (NASEM, 2016), the need for a midcourse assessment (NASEM, 2016, 2018), climate change (NASEM, 2016, 2023; NRC, 2014), invasive species (NRC, 2014), ecosystem services (NRC, 2010), and water quality and quantity challenges (NASEM, 2023; NRC, 2010) and trajectories (NRC, 2012). Past reports have also discussed various aspects of the CERP monitoring and assessment plan (NRC, 2004a, 2008, 2010, 2012, 2014), including project-level monitoring (NASEM, 2018).

The full committee met 12 times for information gathering using a combination of virtual and hybrid meeting formats during the course of this review, and received briefings at these meetings from agencies, organizations, and individuals involved in the restoration as well as from the public (see Acknowledgments). The committee also participated in two field trips and held additional meetings in closed session to achieve consensus on this report. In addition to information received during the meetings, the committee based its assessment of progress on information in relevant CERP and non-CERP restoration documents. The committee's conclusions and recommendations were also informed by a review of relevant scientific literature and the experience and knowledge of the committee members in their fields of expertise. The committee was unable to consider in significant detail new materials received after May 1, 2024.

REPORT ORGANIZATION

In Chapter 2, the committee analyzes the natural system restoration progress associated with the CERP and systemwide operational changes, along with programmatic factors, planning efforts, and other issues that affect future progress. In Chapters 3-5, the committee discusses different ways in which science can better inform restoration decision making. In Chapter 3, the committee reviews challenges and opportunities to better apply Indigenous Knowledge in restoration planning and management. In Chapter 4, the committee discusses opportunities to better incorporate climate change information and tools in restoration planning and implementation. Finally, in Chapter 5, the committee reviews the implementation of adaptive management and processes to incorporate new information in restoration decision making to improve restoration outcomes.

2

Restoration Progress

This committee is charged to discuss restoration accomplishments and assess "the progress toward achieving the natural system restoration goals of the Comprehensive Everglades Restoration Plan [CERP]" (see Chapter 1 for the statement of task and Appendix A for a discussion of restoration goals). In this chapter, the committee updates the National Academies' previous assessments of CERP and related non-CERP restoration projects (NASEM, 2016, 2018, 2021, 2023; NRC, 2007, 2008, 2010, 2012, 2014). The committee also discusses programmatic and implementation progress and the ecosystem benefits resulting from the progress to date.

PROGRAMMATIC PROGRESS

To assess programmatic progress, the committee reviewed primary issues that influence CERP progress toward its overall goals of ecosystem restoration. These issues, described in the following sections, relate to project authorization, funding, and project scheduling.

Project Authorization

Once project planning is complete, CERP projects with costs exceeding $25 million must be individually authorized by Congress before they can receive federal appropriations. Water Resources Development Acts (WRDAs) have served as the mechanism to congressionally authorize U.S. Army Corps of Engineers

(USACE) projects. In the 20 years since the CERP was launched in WRDA 2000, six WRDA bills have been enacted:

- WRDA 2007 (Public Law 110-114), which authorized Indian River Lagoon (IRL)-South, Picayune Strand Restoration, and the Site 1 Impoundment projects;
- Water Resources Reform and Development Act (WRRDA) 2014 (Public Law 113-121), which authorized four additional projects (C-43 Reservoir, C-111 Spreader Canal [Western], Biscayne Bay Coastal Wetlands [BBCW Phase 1], and Broward County Water Preserve Areas [WPAs]);
- WRDA 2016 (Title I of the Water Infrastructure Improvements for the Nation Act [WIIN Act]; Public Law 114-322), which includes authorization for the $1.9 billion Central Everglades Planning Project (CEPP);
- WRDA 2018 (Public Law 115-270), which authorized the CEPP post-authorization change that included the 240,000 acre-foot (AF) Everglades Agricultural Area (EAA) Storage Reservoir;
- WRDA 2020 (Public Law 116-260), which authorized the Loxahatchee Watershed Restoration Project and combined the EAA Storage Reservoir and the CEPP into a single project; and
- WRDA 2022 (Public Law 117-263), which included no new CERP project authorizations but included additional authorization for IRL-South and authorized expedited completion of the EAA Reservoir.

Authorized CERP projects are sometimes classified by the WRDA bills in which they were authorized—Generation 1 (WRDA 2007), Generation 2 (WRDA 2014), Generation 3 (WRDA 2016 and 2018), and Generation 4 (WRDA 2020)—with the Melaleuca Eradication Project, which was authorized under programmatic authority, included in Generation 1. The occurrence of WRDAs every 2 years (since 2014) has ensured that the authorization process does not delay CERP restoration progress.

Funding

Within the past 2 years, authorized and requested funding—from fiscal years (FYs) 2023 and 2024, respectively—have continued at record-high levels. In FY 2023, both the state and federal agencies allocated more than $400 million in CERP funding—a first in the history of Everglades restoration (Figure 2-1). These allocations continue the historic pace of funding from FY 2022, when $1.1 billion was appropriated for the Everglades as part of the Infrastructure Investment and Jobs Act in addition to the annual USACE budget. In FY 2024, the USACE has budgeted

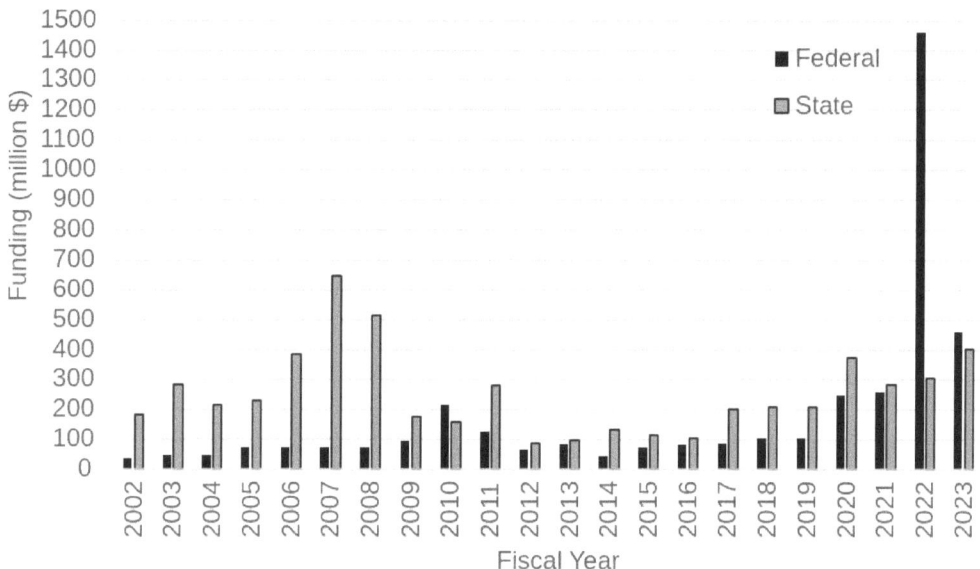

FIGURE 2-1 Federal and state funding for the CERP.

SOURCE: Data from SFERTF, 2024.

$413 million for the CERP,[1] and the FY 2025 President's Budget includes a request for $444 million for this program (OMB, 2024). In FY 2024, the State of Florida budgeted $470 million for the CERP (State of Florida, 2023), and the governor has recently signed the budget for FY 2025 with $664 million for the CERP (State of Florida, 2024). Continuation of these record-high budgets will help sustain a rapid pace of project implementation, which will expedite natural system restoration progress.

Through FY 2023 the federal government and the State of Florida have each spent $2.6 billion on CERP planning and construction (USACE, 2023d). Current projections estimate that CERP completion will total $23.2 billion (in FY 2020 dollars; USACE and DOI, 2020), although future project authorizations and modifications may affect that estimated total.

Project Scheduling and Prioritization

The anticipated future progress of CERP projects and the relationships among all the federally funded South Florida ecosystem restoration projects and some

---

[1] See https://www.usace.army.mil/Missions/Civil-Works/Budget/#Work-Plans.

highly relevant state-funded projects are depicted in the Integrated Delivery Schedule (IDS). The IDS is not an action or decision document but rather a useful communication tool that provides information to decision makers across agencies to guide planning, design, construction sequencing, and budgeting. The schedule is developed by the USACE and the South Florida Water Management District (SFWMD) in consultation with the U.S. Department of the Interior (DOI), the South Florida Ecosystem Restoration Task Force, and the many CERP constituencies. The IDS replaced the Master Implementation Sequencing Plan, initially developed for the CERP, as required by the Programmatic Regulations (33 CFR §385.31).

Updated versions of the IDS were released in November 2022 (USACE, 2022a) and November 2023 (USACE, 2023e). The 2023 IDS update provides a forecasted project planning, design, and construction schedule for the next 10 years (through 2034), based on optimal funding scenarios. Essentially, the IDS provides the agency's vision of the appropriate implementation plan based on project dependencies and construction capacity, assuming that funding is not a constraint. All authorized projects are included in the schedule, but not all components are listed if the design has not been completed (e.g., IRL-South natural storage areas, muck removal, and habitat creation).

The past 2 years of historically large funding levels enabled the program to largely keep pace with the aggressive schedule set forth in the 2022 IDS, with no delays on the completion of Generation 1 projects and some delays in Generation 2 and 3 projects, but an expedited plan for completion of the EAA Reservoir. The 2023 IDS assumes continuation of these historically high funding levels, with estimated project costs of $1.5 to $2.0 billion per year between 2024 and 2027. If funding levels return to pre-2023 levels, the pace of implementation will be substantially slower than that shown in the IDS and will require CERP leadership to decide whether to prioritize the rapid implementation of some projects over others to expedite overall ecosystem benefits or substantially delay the completion of all projects.

NATURAL SYSTEM RESTORATION PROGRESS

In the following sections, the committee focuses on recent information on natural system restoration benefits emerging from the implementation of CERP and major non-CERP projects. The implementation status of CERP projects is shown in Table 2-1, and pending unplanned projects are listed in Table 2-2, including Southern Everglades, the last major CERP planning effort and a critical element of restoration for Everglades National Park and Florida Bay.

TABLE 2-1 CERP Project Implementation Status as of May 2024

Project or Component Name	Yellow Book (1999) Estimated Completion	IDS 2023 Estimated Completion	Project Implementation Report Status	Authorization Status	Construction Status
GENERATION 1 CERP PROJECTS					
Picayune Strand Restoration (Fig. 2-2, No. 1)	2005	2005	Submitted to Congress, 2005	Authorized in WRDA 2007	Ongoing
Site 1 Impoundment (Fig. 2-2, No. 2)	2007		Submitted to Congress, 2006	Authorized in WRDA 2007	
– Phase 1		Completed			Completed, 2016
– Phase 2		Not specified		Requires authorization	Project on hold
Indian River Lagoon-South (Fig. 2-2, Nos. 3 and 4)			Submitted to Congress, 2004	Authorized in WRDA 2007	
– C-44 Reservoir/STA (Fig. 2-2, No. 3)	2007	Completed			Completed, 2021
– C-23/C-24 North and South Reservoirs (Fig. 2-2, No. 4)	2010	2030			Not begun
– C-23/24 STA (Fig. 2-2, No. 4)	2010	2025			Ongoing
– C-25 Reservoir/STA (Fig. 2-2, No. 4)	2010	2028			Not begun
– Natural Lands	NA	Not specified			Not begun
Melaleuca Eradication and Other Exotic Plants (Fig. 2-2, No. 5)	2011	Completed	Final June 2010	Prog. Authority WRDA 2000	Completed 2013, operations ongoing
GENERATION 2 CERP PROJECTS					
C-111 Spreader Canal (Western) Project (Fig. 2-2, No. 6)	2008	Final component not specified	Submitted to Congress, 2012	Authorized in WRRDA 2014	Mostly completed in 2012; S-198 construction on hold
Biscayne Bay Coastal Wetlands (Phase 1) (Fig. 2-2, No. 7)	2018	2025	Submitted to Congress, 2012	Authorized in WRRDA 2014	Ongoing
C-43 Basin Storage: West Basin Storage Reservoir (Fig. 2-2, No. 8)	2012	2025	Submitted to Congress, 2011	Authorized in WRRDA 2014	Ongoing

continued

TABLE 2-1 Continued

Project or Component Name	Yellow Book (1999) Estimated Completion	IDS 2023 Estimated Completion	Project Implementation Report Status	Authorization Status	Construction Status
Broward County WPAs (Fig. 2-2, No. 9)			Submitted to Congress, 2012	Authorized in WRRDA 2014	
– C-9 Impoundment	2007	2034			Not begun
– C-11 Impoundment	2008	2031			Ongoing
– WCA-3A & -3B Levee Seepage Management	2008	2031			Not begun
GENERATION 3 CERP PROJECTS					
Central Everglades Planning Project (CEPP) North (Fig. 2-2, No. 10)	NA	2030	Submitted to Congress, 2015	Authorized in WRDA 2016	Ongoing
CEPP South (Fig. 2-2, No. 11)	NA	2031			Ongoing
CEPP New Water (Fig. 2-2, No. 13)	NA	2024			Completed 2024
CEPP EAA (Fig. 2-2, No. 12)	NA		Submitted to Congress, 2018	Authorized in WRDA 2018, 2020	Ongoing
– EAA Reservoir and Pump Station		2030			
– EAA A-2 STA		2023			
GENERATION 4 CERP PROJECTS					
Loxahatchee River Watershed (Fig. 2-2, No. 14)	2013	2032	Submitted to Congress, 2020	Authorized in WRDA 2020	Not begun
CERP PROJECTS IN PLANNING					
Lake Okeechobee Watershed (Fig. 2-2, No. 15)	2009–2020	NA	Third revised draft, Jun. 2022	Requires authorization	NA
Western Everglades (Fig. 2-2, No.16)	2008–2016	NA	Draft Dec. 2023	Expected 2024	NA
Biscayne Bay and Southeastern Everglades Ecosystem (Fig. 2-2, No.17)	2008–2020	NA	In development	Expected 2026	NA

NOTES: Does not include non-CERP foundation projects. NA = not applicable.
SOURCES: Data from NASEM, 2023; Parrott, 2024; USACE, 2023e; Velez, 2024; G. Ralph, USACE, personal communication, 2024.

TABLE 2-2 CERP Projects or Components That Have Not Been Addressed in Prior or Ongoing CERP Planning Initiatives as of June 2024

Project or Component Name	Estimated Financial Requirement	Status
PENDING CERP PLANNING EFFORTS		
Southern Everglades: Per USACE and DOI (2020) includes – WCA-3 Decompartmentalization (QQ) – Dade Broward Levee/Pensuco Wetlands (BB) – Broward Co. Secondary Canal System (CC) – Flows to eastern Water Conservation Area (EEE) – Lake Okeechobee ASR (GG) – Central Lake Belt Storage Area (S) – Bird Drive Recharge Basin (U) – Divert WCA-2 Flows to Central Lake Belt Storage (YY) – Divert WCA-3 Flows to Central Lake Belt Storage (ZZ)	Not available until project planning completed	Planning process anticipated 2025–2028
PENDING MAJOR UNPLANNED CERP COMPONENTS		
C-43 Basin ASR (D Phase 2)	$483,000,000	Not yet begun
L-8 Basin ASR (K Part 2) and **C-51 Regional Groundwater ASR (LL)**	$387,000,000	On hold
Site 1 Impoundment ASR (M Phase 2)	$234,000,000	Inactive after Hillsboro ASR Pilot
Palm Beach Agricultural Reserve Reservoir and ASR (VV)	$211,000,000	Not yet begun
Caloosahatchee Backpumping with Stormwater Treatment (DDD)	$136,000,000	Not yet begun
Southern CREW (OPE)	$28,700,000	On hold
Florida Keys Tidal Restoration (OPE)	$23,100,000	Suspended
Loxahatchee National Wildlife Refuge Internal Canal Structures (KK)	$17,600,000	On hold
Henderson Creek – Belle Meade Restoration (OPE)	$10,800,000	On hold
Comprehensive Integrated Water Quality Plan (CIWQP)	$8,300,000	On hold
Florida Bay Florida Keys Feasibility Study (FBFKFS)	$6,500,000	Suspended in 2007. The project is planned for the future

NOTES: Remaining unplanned CERP projects include all projects more than $5 million (2019 dollars) as reported in USACE and DOI (2020), for which the components have not been incorporated in other planning efforts or formally removed from the CERP. Letters in parentheses represent project component code from Yellow Book. Estimated financial requirement derived from SFERTF (2023).
SOURCES: Data from SFERTF, 2023; USACE, 2023e; USACE and DOI, 2020.

The discussions of restoration progress that follow are organized into four major sections based on implementation status.

1. **Operating CERP projects (or project increments)**
   - Picayune Strand Restoration Project
   - Melaleuca Eradication
   - CEPP New Water
   - C-111 Spreader Canal
   - BBCW
2. **Regional operations plans**
   - Kissimmee Headwaters Revitalization Schedule
   - Lake Okeechobee System Operating Manual (LOSOM)
   - Combined Operational Plan (COP)
3. **Authorized CERP projects/components not yet operating (or constructed)**
   - CEPP South, CEPP North, and the EAA Reservoir
   - C-43
   - IRL-South
   - Loxahatchee River Watershed
   - C-11 Impoundment
4. **CERP projects in planning**
   - Western Everglades Restoration Project (WERP)
   - Lake Okeechobee Watershed Restoration Project (LOWRP)
   - Biscayne Bay and Southeastern Everglades Ecosystem Restoration (BBSEER)

The committee's previous report (NASEM, 2023) contains additional descriptions of the projects and progress through mid-2022. The South Florida Environmental Report (SFWMD, 2024a) and the 2023 Integrated Financial Plan (SFERTF, 2023) also provide detailed information about implementation and restoration progress.

CERP Projects with Recent Information on Restoration Benefits Delivered

The committee's review in this section is based on reported monitoring data to date for CERP projects for which construction has begun, with emphasis on progress and new information gained during the past 2 years.

Picayune Strand Restoration Project

The Picayune Strand Restoration Project (Figure 2-2, No. 1) was the first CERP project under construction. The 55,000 acre (86 mi^2) Picayune Strand area in southwest Florida was drained for an intended real estate development,

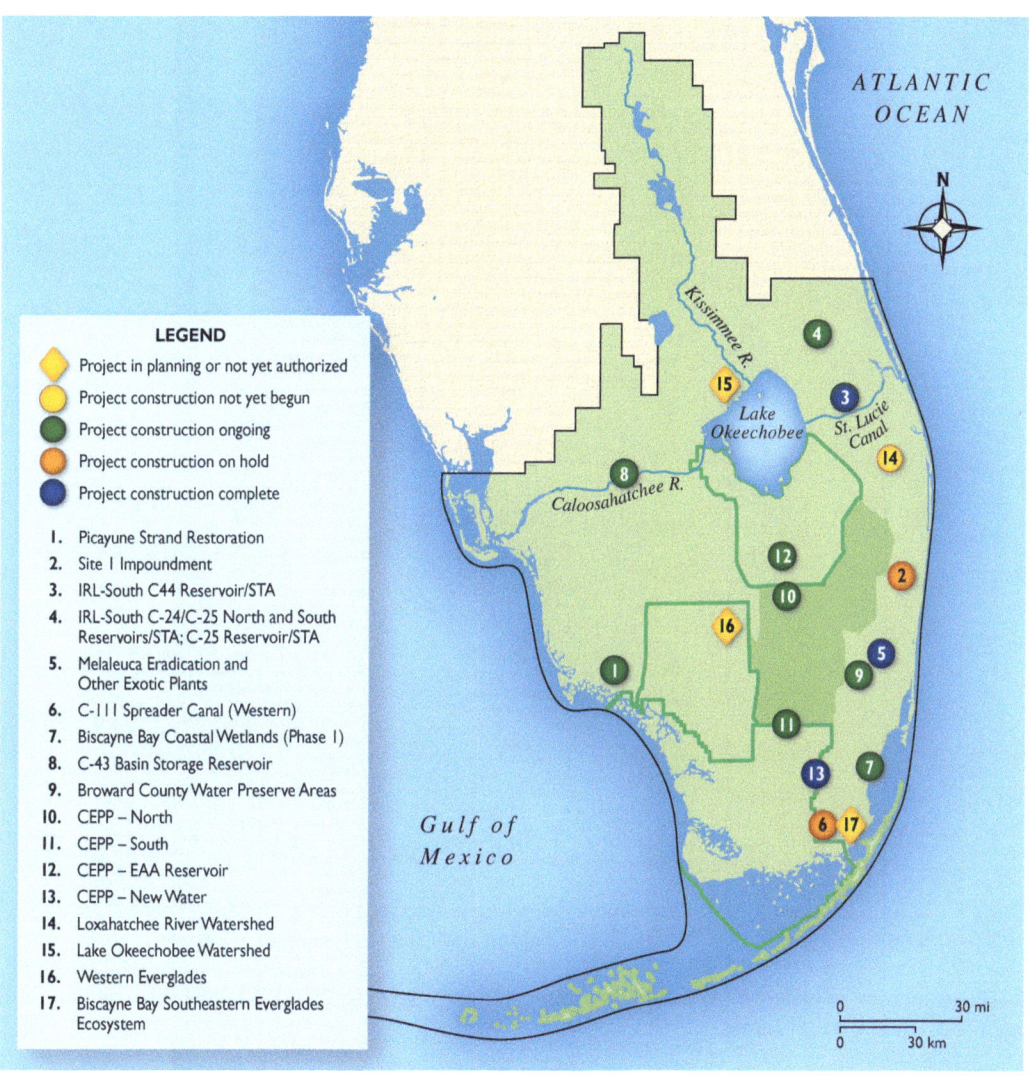

FIGURE 2-2 Locations and status of CERP projects and pilot projects.

SOURCE: Map by International Mapping.

Golden Gate Estates-South, which was abandoned before completion. Construction of drainage canals and an extensive road network drained a large area of wetlands, reduced sheet flow to the south into the Ten Thousand Islands National Wildlife Refuge, and altered regional groundwater flow into surrounding areas (Figure 2-3). Restoring the predrainage hydrology should bring multiple

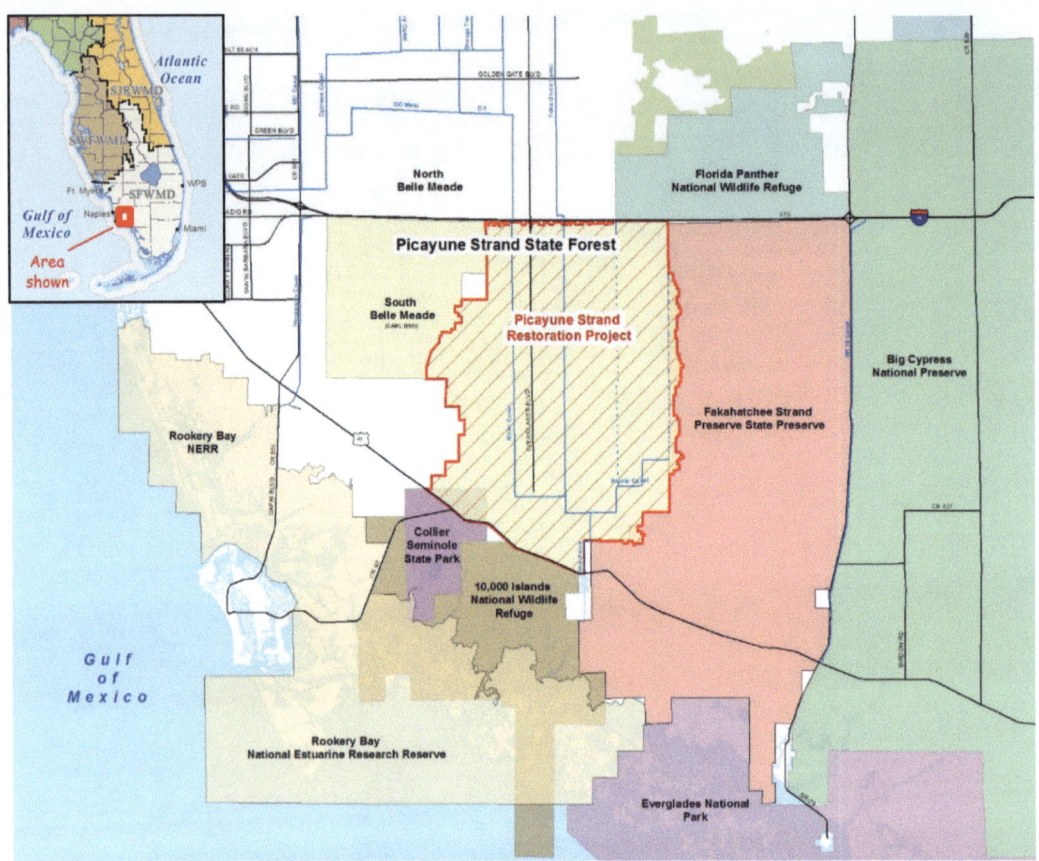

FIGURE 2-3 The Picayune Strand Restoration Project area is surrounded by several other natural areas, including Ten Thousand Islands National Wildlife Refuge and Fakahatchee Strand Preserve State Park.

NOTE: Restoration of water levels within the project footprint will enhance the hydrologic conditions in these surrounding natural areas.

SOURCE: Chuirazzi et al., 2018.

ecological and environmental benefits, including an increase in the spatial extent of wetlands, decreased frequency and intensity of forest fires, and increased habitat for endangered species such as the wood stork and Florida panther. The project is also expected to improve groundwater recharge to the City of Naples' eastern Golden Gate well field, as well as coastal estuarine salinities affected by freshwater point discharges from the Faka Union Canal (RECOVER, 2019).

Project components include plugging drainage canals, degrading roads, and removing logging trams (Figure 2-4; USACE and SFWMD, 2004). Construction has

FIGURE 2-4 Photos looking north from 100th Avenue toward the Merritt Canal: (left) in October 2010, prior to canal filling and (right) in May 2024, 9 years after canal filling.

SOURCE: M. Duever, consultant to the SFWMD, personal communication, 2024.

---

occurred in stages starting with the easternmost portion of the area and proceeding west (Figure 2-5, Table 2-3). The Eastern Stair-step canal was plugged in summer 2021, and 8.4 miles of the Faka Union Canal were plugged by May 2024. The Miller Canal, the westernmost canal, is scheduled to be plugged in late 2024 and 2025 after completion of the Southwest Protection Feature, a levee on the southwest edge of the project intended to reduce flood risk to the agricultural lands to the west of the project. Because of the staged plugging of drainage canals, the degree of hydrologic restoration varies both spatially and temporally. Prior to the 2019–2021 construction, only the northeast corner was considered to have full hydrologic restoration; as of May 2024, roughly half of the project footprint is considered to have full hydrologic restoration, with considerable additional hydrologic restoration expected by the end of 2025 (Figure 2-5).

Recent data on natural system restoration progress. NASEM (2018, 2023) provided a comprehensive review of the hydrologic and vegetation monitoring program. Ongoing groundwater monitoring since release of the committee's last report suggests that the area south of the Merritt Pump Station continues to show natural hydrological patterns, with target hydroperiods attained for cypress habitat (Figure 2-6). No additional vegetation data are being collected until 2025, with analysis expected in 2026. Thus, the remainder of this section focuses on information obtained during the past 2 years on invertebrate and vertebrate survey data in relation to hydrologic and vegetation outcomes from restoration efforts.

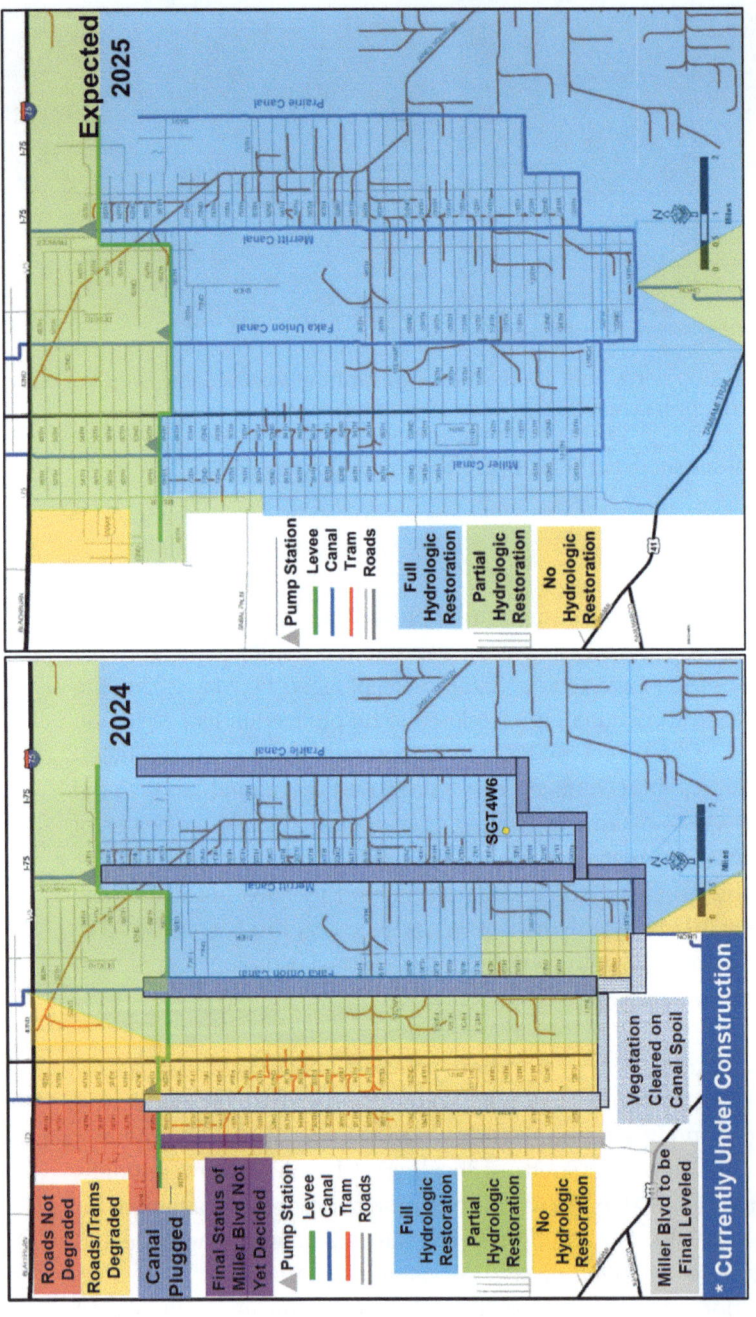

FIGURE 2-5 Maps showing areas of partial hydrologic restoration and full hydrologic restoration as of May 2024 (left) and expected areas of partial hydrologic restoration and full hydrologic restoration by December 2025 once all canal plugging is complete (right). Since 2022, the fully restored area (shaded blue) in Picayune Strand has expanded westward based on additional plugging of the Faka Union Canal.

SOURCE: M. Duever, consultant to the SFWMD, personal communication, 2024.

TABLE 2-3 Phases and Progress of the Picayune Strand Project

Component	Lead Agency	Road Removal (mi)	Logging Tram Removal	Canals to Be Plugged (mi)	Description	Project Phase Status
Tamiami Trail Culverts	State	NA	NA	NA	17 culverts constructed	Completed in 2007
Prairie Canal Phase	State (expedited)	64	30	7	Hydrologic restoration of 11,000 acres in Picayune Strand and 9,000 acres in Fakahatchee Strand State Preserve Park	Plugging and road removal completed in 2007; logging trams removed in 2012
Merritt Canal Phase	Federal	65	16	8.5	Merritt pump station, spreader basin, and tie-back levee constructed	Completed in 2015; pump station transferred to the SFWMD in 2016
Faka Union Canal Phase	Federal	81	11	8.4	Faka Union pump station, spreader basin, and tie-back levee constructed	Pump station completed in 2017; 8.4 miles canal plugging completed by May 2024
Miller Canal Phase	Federal/State	77	11	13	Construct pump station, spreader basin, tie-back levee, and private lands drainage canal; remove Western Stair-step canals	Miller pump station completed June 2019. Partial road removal completed September 2022; canal plugging to be completed 2025
Manatee Mitigation Feature	State	0	0	0	Construct warm water refugium to mitigate habitat loss	Completed in 2016
Southwestern Protection Feature	Federal	0	0	0	Construct 7-mile levee, canal, and water control structures for flood protection of adjacent lands	Construction completion estimated in 2024
Eastern Stair-step Canals	Federal	0	0	5.2		Plugging completed in June 2021

SOURCES: NASEM, 2023; data updates from M. Duever, consultant to the SFWMD, personal communication, 2024.

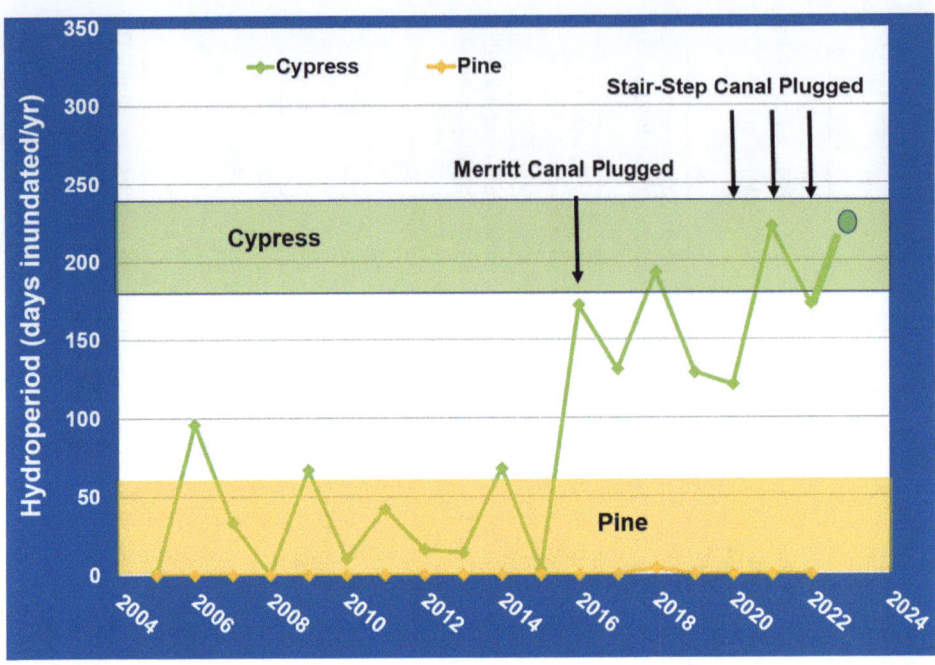

FIGURE 2-6 Changing hydroperiods relative to the target hydroperiods for cypress and pine habitats at site SGT4W6, located between the Prairie and Merritt Canals and near the Stair-step canals, filled in 2007, 2015, and 2021, respectively. The location for this site is shown on Figure 2-5.

SOURCE: Duever, 2023.

There is some early evidence that animal communities are responding to the changes in hydrology and vegetation enabled by restoration. In comparison to baseline data on the relative abundance and species diversity of aquatic macroinvertebrates collected in 2005–2007 (Bartoszek et al., 2007; Ceilley, 2008), macroinvertebrate assemblages by 2019 in some of Picayune Strand's cypress habitats had started to exhibit shifts toward the species composition found in reference wetland communities. These shifts included an increase in species richness and recolonization of sites by longer hydroperiod (wetter) indicator species such as freshwater sponges, limpets, and crayfish (Ceilley et al., 2020). Additional monitoring in 2021–2022 revealed similar patterns, including evidence of further convergence of restored sites toward reference conditions since the 2019 macroinvertebrate sampling (Figures 2-7 and 2-8; Ceilley, 2022; Gaglia, 2022). Data on how fish and anuran (frogs and toads) communities are responding to restoration activities in Picayune Strand are limited compared to data on macroinvertebrates but yield important insights for future monitoring of

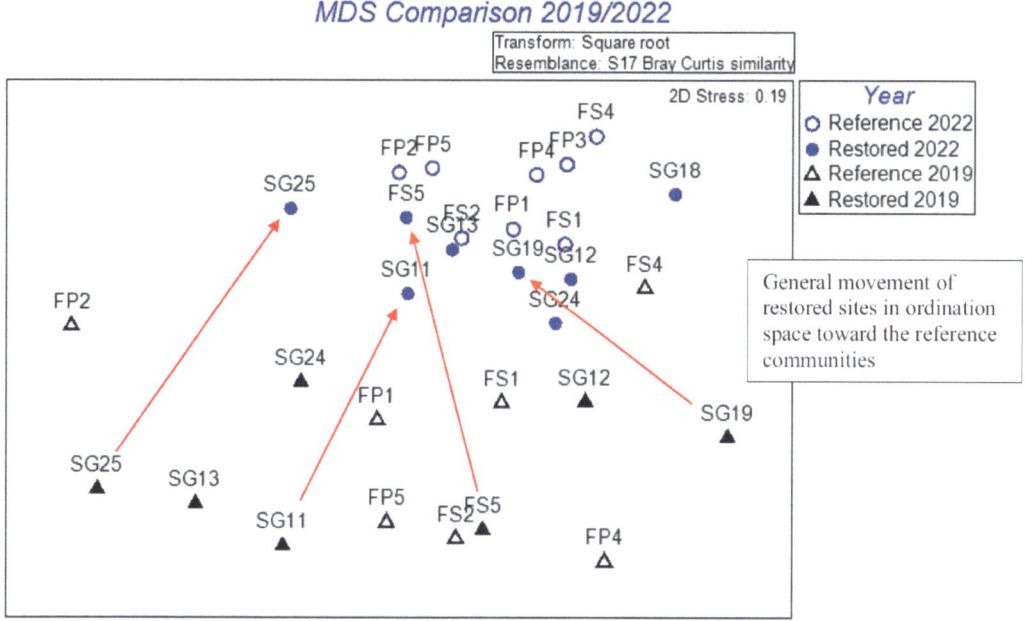

FIGURE 2-7 Multidimensional scaling ordination of species richness of combined macroinvertebrate communities comparing all sites in 2019 to 2022.

NOTE: Directional arrows show that restored sites are moving in ordination space toward reference habitats.

SOURCE: Gaglia, 2022.

these taxonomic groups. As of 2019, the spatial extent of successful sampling remained constrained by short hydroperiod conditions not conducive to supporting fish. Moreover, Ceilley et al. (2020) reported abundant capture of the African jewelfish, a nonnative, invasive species, which may complicate the ability to track native fish community recovery. Anuran diversity and abundance surveys have thus far been limited to monitoring of artificial structures (i.e., PVC pipes that treefrogs use for shelter). The results have primarily yielded invasive treefrogs to date; more than 90 percent of frogs sampled from 2005 to 2019 in Picayune Strand have been Cuban treefrogs (Ceilley et al., 2020; Clark, 2020), obfuscating the ability to discern how native species are responding to restoration. Although this sampling technique has benefits such as ease of standardized, repeated sampling over time, it has several limitations. For example, only refuge-seeking species such as treefrogs typically utilize these artificial habitats, leaving the presence of other species (e.g., Ranid frogs) in the anuran community unknown. Adopting the use of automated audio recording technology (e.g., "frog loggers";

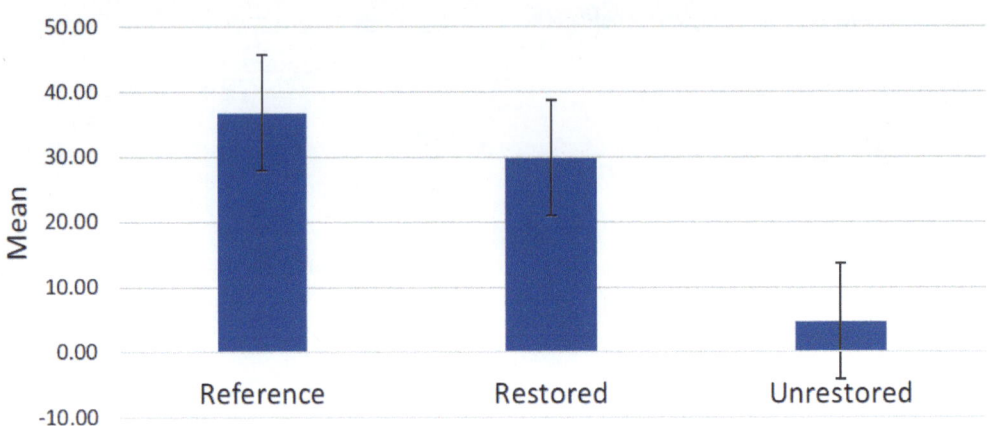

FIGURE 2-8 Mean macroinvertebrate species richness for reference, restored, and unrestored sites, with data pooled among all three sampling periods. Error bars represent the 95% confidence interval around the mean.

SOURCE: Gaglia, 2022.

De Solla et al., 2006; Dorcas et al., 2009; Measey et al., 2017; Stevens et al., 2002), with the possibility of employing artificial intelligence to support analyses of recordings (Gan et al., 2021; Lapp et al., 2021), would provide a more comprehensive assessment of the species of amphibians using restored habitats and might help to overcome possible sampling biases toward a single invasive species. Automated audio recordings offer the additional benefit of documenting broader changes to the soundscape in response to restoration activities. For example, ecoacoustics are increasingly used to document bird diversity and even underwater fauna (Desjonquères et al., 2020), as well as responses of ecosystems to restoration (Ramesh et al., 2023) and the invasion of nonnative species (Barney et al., 2024; Hopkins et al., 2022; Ribeiro et al., 2022).

The committee's overall conclusion is that restoration of hydrology in the Picayune Strand Restoration Project appears to already be generating benefits to the local flora and fauna, with vegetation and macroinvertebrate communities responding favorably. Additional longitudinal monitoring will be needed to continue documentation of recovery, especially given the magnitude of seasonal and inter-annual variation. Sampling methods for some species (e.g., amphibians) need to evolve to generate a clearer picture of what species are responding to restored hydrology. Modern passive acoustics are one example of a sampling method that could have far-reaching benefits for Everglades restoration. Invasive species will remain a challenge, and their management will be key to the reestablishment of some native species.

Recent issues affecting progress. Newly discovered, unanticipated impacts of the project on endangered red-cockaded woodpeckers (RCWs, *Dryobates borealis*) will likely require design modifications in the vicinity of the Miller Canal. To date, impacts on RCWs have not been an important consideration because the species does not occur in the original project area. The U.S. Fish and Wildlife Service (FWS) Biological Opinion included in the Project Implementation Report (PIR) (USACE and SFWMD, 2004) thus focused on how the project would impact the species' potential habitats—mesic and hydric pine flatwoods—in the project footprint and concluded that the project would not adversely affect the species. However, new analyses completed in early 2024 have revealed that the project will likely harm RCW habitat in the adjacent South Belle Meade tract to the west of the original project area (Figure 2-9).

Adverse impacts of the project on RCWs in Belle Meade were not totally unanticipated, because analyses included in the PIR estimated that the project would result in a roughly 14 percent loss each of mesic and hydric pine flatwoods in Belle Meade (USACE and SFWMD, 2004). At the time, the RCW population in Belle Meade was very small, having been extirpated in 2000 and reestablished in 2001. In 2004, it consisted of only four family groups (Folk, 2018). Therefore, there was little concern about the impact of the project on the Belle Meade RCWs, but the population continued to expand, reaching a stable population size of 14–15 family groups by 2016 (Folk, 2018; Mangione and Spickler, 2022). Potential impacts on the RCW population became an issue in 2019 when new modeling associated with National Environmental Policy Act analysis for the Southwest Protection Feature redesign indicated that the project would create a hydrologic flow-way running from the Miller pump station through the southeastern portion of South Belle Meade (Figure 2-9). The resulting increase in hydroperiod was projected to convert a significant portion of the pine habitat currently occupied by RCWs to wet prairie and marsh that is unsuitable for RCWs.

The situation is complicated by the fact that unlike the original project area, in which development of the failed Golden Gates Estates-South left the landscape in need of restored hydrology, no such need exists in South Belle Meade, which is part of Picayune Strand State Forest. Although RCW habitat in Belle Meade has suffered from wildfires and a hurricane in recent years, hydrologically it is otherwise in good condition (J. Spickler, Florida Fish and Wildlife Conservation Commission, and M. White, Florida Forest Service, personal communication, 2024). Hence, the Picayune Strand Restoration Project does not appear to provide any benefits to RCWs in South Belle Meade that could compensate for identified negative effects.

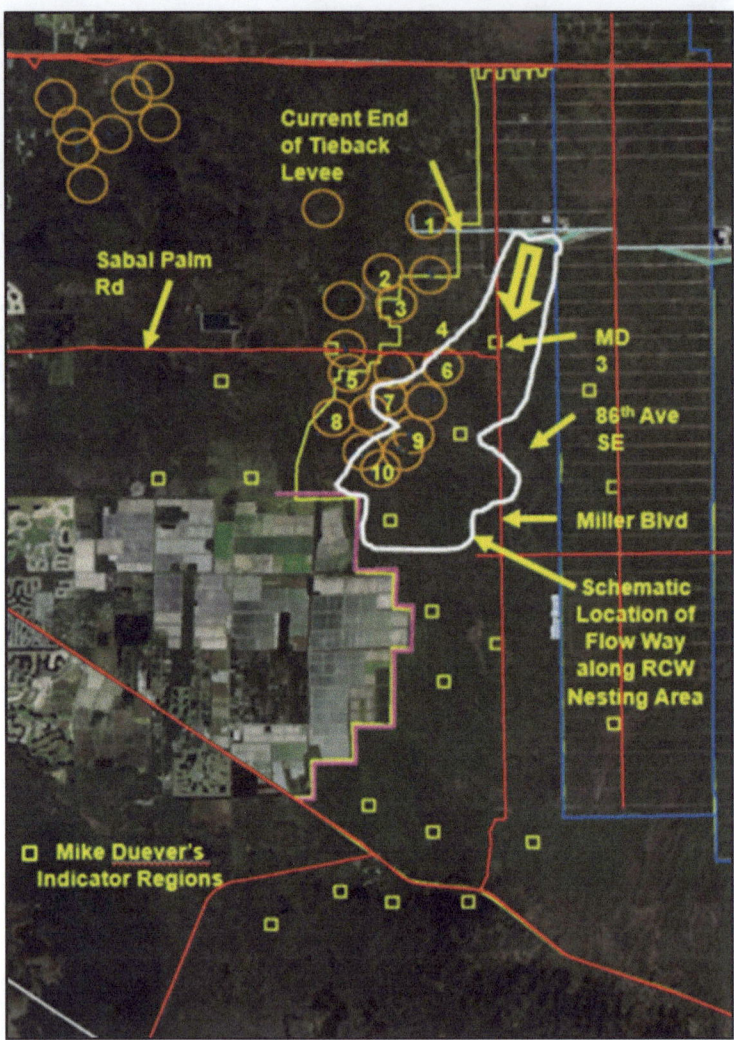

FIGURE 2-9 Potential impacts of surface flows from Picayune Strand on RCW territories in South Belle Meade (orange circles). Pink line indicates Southwest Protection Feature, white area the projected flow-way from Miller pump station to the Feature, and yellow line the current western boundary of the Picayune Strand Restoration Project footprint.

SOURCE: USACE et al., 2023b.

In response to the projected adverse impacts on RCW habitat, the Project Delivery Team is proposing to leave a section of Miller Boulevard from 64th Avenue to 80th Avenue to act as a berm. This option does not require new features or significantly increase costs, nor is its implementation likely to delay construction (A. McKenzie, SFWMD, personal communication, 2024). Whether resolving this issue delays completion of the final stage of construction of the Picayune Strand Restoration Project instead depends on the length of time it takes to move the selected option through the approval process. This issue is discussed further in Chapter 5 in the context of adaptive management.

Melaleuca Eradication and Other Exotic Plants

CERP funds were used to construct the Biological Control Rearing Annex, adjacent to the U.S. Department of Agriculture's Invasive Plant Research Laboratory (IPRL) in Davie, Florida, and CERP operational funding ($660,000 per year) supports biocontrol efforts for invasive plants in the Everglades (A. Dray, USDA, personal communication, 2024). The IPRL evaluates potential insect or fungal natural enemies in the native range of an invasive plant, obtains permits to export and import those species, raises them in quarantined conditions, evaluates how best to produce large numbers and release them to the wild, obtains permits for those releases, and then monitors the success of the releases. In general, biological control measures complement physical control measures (e.g., cutting or herbicide application) (Dray et al., 2023).

Five particularly problematic invasive plants that are the focus of recent and ongoing management and biocontrol in the South Florida ecosystem are listed in Table 2-4. Figure 2-10 shows recent information on the locations, numbers of release events, and numbers of released insects for two major biocontrol species. Melaleuca was once the most vigorously managed invasive plant, but integrated control measures, including biocontrol by a weevil, a sap-sucking psyllid, and a midge, have reduced the area dominated by Melaleuca by approximately 75 percent (Smith, 2022). In South Florida, Melaleuca control is considered to be in maintenance mode and is no longer targeted by active control methods except in areas where other invasive species are being controlled. The biological control agents drastically reduce establishment of seedlings except in persistently wet areas where the weevil—the most effective biocontrol agent—does not pupate (Smith, 2022). IPRL identified another midge species from the Melaleuca native range, the tip-galling midge, *Lophodiplosis indentata*, that was able to complete its life cycle in wet areas. *Lophodiplosis indentata* was granted a release permit in April 2022 and was introduced at one site in June 2023. A monitoring grid has been set up to confirm establishment of a field population and detect spread from the release site (Dray et al., 2023).

TABLE 2-4 Biocontrol Agent Rearing for Invasive Plant Species Control from 2019 to 2024

Invasive Species	Biocontrol Agents in Use
Melaleuca (*Melaleuca quinquenervia*)	tip-galling midge (*Lophodiplosis indentata*; in testing)
Brazilian pepper (*Schinus terebinthifolia*)	Brazilian peppertree thrips (*Pseudophilothrips ichini*)
water hyacinth (*Eichhornia crassipes*)	water hyacinth planthopper (*Megamelus scutellaris*) has been used but is no longer needed to maintain control
Old World Climbing Fern (*Lygodium microphyllum*)	Lygodium mite (*Floracarus perrepae*) Lygodium defoliator moth (*Neomusotima conspurcatalis*)
air potato (*Dioscorea bulbifera*)	air potato beetle (*Lileoceris cheni*) has been used but is no longer needed because control has been reached

SOURCE: Data from Dray et al., 2022.

Brazilian pepper is one of the most invasive upland shrubs in Florida and is problematic elsewhere. It has been primarily controlled by chemical and mechanical means, but biocontrol is now provided by two insect species. The Brazilian pepper thrips, *Pseudophilothrips ichini*, was first released in July 2019 and is currently being mass reared and distributed. Monitoring has demonstrated that it has established persistent populations in the field and is spreading into new areas (Dray et al., 2023). A second species, a leaf galler (*Calophya latiforceps*), was approved for release in 2019 but as of 2022 has not been released because the rearing colony at IPRL was lost during the pandemic and has not yet been replaced by collecting new insects in Brazil (Cuda et al., 2023).

Water hyacinth is an extremely invasive aquatic nuisance. Control efforts using herbicides started in the 1940s. Frequent repeated herbicide sprays are required because water hyacinth rapidly regrows after being sprayed. Biocontrol was started in the 1970s in an attempt to provide more sustainable biological control (Goode et al., 2021). The three biocontrol insect species introduced in the 1970s effectively reduced biomass of water hyacinth but did not substantially reduce cover (Tipping et al., 2014), and immature stages were killed by herbicide use. Since 2007, IPRL has used CERP funding to identify biocontrol species that are less effected by herbicide use. The water hyacinth planthopper, *Megamelus scutellaris*, was released starting in 2010. In experimental conditions, the combination of biocontrol and herbicide use reduces water hyacinth biomass to zero or near-zero and reduces regrowth, which increases the time between required herbicide sprays (Tipping et al., 2017). *Megamelus scutellaris* is less effective in the field because dispersal of the planthopper makes it difficult to maintain high densities on water hyacinth plants.

The Old World Climbing Fern is one of the worst invasive plants in moist areas of central and southern Florida. First reported as naturalized in 1965, it spreads horizontally and vertically to form thick mats covering other vegetation.

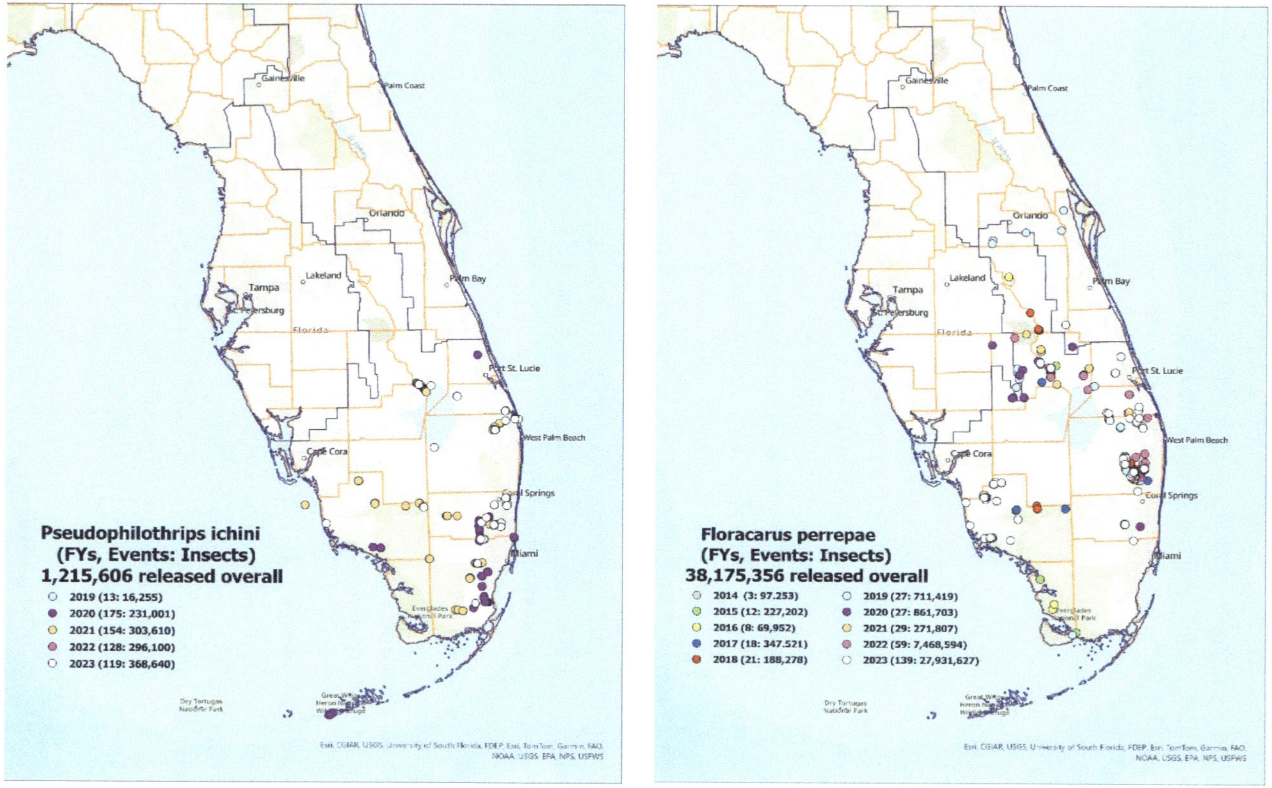

FIGURE 2-10 CERP release locations, events, and quantity of insects released for two major biocontrol species: *Pseudophilothrips ichini* (left) and *Floracarus perrepae* (right), which target Brazilian Pepper and Old World Climbing Fern, respectively.

SOURCE: Dray et al., 2023.

It is a difficult plant to control because it is widespread, it covers large areas, and herbicide use is expensive. Two species, brown lygodium moth, *Neomusotima conspurcatalis,* and the lygodium mite, *Floracarus perrepae*, were released in 2008, but in the field, neither has a consistent effect on Lygodium spread (Walker et al., 2024). The brown lygodium moth is no longer being reared and released because of disease issues and labor-intensive maintenance (Dray et al., 2023). IPRL is evaluating the suitability of additional natural enemies as biocontrol agents (Walker et al., 2024).

The air potato vine is invasive across the southeastern United States. Like the Old World Climbing Fern, it spreads horizontally and vertically and covers extant vegetation. Large-scale field releases of a beetle, *Lilioceris cheni*, started in 2012 (Rayamajhi et al., 2019) and continued until 2019. Field monitoring shows that greater than 90 percent of air potato vines are damaged by *L. cheni* and vegetative reproduction is greatly reduced, to the extent that additional releases and herbicidal measures are not needed. The plant is no longer considered a priority invasive species by land managers (Overholt et al., 2016; Rayamajhi et al., 2019). Rearing and release of *L. cheni* was terminated in 2022 (Dray et al., 2021).

Evaluating the success of biocontrol measures is complicated because biocontrol has the potential to impact large areas but usually has a chronic, not an acute, effect that acts over long time frames. Current species-specific monitoring has focused on assessing population persistence in and dispersal from introduction sites (Dray et al., 2022). There are examples of substantial impact in limited areas (Figure 2-11), but landscape-scale effects are more difficult to monitor. In general, biocontrol has generally been viewed as a success for Melaleuca and air potato vine but has been less effective for other species for various reasons.

Vegetation mapping using remote sensing is the best way to assess large-scale effects of invasive species management, using trained staff to identify species-specific signatures in remote-sensed imagery. The recently completed vegetation map of Everglades National Park, using color-infrared aerial photography taken in 2009 (Ruiz et al., 2021), includes maps of locations and cover of Brazilian pepper, Melaleuca, and Lygodium, although Lygodium was difficult to detect. Because these maps are based on aerial photography, they only show areas where an invasive species is apparent in the canopy, but they provide a baseline landscape-scale assessment. An updated vegetation map based on more recent imagery would provide a useful means to assess large-scale progress in the control of invasive plants in the Everglades.

Central Everglades Planning Project

The CEPP is the keystone project in the restoration of the central heart of the Everglades. It will provide the means to send additional water south through

FIGURE 2-11 Lygodium "brownout" in the Kissimmee River Valley caused by *Neomusotima conspurcatalis* during fall 2018.

NOTE: The helicopter shadow helps delineate the extent of the damage.

SOURCE: Dray et al., 2022.

---

the Water Conservation Areas (WCAs) and Everglades National Park to Florida Bay, reduce harmful discharges to the northern estuaries, and improve the timing and distribution of flow in the central Everglades (USACE, 2024a). The CEPP continues to progress at the rapid pace that has characterized it throughout its development, authorization, and implementation (NRC, 2014). The CEPP is a complex project with four phases, each with multiple components: CEPP South, CEPP North, CEPP New Water, and CEPP EAA. Its many parts include improvements in seepage management; improvements in conveyance through filling of canals, levee removal, and addition of new structures such as pump stations and gated spillways; a large water storage feature (the EAA Reservoir); and construction of the A-2 stormwater treatment area (STA) to ensure that the new inflows comply with existing water quality requirements (Figure 2-12). Although all of these project components are under active construction, in this section the committee discusses CEPP New Water, which is now operating with evidence of restoration benefits. The other CEPP project components are discussed in a later section under projects under construction.

CEPP New Water. CEPP New Water consists of a single project—a partial depth seepage barrier. Its objective is to reduce seepage from Northeast Shark River Slough into the 8.5 Square Mile Area to the east (Figure 2-13), reducing flood control constraints on operations to allow more existing water to flow south. Previously, the Limestone Products Association constructed 5 miles of seepage barrier (35 feet deep) south of Tamiami Trail along the L-31N levee (also known as the L-31N Rock Miners Seepage Wall) to the north of the CEPP New Water project, and in December 2022 the SFWMD completed construction of a 2.3-mile, 63-foot-deep seepage barrier adjacent to the 8.5 Square Mile Area immediately south of the project (Figure 2-13). CEPP New Water created an additional 5.0 miles of seepage barrier (55–65 feet deep) running from the SFWMD seepage barrier north to the L-31N Canal. There will be a 1.6-mile gap between this new seepage barrier and the existing Rock Miners Seepage Wall to the north (Figure 2-13). CEPP New Water was completed in early 2024 (Parrott, 2024).

It is already evident that the two new seepage barriers will greatly reduce, and perhaps even eliminate, flood control constraints imposed by the 8.5 Square Mile Area/Las Palmas community within the CEPP footprint. Prior to the COP, these constraints for decades completely stymied every attempt to restore the historic distribution of flow between Northeast and Western Shark River Slough (NASEM, 2021). Although the COP has been highly successful in restoring historic flow patterns, it has not been immune to these constraints, experiencing four events in which flood control constraints affected operations in water year (WY) 2021 and two more in WY 2022 (NASEM, 2023). With the completion of the SFWMD seepage barrier and the CEPP New Water curtain wall, there have been no such events since (see also *The Combined Operational Plan* below) (B. Mills, SFWMD, personal communication, 2024). The change in hydrology can be seen in Figure 2-14, which shows that water levels within the 8.5 Square Mile Area now remain up to 3 feet lower than water levels in the adjacent slough. The importance of eliminating flood control constraints emanating from the 8.5 Square Mile Area on water management to restoration in the central Everglades cannot be overstated and paves the way for additional water deliveries in the CEPP.

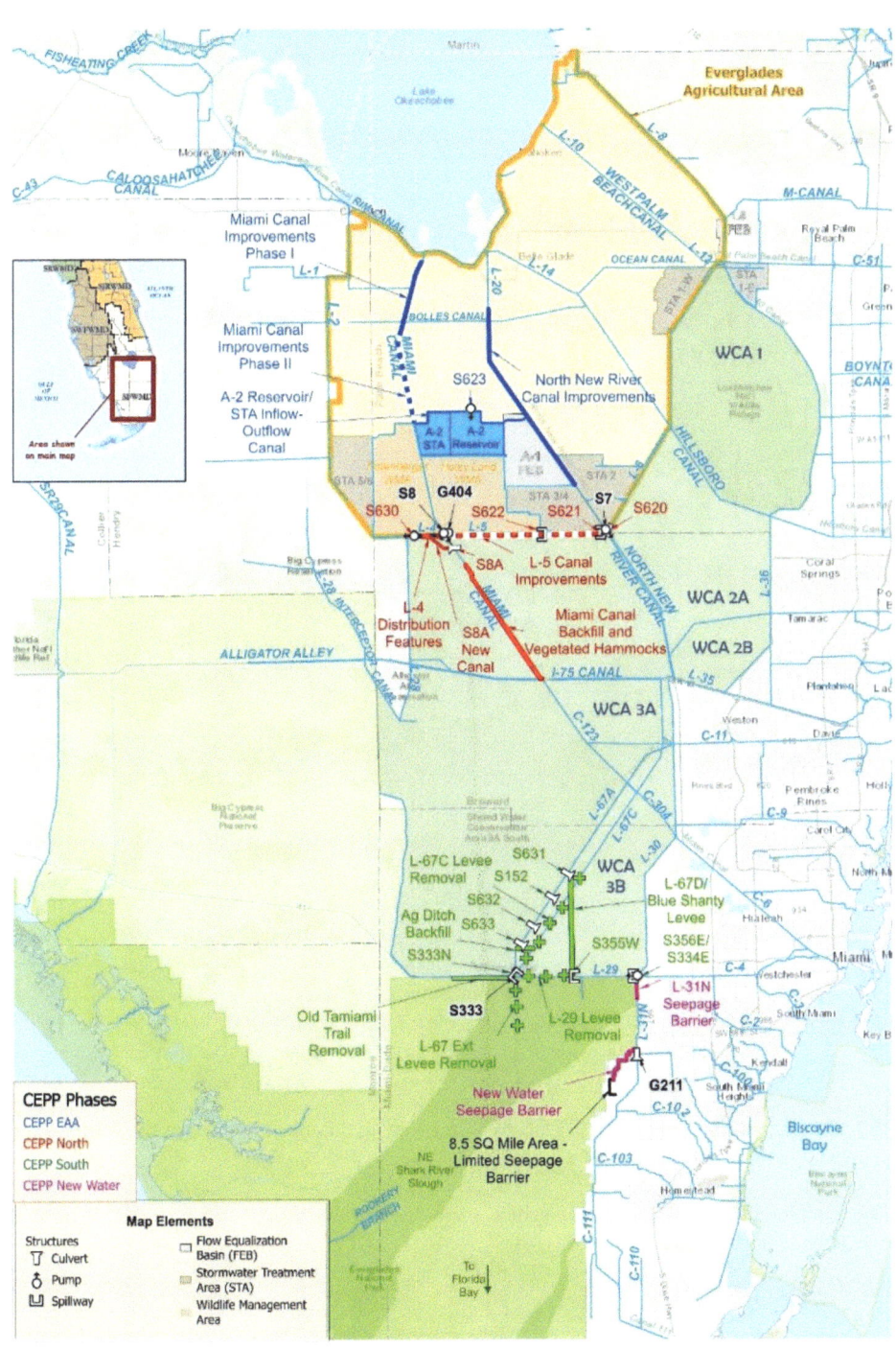

FIGURE 2-12 CEPP features.

SOURCE: USACE and SFWMD, 2024.

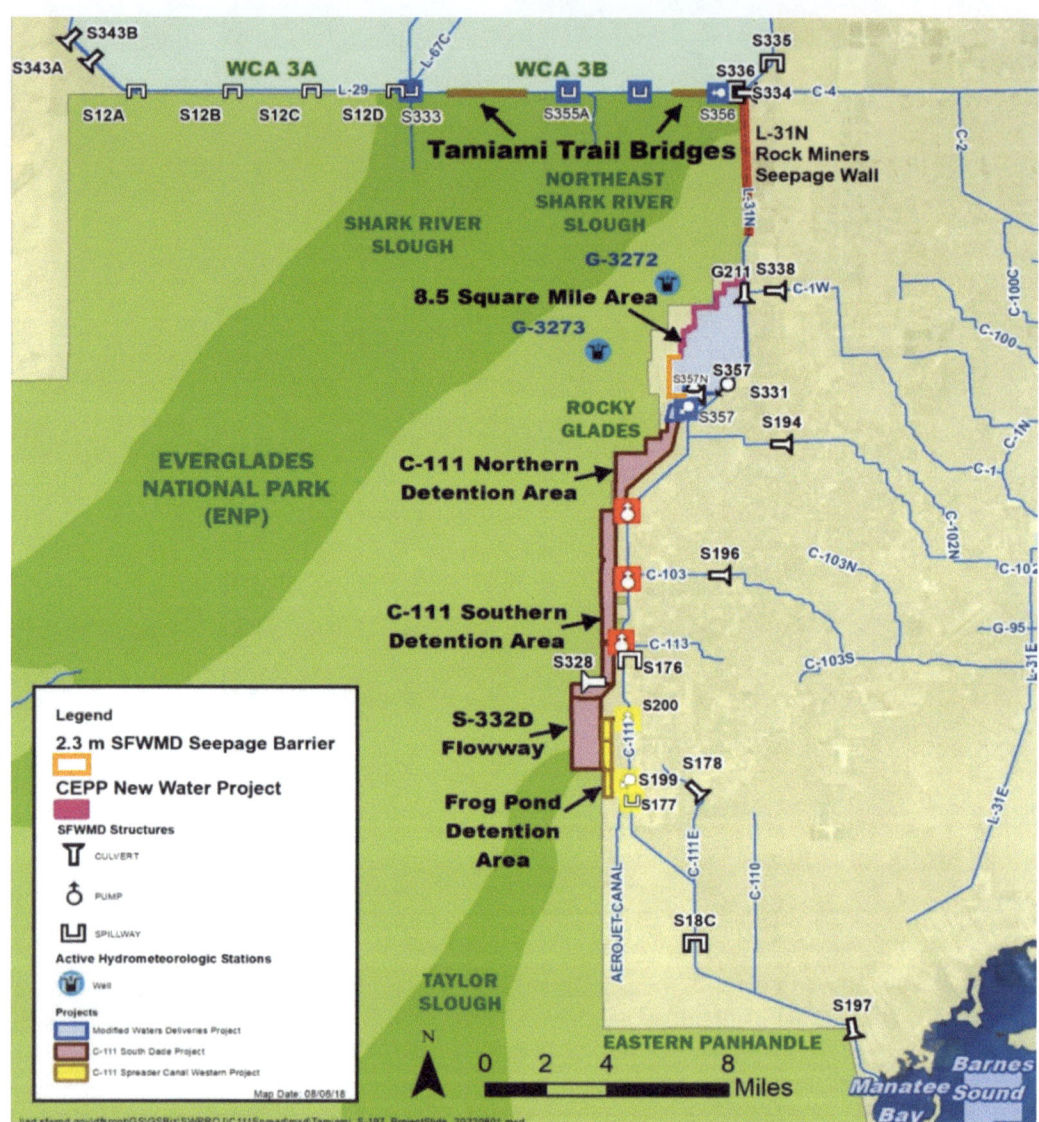

FIGURE 2-13 Location of the 4.9-mile CEPP New Water seepage barrier, adjacent to the 2.3-mile seepage barrier currently under construction by the SFWMD.

NOTE: The combined seepage barrier along the 8.5 Square Mile Area would end 1.6 miles from the end of the 5-mile Rock Miners Seepage Wall.

SOURCE: Modified from Reynolds, 2022.

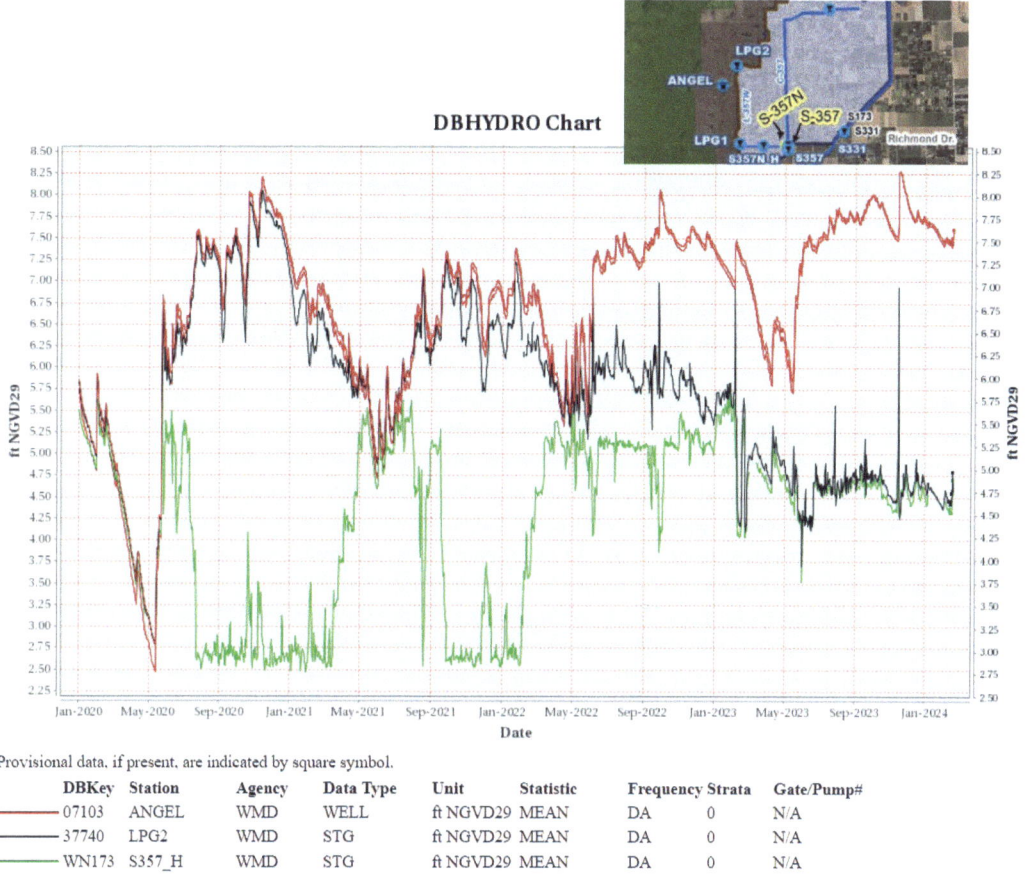

FIGURE 2-14 Water levels in three monitoring wells in and adjacent to the 8.5 Square Mile Area show the impact of the two new seepage barriers. Inset map shows well locations relative to the 8.5 Square Mile Area.

NOTES: Since the start of construction of the CEPP New Water seepage barrier in 2022, Angel's well (red line), located outside the 8.5 Square Mile Area, has exhibited water levels up to 3 feet higher than those of a well just inside the new seepage barrier (LPG [black line]). In September 2022, the SFWMD's 2.3-mile seepage barrier was completed, which together with a previously constructed interior seepage drainage system has managed flooding in the southern end of the 8.5 Square Mile Area (green line).

SOURCE: B. Mills, SFWMD, personal communication, 2024.

C-111 Spreader Canal Western Project

The C-111 Canal, originally designed to provide flood protection in Dade County, spurred agricultural development on lands to the east while draining water from the Southern Glades and Taylor Slough in Everglades National Park. A principal source of the freshwater in the canal is seepage from Everglades

National Park. Because seepage drains water from the park and alters the flow pattern of Taylor Slough, the C-111 Canal has had detrimental ecological and environmental effects on Taylor Slough and Florida Bay. The C-111 Canal also discharges large volumes of freshwater through the S-197 structure into Manatee Bay and Barnes Sound, while reducing overland flows that entered the central zone of Florida Bay, altering the natural salinity regime and ecology of those waters.

The construction of the C-111 Spreader Canal project (Figure 2-2, No. 6) was envisioned in two phases—the eastern and western projects. Planning for the features and objectives of the C-111 Spreader Canal Eastern project are part of the BBSEER project, discussed later in this chapter. The western project (Figure 2-15) was designed to help restore the quantity, timing, and distribution of water delivered to Florida Bay via Taylor Slough; improve hydropatterns within the Southern Glades; and lower coastal-zone salinities in central and eastern Florida Bay. The project was largely completed in February 2012 through expedited investment by the SFWMD, and operations began in June 2012. Construction of the final

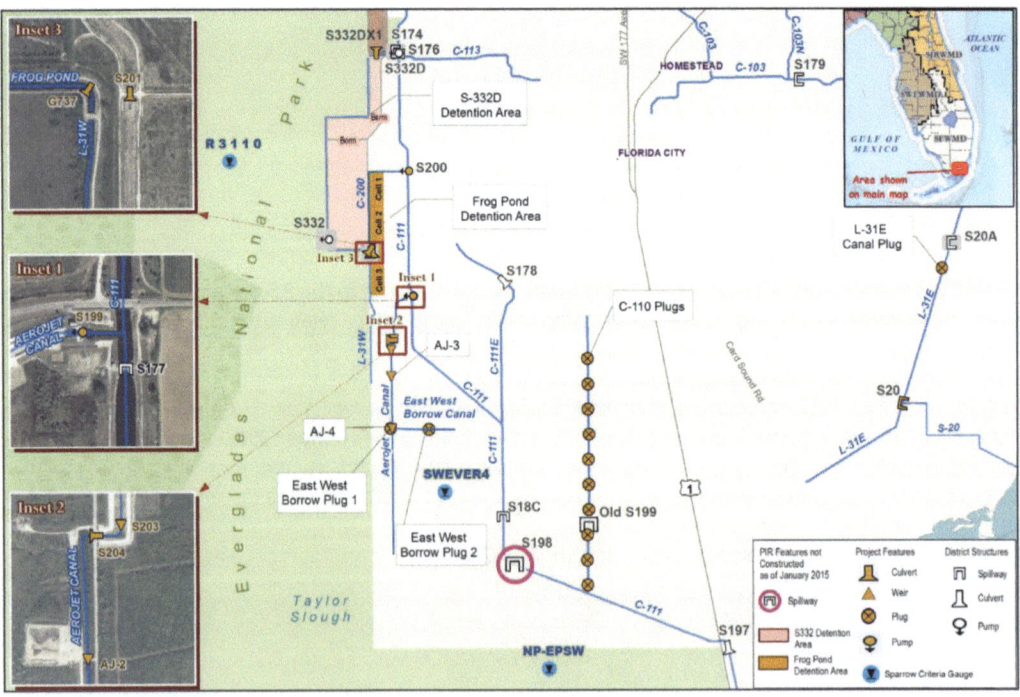

FIGURE 2-15 C-111 Spreader Canal (Western) Project features.

SOURCE: Qiu et al., 2018.

project component—the S-198 Spillway (Figure 2-15)—is currently on hold. The SFWMD reports that it "may be implemented" if it is "anticipated to increase restoration and can be implemented without adversely impacting pre-project levels of flood protection" (Gottlieb et al., 2024); it remains unclear whether it will be constructed, because it is not included in the next 10-year schedule in the 2023 IDS (USACE, 2023e).

The C-111 Spreader Canal Western project pumps excess water from the canal into the 600-acre Frog Pond Detention Area and into the Aerojet Canal impoundment (Figure 2-15), thereby creating a 6-mile hydraulic ridge along the eastern boundary of Everglades National Park to reduce seepage from the park and improve the hydrologic conditions of Taylor Slough. Rather than a persistent feature, the hydraulic ridge is present and functions only when water is available to fill the detention area. The project is also intended to contribute to improved distribution of flows in the Southern Glades through emplacement of earthen plugs along the C-110 Canal and through modified operations of structures located principally along the southern segment of the C-111 Canal.

Annual SFWMD reports show prior year data on flow and stages (e.g., Gottlieb et al., 2024), but no recent long-term analysis is available on the hydrological or ecological outcomes of the project relative to expectations or baseline conditions. Gottlieb et al. (2024) report, "Within Taylor Slough, dry-season water levels have risen, and dry season tidal creek flows have increased" since operations began, although the data analysis is not presented. Given the overlap in downstream footprint of the project with other CERP and non-CERP projects, the SFWMD concluded that potential interactions will make it difficult to tease out specific direct benefits of the project (Gottlieb et al., 2024). The project features are operated as part of the COP, which outlines the integrated operations of CERP and non-CERP projects in the region. The benefits of the COP are discussed later in this section.

Biscayne Bay Coastal Wetlands, Phase 1

Historically, Biscayne Bay received freshwater from overland flow passing through the coastal ridge and wetlands, and from extensive groundwater seepage. As a consequence of historical hydrologic alteration and development, freshwater delivery to Biscayne Bay has been greatly reduced, particularly in the dry season, resulting in loss of wetlands and an increase in salinity along the western margin of the bay. At the same time, controlled freshwater pulse discharges as point sources create altered flow, salinity, and nutrient inputs into the bay. Freshwater wetlands in the Southern Everglades have been reduced in area, altered, and degraded because of water management practices, land development, and sea-level rise, and much of the Model Lands, Southern Glades, and South Dade Wetlands are drained. These factors have contributed to landward expansion of saltwater and mangrove wetlands, including low-productivity, sparsely vegetated dwarf mangroves, and invasive exotic vegetation. The BBCW Project (Figure 2-2, No. 7) was developed to address these issues.

The primary goal of the BBCW Project is to reduce near-shore salinity and improve the ecological condition of wetlands, tidal creeks, and other habitats by increasing freshwater flows to Biscayne Bay and Biscayne National Park. The full BBCW Project, as outlined in the Yellow Book (USACE and SFWMD, 1999), envisioned restoration of wetland hydroperiods to 11,300 acres of the total 22,500 acres of wetlands. The footprint of BBCW Project Phase 1 is small. Its goals are to restore about 400 acres of freshwater wetlands and redistribute existing surface water to another approximately 2,000 acres in three geographically distinct components: the Deering Estate component, just north of the Biscayne Bay National Park, and the L-31E Flow-way and the Cutler Wetlands components, portions of which are within Biscayne National Park (Figure 2-16). The Deering Estate component was completed in 2012 and is operational. L-31E Flow-way is under construction, with some incremental benefits and completion expected in 2025 (Charkhian, 2023). No restoration progress is expected from the Cutler Wetlands component, which is only recently under construction, with completion estimated in 2025 (USACE, 2023e). The objectives of BBCW Project Phase 2 are being addressed through the BBSEER planning process, discussed later in the chapter.

Deering Estate. The goal of the Deering Estate component is to rehydrate the wetland region east of Old Cutler Road, reduce point source discharges at S-123, and restore a more natural freshwater flow regime. The hydrologic objective was to redirect up to 100 cfs of water from the C-100A Spur Canal, via the S-700 pump station, to the coastal wetlands using nighttime pumping (see Chapter 5, Figure 5-8), thereby reducing point source freshwater discharges. However, in WY 2019 the SFWMD moved to continuous pumping at a minimum rate of 25 cfs to alleviate the hydrologic flashiness that occurred with intermittent pulsed releases (Charkhian, 2023). This change improved the hydration and increased the hydroperiod in the remnant wetlands over approximately 19 acres in this project area underlain by extremely porous limestone, which significantly reduced the groundwater and surface-water salinity in the wetland. Annual vegetation monitoring between 2019 and 2022 noted increasing numbers of willows and royal palm in the wetland sloughs (Charkhian and Niemeyer, 2023).

L-31E Flow-way. The goal of the L-31E Flow-way component is to improve habitat conditions by diverting water that would normally be released through the L-31E Canal to the adjacent coastal wetlands via 10 newly constructed culverts, thereby lowering near-shore salinities. These culverts have been in place since 2010, but without completion of the associated pumps, it is difficult to maintain

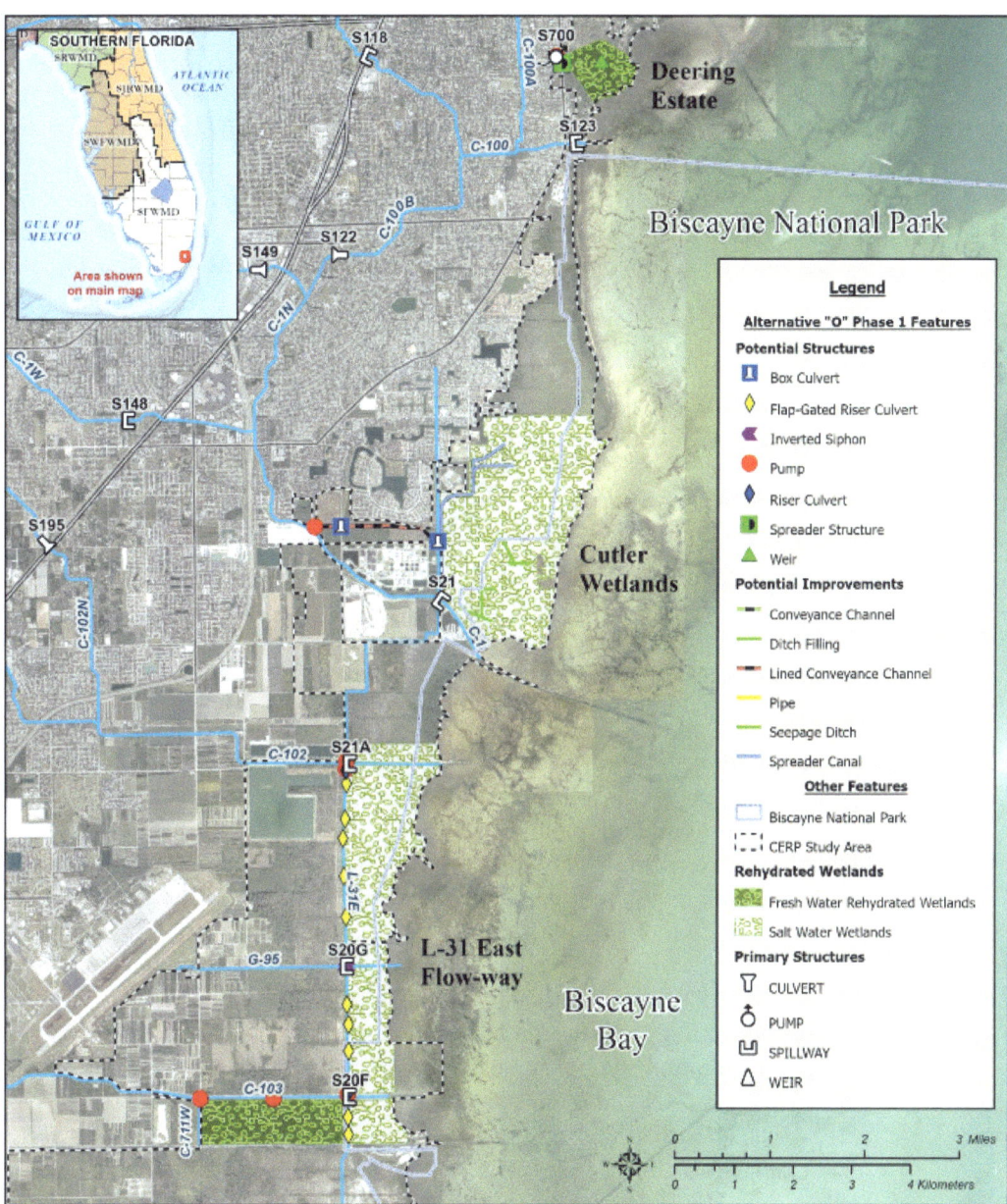

FIGURE 2-16 Biscayne Bay Phase 1 coastal wetlands project locations.

SOURCE: Charkhian, 2024.

the canal stage high enough (stage target level is 2.2 feet National Geodetic Vertical Datum [NGVD]) to promote outflow through the culverts. The USACE is expected to finish construction of the L-31E Flow-way component in 2025, which will include a total of five pump stations (USACE, 2023e).

Several coastal wetland vegetation transects downstream from the L-31E culverts were sampled in 2022 and 2023. All transects remained dominated by a mangrove overstory, and a large decrease in mangrove with freshwater inflows has not been observed. New establishment of herbaceous species was limited and largely consisted of salt-tolerant species. The presence or increase of other freshwater species such as sawgrass outside of a few meters of established culverts was not observed (Charkhian, 2024). Charkhian (2024) stated,

> As future deliveries of fresh water to the wetland are anticipated to substantially increase in the next two years, future monitoring efforts will determine whether this early lack of response is a result of the limited volumes of fresh water that have been delivered till now, or the strength of the opposing influences represented by rising sea level in Biscayne Bay, irrespective of any changes in freshwater delivery.

Regional Operations Plans

Kissimmee River Headwaters Revitalization Schedule

The Kissimmee Basin includes more than two dozen lakes in the Kissimmee Chain of Lakes, their tributary streams and associated marshes, and the Kissimmee River and floodplain (Figure 2-17). The Kissimmee River Restoration Project was authorized in 1992 with the goal of restoring more than 40 mi^2 (or one-third) of the river-floodplain ecosystem and 44 miles of the river channel. Project features, which included backfilling 22 miles of the C-38 Canal, removing water control structures, and reconnecting remnant river segments, were completed in 2021, setting the stage for implementation of the new Headwaters Revitalization Schedule, which is beginning phased implementation in 2024.

Once fully implemented, anticipated in 2026, the new stage regulation schedule for the S-65 water control structure will allow water levels to rise up to 1.5 feet higher than the current S-65 schedule and will increase the water storage capacity of the Upper Kissimmee Basin by approximately 100,000 AF (Koebel et al., 2024). This increased capacity will allow releases to more closely approximate the historic flows needed for restoration of the Kissimmee River and its floodplain wetlands and is also expected to improve littoral zone conditions in Headwaters Lakes.

The ultimate goal of hydrologic restoration for the entire Kissimmee River is to restore the single, continuous floodplain inundation event that occurred

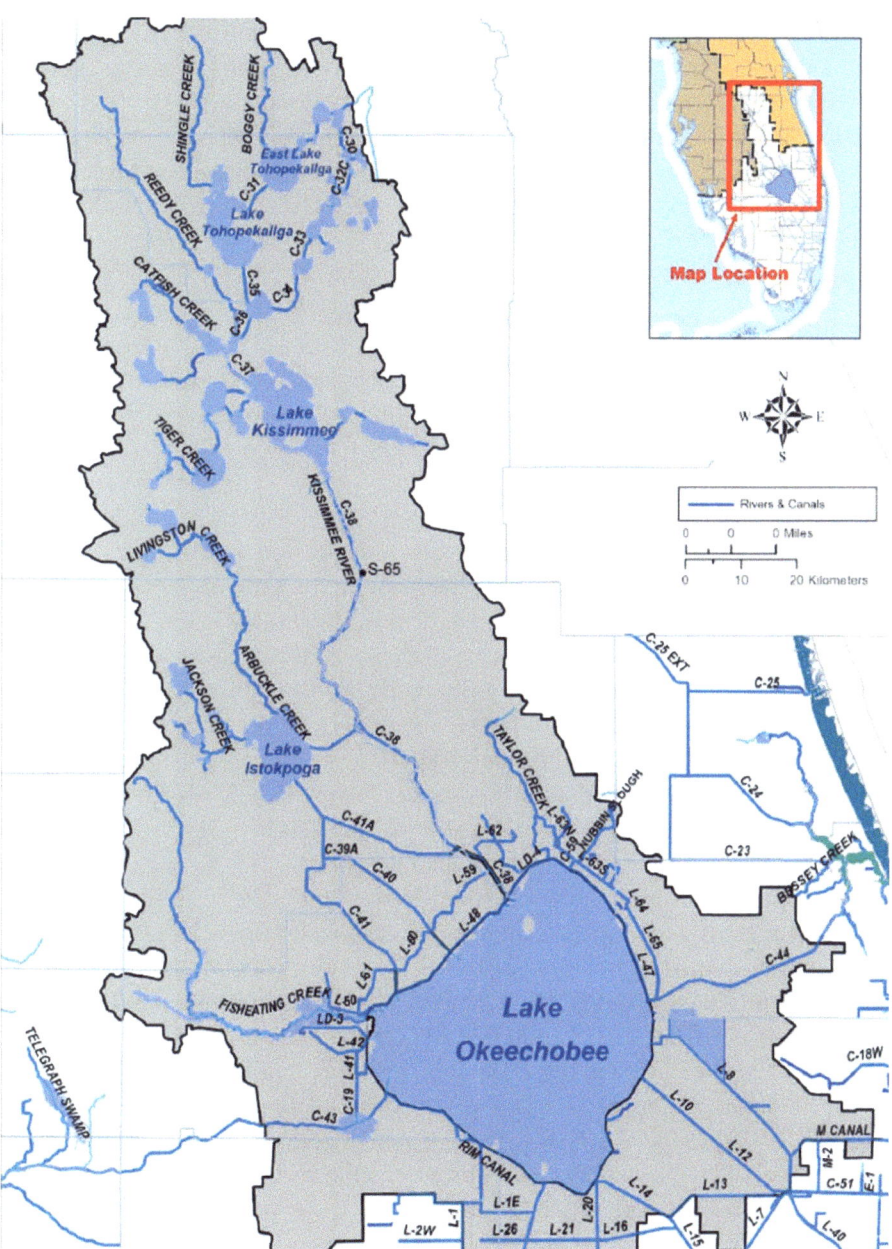

FIGURE 2-17 Major hydrologic features in the Lake Okeechobee Watershed.

SOURCE: Betts et al., 2024.

most years prior to channelization that typically began late in the wet season, continued well into the dry season, and extended throughout all seasons in some years. Floodplain inundation supported wetland vegetation along the Kissimmee River floodplain and provided important foraging habitat for wading birds and waterfowl and nursery areas for native fish. Modeling results project that floodplain inundation will improve with the 1,400-cfs discharge plans while also improving conditions to the Headwaters Lakes (Koebel et al., 2024).

Lake Okeechobee System Operating Manual

Management of Lake Okeechobee is challenged by its fast inputs and limited managed outflow, allowing high precipitation events to raise water levels quickly. Prolonged high-water levels result in the loss of submerged aquatic vegetation (SAV) in the littoral zone (because of low light transmissivity) and extreme high-water levels can pose dike safety issues, thereby requiring releases to the northern estuaries. These releases alter salinity and increase turbidity and nutrient loads, which can negatively affect SAV in the estuaries and exacerbate harmful algal blooms. In contrast, extended periods of low precipitation can lower water levels in the lake, leading to expansion of invasive species in the littoral zone and threats to water supply deliveries to utilities, agricultural producers, and residential populations (see NASEM, 2018, for in depth discussions of Lake Okeechobee and the northern estuaries).

In 2022, after extended stakeholder input, the USACE released a draft of LOSOM as the updated plan to manage the water levels in the lake after completion of the Herbert Hoover Dike rehabilitation efforts (USACE, 2022b). Zone D of the proposed operating schedule provided considerable flexibility compared to previous plans as to when to hold or release water from the lake, but with this flexibility comes uncertainty regarding the criteria for operational decisions within Zone D (Figure 2-18). LOSOM also resulted in substantially less water storage (between 460,000 and 800,000 AF) compared to the operations in place when the CERP was originally designed (for more information, see NASEM, 2023). Implementation of LOSOM was scheduled for March 2023 but was delayed pending final agency review after recent consultation and a biological opinion from the National Oceanic and Atmospheric Administration's National Marine Fisheries Service (NMFS, 2023).

The LOSOM Final Environmental Impact Statement (EIS; USACE, 2024b) was published in May 2024. The EIS reemphasizes the USACE operational strategy to balance the needs of managing flood risk, water supply, navigation, recreation, and the ecological health of fish and wildlife. The committee acknowledges the greater flexibility in LOSOM to make releases to improve water supply and enhance fish and wildlife, as well as use whole-system behavior to make

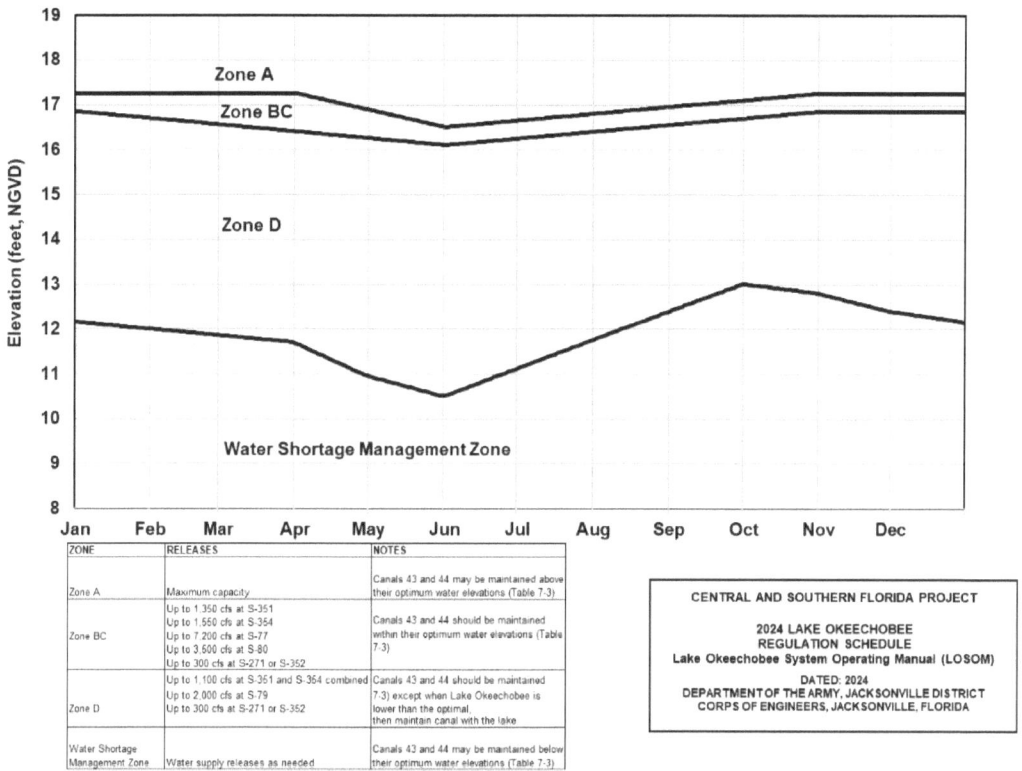

FIGURE 2-18 LOSOM water control plan.

NOTES: If lake level exceeds the top line bordering Zone A, water must be released to the northern estuaries at maximum discharge capacity. In Zone BC, discharge levels are high but below maximum (≤7,200 cfs to the Caloosahatchee River Estuary and ≤3,500 cfs to the St. Lucie Estuary) to lower lake levels. Within Zone D, no water is discharged to the St. Lucie Estuary and up to 2,000 cfs may be discharged to the Caloosahatchee River Estuary.

SOURCE: USACE, 2024b.

real-time decisions, because of the latitude in Zone D. NASEM (2023) recommended that the USACE implement a process for periodic multi-stakeholder review of Lake Okeechobee operations relative to the objectives of LOSOM to build confidence that the flexibility of the new operational schedule is being used as designed and to support learning to enhance future decision making. USACE (2024b) noted that it plans to hold three seasonal periodic scientist call meetings per year for agency and stakeholder communication and input.

The Combined Operational Plan

The COP is a comprehensive, integrated water control plan that defines the operations of the constructed features of the Modified Water Deliveries to Everglades National Park (Mod Waters) and C-111 South Dade projects (Figure 2-13) and other water management structures in the region. These non-CERP projects whose capabilities it incorporates are considered foundation projects for the CERP because they alter the delivery and flow of existing water in ways that are critical to the CERP's capacity to deliver additional flow volumes and restoration benefits. In addition, completion of Mod Waters and its operations plan was required before federal funding could be appropriated to begin construction of the CEPP. Therefore, the implementation of the COP marks not only the largest step by far toward restoring the hydrology and ecology of the central Everglades yet achieved but also the beginning of the next phase of the restoration of the heart of the Everglades embodied in the CEPP.

Two features of Mod Waters and other related projects are especially critical to the capacity of the COP to make significant changes to the hydrology of the central Everglades. First, raising the Tamiami Trail and bridging extensive portions of it enables increased flows into Northeast Shark River Slough and Everglades National Park, and much more as sheet flow (Box 2-1). Second, seepage management and flood mitigation features along the eastern boundary of Everglades National Park (Figure 2-13) reduced flood risk management constraints that limited flows into Northeast Shark River Slough. The C-111 South Dade Project improved seepage management along the eastern boundary of Everglades National Park further south, enabling more flow through Taylor Slough to Florida Bay, while continuing to honor flood risk management constraints for the agricultural lands east of the park (USACE, 2022c). The C-111 Spreader Canal Western project extends this hydraulic ridge southward, providing additional restoration benefits to Taylor Slough.

The ongoing ecological degradation of the central Everglades (see NRC, 2012 for a review) has long been a major concern motivating restoration efforts, and management of water in this area is a source of controversy. The COP is the latest in a series of water management plans that attempts to address the issues in this region. The development of the COP was informed by data gathered during a period of incremental operational testing, beginning in 2015. Thus, the hydrologic and ecological changes discussed in this section reflect incremental operational changes from 2015 to 2020 and full implementation of the COP in September 2020, which were made possible by the new infrastructure available from the Mod Waters, C-111 South Dade, and Tamiami Trail Next Steps projects.

A comprehensive assessment of observed COP benefits can be found in the COP Biennial Report (USACE et al., 2023a), which summarizes COP operations,

BOX 2-1
Tamiami Trail Next Steps Project

The 10.7 miles of the Tamiami Trail between the L-31N and L-67 extension levees have been an impediment to surface flow from WCA-3 into Northeast Shark River Slough located in Everglades National Park since the completion of this highway in 1928. Reducing the impact of this barrier has been an important component of the restoration. The Modified Water Deliveries Project created a 1-mile bridge, completed in 2013, near the eastern end of this portion of the Trail. Phase 1 of the Tamiami Trail Next Steps project, completed in 2019, addressed an additional 2.6 miles of highway with 2.3 miles of bridging at the western end of the 10.7 miles of the Trail (Figure 2-19). The Tamiami Trail Next Steps Phase 2 project will reduce the impact of the remaining 6.7 miles of highway on sheet flow through the addition of six new 60-foot-wide slab bridges, improvements to seven culverts, and raising of the highway in unbridged segments. Phase 2 construction began in 2021 and is projected to be completed in 2026 (USACE, 2023e). With the completion of this project, the entire 10.7 miles of the Trail will have been modified, which will enable raising of the maximum water level in the L-29 Canal to 9.7 feet to accommodate the CEPP, which will have a profound effect on capacity to manage water moving through WCA-3 into Everglades National Park. At that time managers anticipate implementing a new water management plan (CEPP 1.0) that will replace the COP and that incorporates these new features.

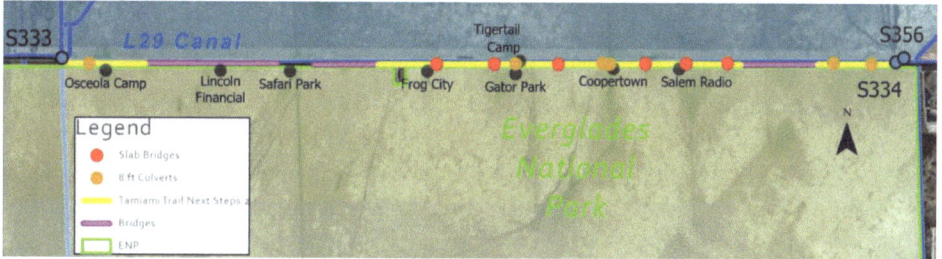

FIGURE 2-19 Tamiami Trail Next Steps Phase 2 bridging and culvert locations, which complement road raising on the remaining unbridged segments (in yellow).

SOURCE: Johnson, 2020.

the USACE decision-making rationale for any deviations, and the hydrologic and ecosystem status for the period September 1, 2020, through April 30, 2022. The COP Biennial Report seeks to address whether the plan is achieving its objectives and whether adjustments are recommended. The report also evaluates the adaptive management uncertainties identified in the design of the COP based on recent monitoring results.

Hydrologic responses. The COP Biennial Report (USACE et al., 2023a) presents an analysis of the hydrologic performance of the COP, employing the performance metrics used in the development of the COP as a reporting and evaluation framework. Monitoring results from before and after COP implementation and simulated COP performance from the Regional Simulation Model for the Glades and Lower East Coast Service Areas (RSM-GL) are compared. This comparison enables assessment of actual COP performance relative to the expectations based on modeling, and thus provides insight on the predictive performance of the modeling in terms of recent historical conditions (termed baseline, WY 2002–2015), which sheds light on the actual improvements achieved on the ground. Although encompassing only approximately 2 years of full COP operations, the report includes both a wet (WY 2021) and dry water year (WY 2022), signaling some early hints at COP performance. The biennial reports will be increasingly informative and important as the period of COP implementation lengthens.

For the current report period, the hydrologic results have been consistent with or better than the expectations based on model simulations and better than historical conditions. Even in the dry year of WY 2022, the water deliveries across Tamiami Trail were well above the typical deliveries prior to 2016 (note that the period after 2016 and prior to the COP consisted of incremental field testing of the Mod Waters features and emergency deviations and is less useful for comparisons) (USACE et al., 2023a). As the committee noted previously, the COP has been remarkably successful so far in not only increasing water deliveries across Tamiami Trail, particularly during the dry season (Figure 2-20), but also restoring the historical distribution of those flows between Western and Northeastern Shark River Slough (Figure 2-21; NASEM, 2023).

Water deliveries to Taylor Slough have also shown improvements from COP implementation to date. The longer hydroperiods have been observed relative to the pre-2016 baseline for most of the Everglades National Park region and especially in Northeastern Shark River Slough, Taylor Slough, and the western and eastern marl prairies. Again, the results compare well with the modeled expectations and generally show improvement relative to the historical baseline. Dry days and risk of soil oxidation within Everglades National Park have been reduced under the COP relative to the baselines even in the dry year, although risk of soil oxidation increased as was anticipated in north-central WCA-3A. In terms of increases in freshwater flows to Florida Bay, the results are modest and reflect the need for additional projects to come online before improvements can be expected there (USACE et al., 2023a). Overall, the early hydrologic results from the COP are promising and provide confidence in the modeling, coordination, and evaluation process used to produce it.

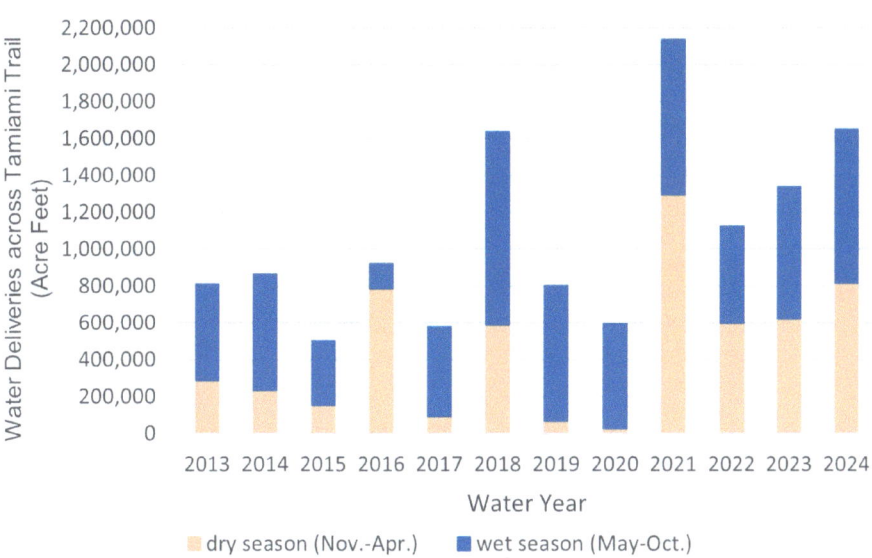

FIGURE 2-20 Water deliveries across Tamiami Trail in AF by water year.

SOURCES: Data from Velez, 2024, and L. Do, USACE, personal communication, 2024.

Ecological responses. To assess ecological performance, the COP Biennial Report (USACE et al., 2023a) employs a similar process as that used to assess hydrological performance. Of the four ecological performance measures monitored—freshwater fish, tree islands, wood stork and wading bird nesting, and slough vegetation—there are, as yet, too few data to definitively assess COP performance for freshwater fish and tree islands. For wood stork and wading bird nesting (see also below), the variation in rainfall over the period of review was deemed too large to evaluate the impact of the COP (USACE et al., 2023a). The fourth performance measure, slough vegetation, however, is highly informative. Coincident with increased hydroperiods in Northeastern Shark River Slough under the incremental field testing and the COP, long hydroperiod species such as sawgrass and slough species such as *Eleocharis* are expanding (USACE et al., 2023a). Nocentini et al. (2024) further document this pattern, including transitions to wetter plant communities along the gradient from dry marl prairie to beaksedge (*Rhynchospora tracyi*) marsh to sawgrass marsh to slough communities, as well as loss of marl prairie species. These results indicate that a predicted ecological benefit of the COP is being achieved.

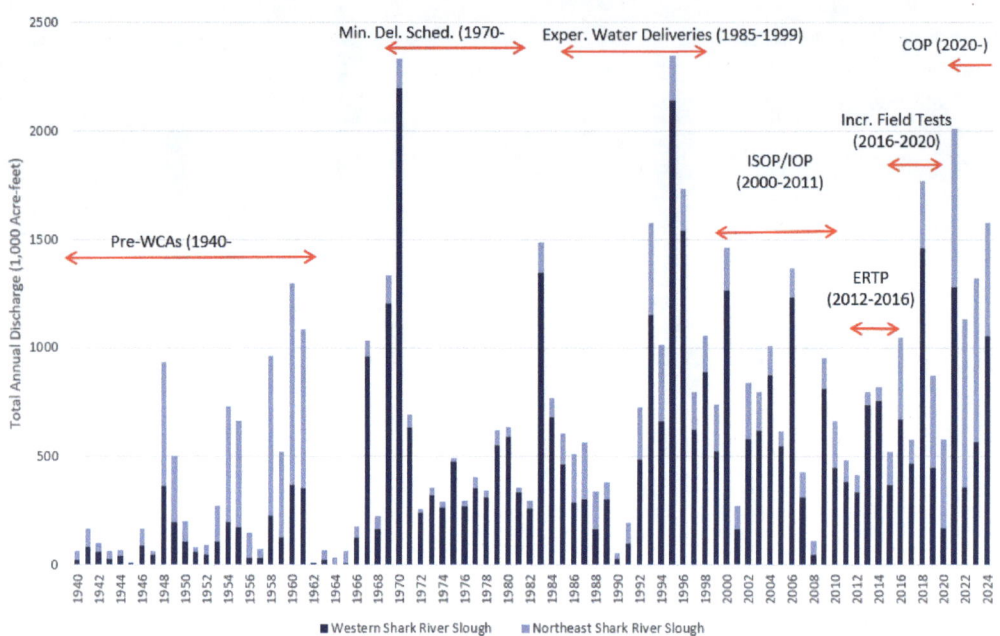

FIGURE 2-21 Changing distribution of flows across Tamiami Trail into Everglades National Park between Western and Northeast Shark River Slough by water management operations plan (Western Shark River Slough shown as dark blue and Northeast Shark River Slough shown as light blue).

NOTES: The red arrows demarcate the water management plans that governed water flows through these pathways from 1940 to 2024. These data highlight the increasing flows into Northeast Shark River Slough in the COP.

SOURCES: R. Johnson, DOI, and E. Stabenau, National Park Service, personal communication, 2024.

Additional ecological monitoring associated with the COP addresses impacts on threatened and endangered species (see Table 2-5 for the committee's summary of results of monitoring for the Cape Sable seaside sparrow [*Ammodramus maritimus mirabilis*], Everglade snail kite [*Rostrhamus sociabilis plumbeus*], and wood storks [*Mycteria americana*]) related not only to performance measures but also to criteria for incidental take, exceedances of which could require reinitiating consultation with the FWS over COP operations.[2] The prior period

---

[2] Several of the criteria require an exceedance in 2 consecutive years to reinitiate consultation. Exceedances may also be attributable to climatic variability rather than operations of the COP. For example, if an exceedance did occur but localized rainfall was the cause of the exceedance, then re-initiation of consultation may not be warranted because the root cause was not attributable to operations (G. Ralph, USACE, personal communication, 2024).

TABLE 2-5 Committee Assessment of COP Effects on Threatened and Endangered Species Relative to 2016–2020 Conditions

Type	Target/Criterion	Baseline Performance (2016–2020)	COP Performance (2021–2023)	Assessment
		Cape Sable Seaside Sparrow		
Take	Population size	Criterion met 4/4 (100%)	Criterion met 3/3 (100%)	Criterion always met, close to exceedance 1/5 baseline and 2/3 COP years, no improvement over baseline
Parameter	Nesting season dry days in 6 subpopulations	Target met 12/30 (40%)	Target met 3/10 (30%)	Target only sometimes met, no improvement over baseline
Parameter	4-year average hydroperiod in 6 subpopulations	Target met 8/30 (27%)	Target met 2/10 (20%)	Target met regularly 1 subpopulation, rarely in 5, no improvement over baseline
		Snail Kite		
Take	Maximum dry season water level (4/15–5/31)	Criterion met 1/5 years	Criterion met 2/3 years	Criterion sometimes met, improvement over baseline
Take	Pre-breeding maximum water level (6/1–12/31)	Criterion met 3/5 years	Criterion met 2/2 years	Criterion usually met, improvement over baseline
Take	Recession rate (maximum stage difference)	Criterion met 18/20 (90%) at 4 gages	Criterion met 6/12 (50%) at 4 gages	Criterion sometimes met, less often compared to baseline
PM	Dry season water level (5/1–6/1)	NA	Target met 56%, 2%, 100% days 2021–2023	Variable success meeting target depending on rainfall regime
PM	Pre-breeding water level	NA	Target met 2/2 years	Target always met
PM	Optimal and suboptimal (but suitable) weekly recession rates during nesting season (1/1–6/1)	NA	Over 3 years, 3% optimal, 30% suboptimal, 67% undesirable	Suboptimal target sometimes met, optimal target rarely met

continued

TABLE 2-5 Continued

Type	Target/Criterion	Baseline Performance (2016–2020)	COP Performance (2021–2023)	Assessment
			Wood Stork	
Take	Nesting season maximum water level in WCA-3A (3/1–5/31)	Criterion met 9/15 (60%) months	Criterion met 8/9 (89%) months	Criterion almost always met, improved over baseline
PM	Optimal and suboptimal (but suitable) weekly recession rates during nesting season (1/1–6/1)	NA	Over 3 years, 5% optimal, 82% suboptimal, 14% undesirable	Suboptimal target almost always met, optimal target rarely met
PM	Maximum undesirable recession rates during nesting season (1/1–6/1)	NA	Target met all weeks 2021 and 2022, met 20/22 weeks 2023	Target almost always met
PM	Optimal and suboptimal (but suitable) water depths in the core foraging areas of active nesting colonies	NA	Generally met in 2021, often not met in 2022, 2023	Target sometimes met
PM	Maximum conversion of short to long hydroperiod wetlands within core foraging areas	NA	Greatly exceeded in 2021, 2022 (no data for 2023)	Target appears unattainable

NOTES: Three types of measure are included: criteria for incidental take (Take), performance measures (PM), and hydrological parameters indicating potential for incidental take of Cape Sable seaside sparrows (Parameter). Results are shown for the incremental testing baseline period (baseline performance) and for full COP operations (COP performance) and summarized in Assessment.
SOURCES: Data from USACE, 2023f, 2023g; USACE et al., 2023a.

of incremental testing (2016–2020) was used as the baseline for comparisons to COP performance (2021–2023), limiting the ability to assess the full benefits of the COP relative to previous management plans (NASEM, 2021). No comparisons to a baseline have been made for performance measures for threatened and endangered species; performance is instead assessed in terms of success in meeting targets.

The monitoring results summarized in Table 2-5 highlight the challenges that the COP faces in meeting precise, complex targets over extended periods, such as water levels and especially recession rates during the nesting season for snail kites and wood storks. The COP and the infrastructure it employs likely lack the capacity to adjust sufficiently to achieve such precise goals under all rainfall regimes, a conclusion evidently reached by FWS in its decision not to reinitiate consultation in response to exceedance of incidental take of snail kites based on recession rates (Table 2-5). Impacts on wood storks and snail kites were not expected to be sufficiently adverse to result in constraints on operations, and no constraints have been placed on operations. For these two species, performance related to incidental take criteria, as anticipated, has instead improved under the COP relative to the 2016–2020 baseline, with the exception of the recession rate criterion for kites.

Results for the Cape Sable seaside sparrow present a very different picture. Simulations of COP performance predicted that the hydrologic targets for the sparrow will not be met (USACE, 2020a), and monitoring confirms this prediction. If anything, performance has declined under the COP relative to baseline (Table 2-5). The evidence is abundant that the COP is reducing short hydroperiod wetlands (see above; see also NASEM, 2023). That the acres of short hydroperiod wetlands within the core foraging areas of stork colonies experiencing long hydroperiods exceeded the target by an order of magnitude in each of the first 2 years of the COP (USACE, 2023g) adds to this evidence. Because of these changes, sparrows are losing habitat in the areas they currently occupy, and to date, they have not colonized new suitable habitat created by the COP. Data as of July 2024 indicate that their estimated population size has fallen below the threshold for incidental take (2,387 sparrows) (M. Meyer, FWS, personal communication, 2024). The future of the sparrow under the COP and subsequently the CEPP is an issue that requires creative solutions, sooner rather than later.

Water quality responses. The 2023 COP Biennial Report highlighted ongoing water quality concerns, noting an exceedance in the 12-month average phosphorus concentration for inflows to Shark River Slough during WY 2019. Since then, exceedances have occurred in WY 2021–2023 (Qiu, 2024a,b).

Uncertainty #16b in the COP adaptive management plan (USACE, 2020a) specifically addressed water quality changes due to COP operations:

> **Water Quality in Northeast Shark River Slough:** Will there be downstream biogeochemical effects associated with modifying inflows and hydrologic conditions in [Everglades National Park] ENP, that result in detrimental effects on nutrient movement, availability, and ecological responses?

During the reporting period, monitoring data were analyzed to investigate the potential to reduce concentrations by shifting flows from the S-333 structure to the S-333N structure and/or making adjustments to the S-12 structures, but the analysis showed that such changes would not reduce total phosphorus (TP) concentrations. Concerns were raised about water quality resulting from S-333 operations, but during the COP biennial reporting period, management options were not implemented. Water quality issues associated with increased restoration flows are discussed in more detail later in the chapter.

Summary of COP results. The initial monitoring results indicate that the COP has met expectations in changing the quantity, distribution, and seasonal patterns of flow and improved conditions relative to the baseline period. Initial results also indicate that these hydrologic changes are producing some of the predicted broader ecological responses that are COP objectives, as illustrated by changes in plant communities and risk of soil oxidation. The COP has achieved much less success with more complex objectives that involve creating fairly precise hydrologic conditions over extended periods of time, as illustrated by recession rates and water levels during the nesting season for threatened and endangered species. Thus, the COP is proving to be what it was intended to be—not a complete solution but rather the first major step toward restoring the central heart of the Everglades. Additional efforts will likely be needed to sustain the Cape Sable seaside sparrow, including but not limited to translocation. The first COP Biennial Report is an effective means of conveying progress toward restoration that should be emulated through the remaining years of the COP and beyond.

CEPP 1.0 operational plan. Lessons learned from the COP can be applied to the next system operating plan, CEPP 1.0, which is currently in planning to adapt operations to recently completed or soon-to-be-completed CERP features. CEPP 1.0 will incorporate CEPP components scheduled for completion by 2025, including CEPP New Water and several CEPP South features.[3] Several non-CERP projects will also be included for the development of this regional operations

---

[3] These features include the S-333N and S-355W Spillways; the S-631, S-632, and S-633 culverts through the L-67A levee; and the S-152 culvert and L-67 levee gap from the Decomp Physical Model.

plan, including the Tamiami Trail Next Steps Phase 2 project (Box 2-1), the 2.3-mile seepage barrier adjacent to the 8.5 Square Mile Area constructed by the SFWMD, the 5.0-mile CEPP New Water seepage barrier (Figure 2-13), and LOSOM (Figure 2-18). Crucially, CEPP 1.0 will raise the constraint on water levels in the L-29 Canal from 8.5 up to 9.7 feet, enabling movement of even higher flows of freshwater through the central Everglades and into Everglades National Park. Thus, CEPP 1.0 represents a major increment in the restoration of historic flow volumes and distribution beyond that achieved by the COP. Currently alternatives are being developed and compared, with the expectation that CEPP 1.0 will be implemented in 2026 (USACE and SFWMD, 2024).

Authorized CERP Projects Not Yet Constructed or Fully Operating

Six CERP projects are actively under construction, of which four—the C-43 Reservoir, IRL-South, the C-11 Impoundment, and the CEPP (including CEPP South, CEPP North, and CEPP EAA)—are too early in their construction to have documented restoration benefits. Additionally, the Loxahatchee River Watershed project has been authorized, but construction has not yet begun. These projects are described briefly here to understand the context of implementation progress.

C-43 Reservoir

Early in the 20th century, the course of the Caloosahatchee River was deepened and straightened, and canals were excavated in the river basin that connected the river to Lake Okeechobee and drained agricultural lands and urban areas. As a result, during prolonged dry periods, freshwater flow to the estuary is greatly reduced, and saline water can migrate far up the river, killing beds of freshwater submerged plants. Conversely, during periods of heavy rainfall, large volumes of nutrient- and sediment-rich freshwater are transported into the Caloosahatchee River estuary, affecting habitat quality for seagrasses, oysters, and other aquatic organisms. The Caloosahatchee River (C-43) West Basin Storage Reservoir (Figure 2-2, No. 8) is a CERP project designed to impound up to 170,000 AF of stormwater runoff from the C-43 drainage basin or from Lake Okeechobee during wet periods (USACE and SFWMD, 2010), hence protecting the estuary from excessive freshwater. During dry periods, this stored water can be released to supplement low river flows to maintain optimal salinity levels in the estuary. Additionally, the SFWMD is planning to construct a non-CERP in-reservoir alum treatment system to reduce phosphorus loading to the Caloosahatchee River estuary to address concerns that storage of high phosphorus water in the reservoir could contribute to algal blooms (J-Tech, 2020). Reservoir construction has been delayed by the need to replace the contractor (A.B. Williams, 2023), although completion is still anticipated in December 2025 (USACE, 2024d), and the SFWMD expects to begin filling the reservoir in late 2024 (Parrott, 2024). It is too soon to realize natural system benefits from this project.

Indian River Lagoon-South

The Indian River Lagoon and the St. Lucie Estuary are biologically diverse estuaries located on the east side of the Florida Peninsula, where ecosystems have been impacted by factors similar to those that have impacted the Caloosahatchee River Estuary—surges of freshwater from Lake Okeechobee and canals in the watershed and nutrient-rich runoff from farmlands and urban areas (USACE, 2022d). The IRL-South Project (Figure 2-2, Nos. 3 and 4) is designed to reverse this damage through improved water management, including the 50,600-AF C-44 storage reservoir, three additional reservoirs (C-23/C-24 South, C-23/C-24 North, and C-25) with a total of 97,000 AF of storage, three new STAs (C-44, C-23/24, C-25), dredging of the St. Lucie River to remove 7.9 million cubic yards of muck, and restoration of 53,000 acres of wetlands (Figure 2-22). The project also involves the restoration of nearly 900 acres of oyster habitats and the creation of 90 acres of artificial habitat for oysters and SAV (USACE, 2022d). Construction was completed on the C-44 STA and C-44 Reservoir in March and December 2021, respectively (USACE, 2024e). The reservoir filling was initiated in 2022, operational testing of the reservoir is ongoing as of mid-2024, and work is under way to repair seepage issues that occur when water depths exceed 10 feet (Booth, 2023). The USACE estimates a 2026 completion of the repairs (Williams, 2024).

Construction of the C-23/24 STA began in 2022, the contracts for construction of the C-23/C-24 North and South Reservoirs are anticipated to be awarded in FY 2024, and the construction award for the C-25 Reservoir is expected in late 2024 (Figure 2-22; USACE, 2024e). Because newly completed features remain in the early stages of operational testing, there is no natural system restoration progress to report.

C-11 Impoundment

The C-11 Impoundment is a major component of the Broward County WPAs project (Table 2-1; Figure 2-2, No. 9), which is designed to reduce seepage from WCA-3. The C-11 Impoundment is expected to have approximately 4,600 AF of storage and is designed to capture urban stormwater runoff from the western C-11 basin and associated nutrient loads, which previously had been pumped into WCA-3 via the S-9 pump station. This stored water is expected to help reduce seepage from WCA-3 while also increasing groundwater recharge

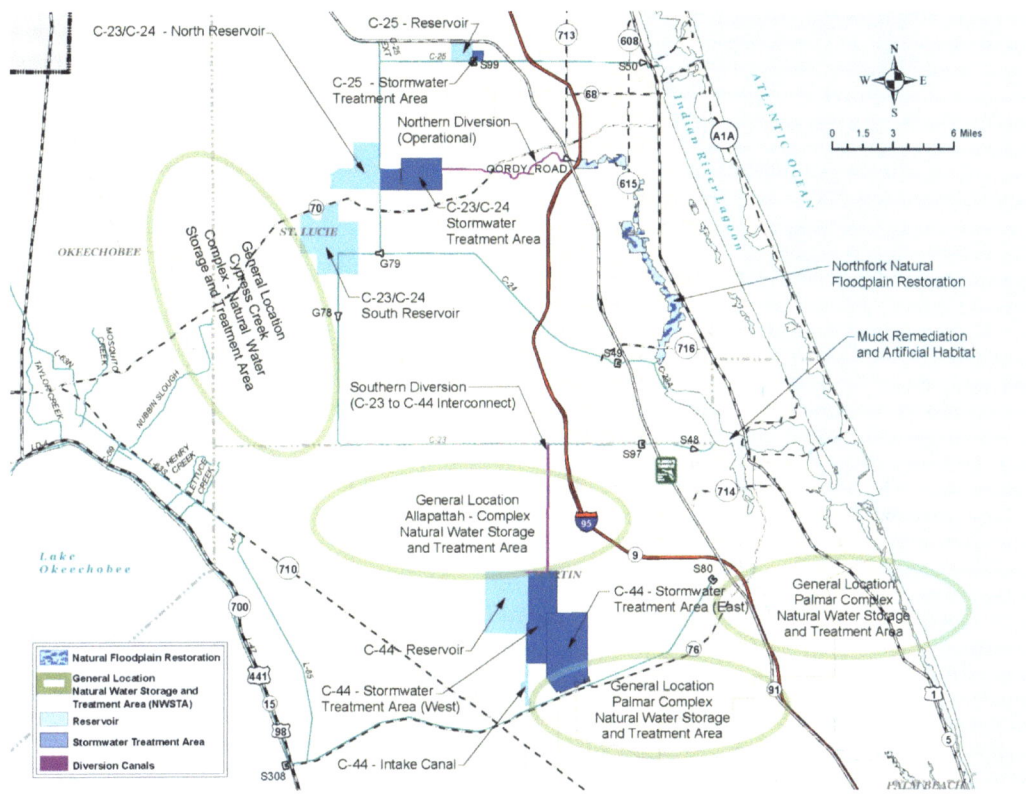

FIGURE 2-22 Features of IRL-South project.

SOURCE: USACE, 2022d.

near the feature, thereby enhancing municipal and agricultural water supply and reducing saltwater intrusion (USACE and SFWMD, 2012). This feature was listed as a dependency for CEPP construction in USACE and SFWMD (2014). Construction began in 2017 on an initial Mitigation Area A berm, and a new contract award for continued construction is expected in September 2024 (G. Ralph, USACE, personal communication, 2024). No natural system restoration benefits are expected from construction to date.

CEPP North

CEPP North (Figure 2-2, No. 10) is designed to improve the distribution of flows into northern WCA-3, which has long been subject to overly dry conditions, to restore its hydrology and ecology. It will also hydrate WCA-2 under high flow conditions. CEPP North includes backfilling of the Miami Canal, as well as several projects designed to improve conveyance (Figure 2-12; USACE and SFWMD, 2014). CEPP North remains in the design phase, and improvements in design based on new information continue to be made (see Chapter 5). As of May 2024, construction was under way on the S-620 gated culvert and the S-8 pump station modifications, and CEPP North is estimated to be completed in 2030 (USACE and SFWMD, 2024).

CEPP EAA

CEPP EAA consists of construction of the EAA (A-2) Reservoir and adjacent A-2 STA (Figure 2-2, No. 12), and several components related to the operation of these new features, such as construction of an inflow pump station and seepage and inflow/outflow canals (Figure 2-12). The objectives of CEPP EAA are to store new water and treat it before moving it south, with projected increases in average annual inflows to the remnant Everglades of 370,000 AF (USACE, 2020b). The principal component of CEPP EAA—the 23-foot-deep, 240,000-AF reservoir—is expected to be completed in 2030. As of May 2024, some features of the A-2 Reservoir are currently under construction, including seepage canal segments and the foundation and seepage cutoff wall, and others are in design or procurement. The A-2 STA is well along in construction and is expected to be completed by December 2024 (USACE and SFWMD, 2024). The timely delivery of the intended benefits of the EAA Reservoir is dependent on the performance of both the existing STAs and the A-2 STA (NASEM, 2023).

CEPP South

The objectives of CEPP South (Figure 2-2, No. 11) are to remove barriers to sheet flow in southern WCA-3A and increase capacity to move more water south (USACE and SFWMD, 2014). It has been progressing at a good pace, and as of May 2024, two components of the project have been completed (S-333N gated spillway, Old Tamiami Trail roadway removal); construction contracts are under way covering construction of three 500-cfs gated structures in the L-67A levee, backfill of the agricultural ditch in WCA-3B, and spoil removal (Figure 2-12). The 2023 IDS estimates completion of most features, including the Blue Shanty Levee, by 2031 if projected funding needs are met.

A more significant issue for the CEPP that may impact restoration progress involves potential ecological impacts from changing phosphorus dynamics associated with increased flows. These issues and their potential causes and solutions are discussed in detail below (see Issues That May Impact Progress: Water Quality later in this chapter), and possible modifications to CEPP South to address these issues are discussed in Chapter 5.

Loxahatchee River Watershed

Alterations of the Loxahatchee River system and watershed over the past century, including dredging, channelization, and drainage, have substantially altered flows in the watershed and have reduced natural water storage of excess waters, resulting in periods of either excessive or limited flows to the Loxahatchee River Estuary. The resulting changes in natural land cover, including up-river migration of mangrove and the displacement of cypress, raised concern, especially in the area designated as a Wild and Scenic River (FDEP, 2010). The Loxahatchee River Watershed Restoration Project (Figure 2-2, No. 14), authorized in WRDA 2020, seeks to capture, store, and redistribute freshwater currently lost to tide; rehydrate natural areas in the headwaters; reduce peak discharges to the estuary; and improve the resilience of estuarine habitats by altering the timing and distribution of water from upstream. Planned components of the project include wetland restoration and hydrologic improvements within the watershed, a single 9,500-AF impoundment, four aquifer storage and recovery (ASR) wells, and several structures related to connectivity in the southern part of the watershed. Together the project components are expected to deliver 98 percent of the wet season restoration flow target and 91 percent of the dry season restoration flow target in the Northwest Fork of the Loxahatchee River (USACE and SFWMD, 2020a). In turn, these flows are expected to limit saltwater penetration in the estuary, conserve the remaining cypress, and promote the recovery of freshwater aquatic vegetation (e.g., *Vallisneria*) and other habitats important for estuarine species such as manatee and oysters. Design is ongoing and construction has not yet begun.

CERP Projects in Planning

Western Everglades Restoration Project

WERP (Figure 2-2, No. 16) is one of the few CERP projects that address ecological degradation in the western Everglades. WERP is intended to reestablish ecological connectivity of wetland and upland habitats, restore hydroperiods and predrainage distributions of sheet flow, restore low-nutrient conditions to reestablish native vegetation, and promote ecosystem resilience. The project footprint covers more than 1,200 mi^2 of the western basin and includes Big Cypress National Preserve and lands of the Florida Seminole and Miccosukee Tribes (Figure 2-23). Planning for WERP started in August 2016 following discussions

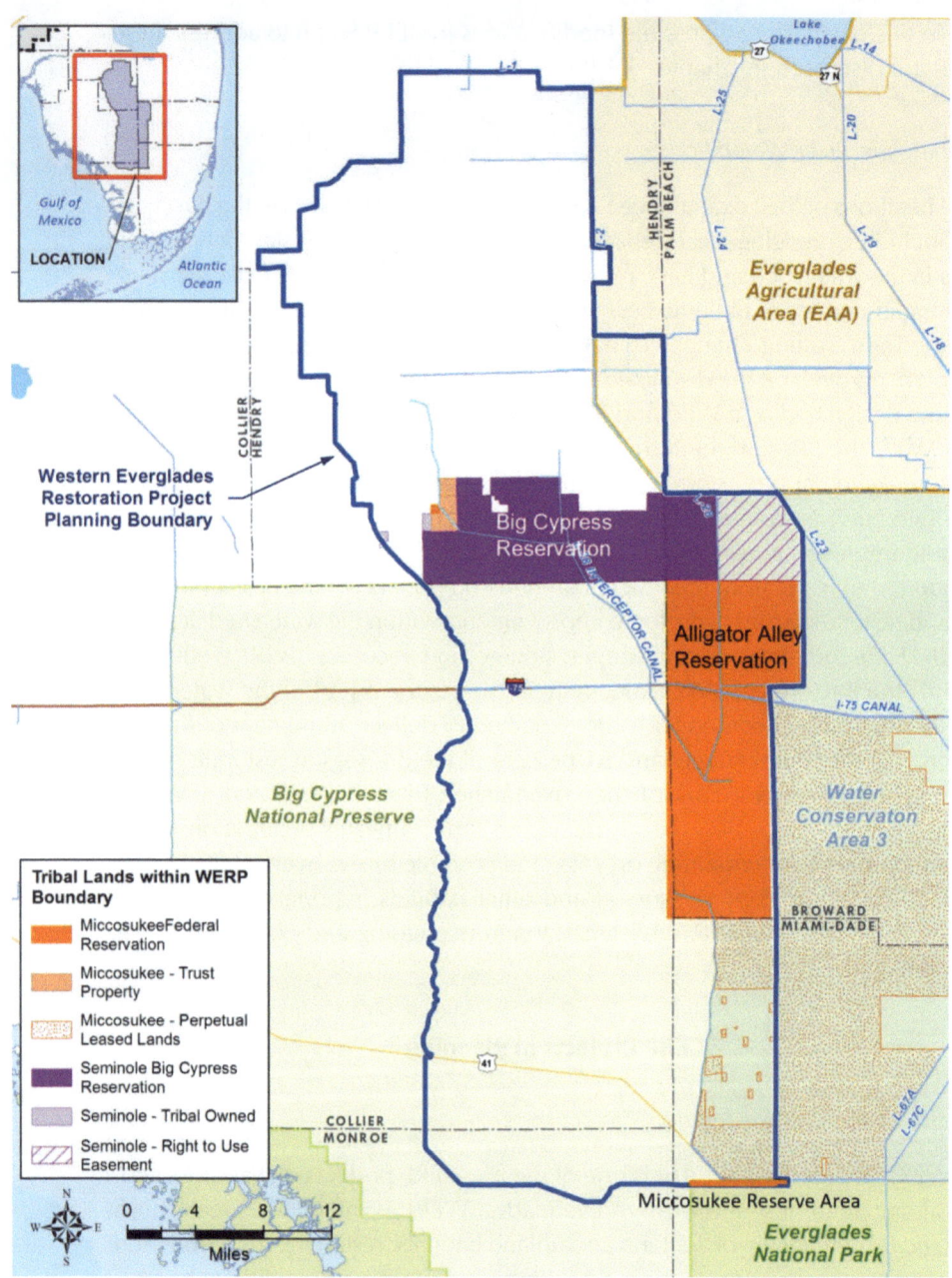

FIGURE 2-23 WERP project area and affected Seminole and Miccosukee Tribal lands.

SOURCE: USACE and SFWMD, 2023a.

by a subgroup of the South Florida Ecosystem Restoration Task Force (Task Force) formed in response to concerns raised by the Seminole Tribe about lack of restoration progress in the western Everglades.

The planning process has been challenging. The initial preferred alternative—Alternative H (Alt-H)— was deemed not cost-effective. The high project costs, together with unmet need for an extensive real estate takings analysis, led to the suspension of WERP planning in 2019. Through strong stakeholder support, the planning was restarted and the project re-scoped, resulting in the Hybrid Revised Alternative (Alt-Hr), which included two STAs that were to cover nearly 7,500 acres and treat runoff entering the northern portion of the project area. Stakeholder objections to the placement of one STA in an area of remnant cypress led to further project revisions. Alternative Hybrid Natural Flow Revised (Alt-HNFR) was ultimately identified as the tentatively selected plan, and the draft PIR was released in December 2023 (USACE and SFWMD, 2023a). The recommended WERP plan is shown in Figure 2-24, although as of August 2024, the final draft PIR was not publicly available. Therefore, the following information is largely derived from the December 2023 draft PIR.

Within the northern portions of the recommended plan (Figure 2-24), several canals will be modified, and a variety of other structures including a weir, spreader canals, and culverts are designed to direct flows into the Seminole Big Cypress Reserve Native Area—an area of great cultural importance to the Seminole Tribe—and the Big Cypress National Preserve to rehydrate these areas. Better hydration in the Big Cypress National Preserve will help reduce the risk and severity of wildfires in the preserve. Further work is planned to restore vegetation in WCA-3A in an area currently dominated by cattail due to water quality impacts near the intersection of the L-28i and L-28N, within the Miccosukee Alligator Alley Reservation. The recommended plan (Figure 2-24) includes one STA to provide treatment for flows from the North Feeder subbasin and areas within the C-139 annex basin, with the objective of improving water quality conditions in the L-28N Canal. Water quality concerns in the West Feeder Basin (where an STA was eliminated in Alt-HNFR) will be addressed using a source-control approach through non-CERP components. Natural removal processes within the Kissimmee Billie Slough are also expected to reduce phosphorus concentrations. Upstream phosphorus levels must be addressed so that WERP does "not cause or contribute to water quality violations in Big Cypress National Preserve, the Seminole Tribe of Florida Reservation or other areas as determined by the regulatory agencies" (USACE and SFWMD, 2023a).

In the southern portion of the project area, portions of the L-28 Tieback and L-28S Canals will be filled; gated control structures will be built on L-28S to increase exchange between Big Cypress and WCA-3A; and culverts will be added beneath 11-Mile Road, US-41, and the Loop Road to enhance hydrologic

FIGURE 2-24 WERP tentatively selected plan.

SOURCE: https://www.saj.usace.army.mil/WERP, accessed August 29, 2024.

connectivity and flows to the southwest, while protecting Tribal tree islands and camps. Overall, WERP features, if implemented as planned, should improve hydration, hydrologic and ecological connectivity, and water quality, which have been longstanding concerns in the WERP study area. The WERP Project Delivery Team has targeted September 2024 for a Chief's Report, with the goal of authorization in WRDA 2024 (Velez, 2024).

The Adaptive Management Plan for WERP (Annex D of USACE and SFWMD, 2023a) identifies nine key uncertainties, monitoring plans, trigger points and or thresholds for management action, and potential management actions (see also Chapter 5). Continuous monitoring and assessment will enable a comparison between what is expected and actual system response, and in turn support an iterative process to guide subsequent management decisions and any needed improvements or adjustment to the current plan. In this regard, it is important that the WERP project team works closely with Miccosukee and Seminole Tribes to continue to refine performance measures and targets/trigger points based on Indigenous Knowledge (see also Chapter 3).

Water quality and WERP implementation dependencies. The 2023 Draft PIR (USACE and SFWMD, 2023a) states that implementation of WERP and its project dependencies will occur in stages over approximately 10 years following authorization, starting with implementation of non-CERP best management practices (BMPs) on privately owned lands in contributing areas. The timely implementation of several WERP features, including modification of the Lard Can and Wingate Mill Canals, backfilling the L-28i and L-28i extension canals, and construction of the L-28i weir and plug, depends on achievement of water quality objectives. In the draft PIR, a "numeric planning placeholder" of 17 µg/L TP was used "for the area influenced by the WERP . . . West Feeder Project components," assessed areawide as an annual geometric mean that must be met at least 2 of every 3 years. The PIR estimates that concentrations of 31–34 µg/L TP at the West Weir (Figure 2-24) would be sufficient to reach this 17 µg/L TP placeholder target in Big Cypress National Preserve, although additional studies are ongoing to establish baseline conditions, investigate appropriate targets, and determine the probable natural treatment likely to be afforded as the water travels through Kissimmee Billie Slough. By comparison, annual flow-weighted mean TP concentrations over the past 10 years at the West Weir have generally ranged between 40 µg/L and 90 µg/L (Figure 2-25; Wang and Mahmoudi, 2024). The proposed implementation schedule estimates a 7-year time frame for non-CERP water quality initiatives to meet these levels (USACE and SFWMD, 2023a).

In essence, much of the implementation of WERP features depends upon achievement of water quality objectives, with a 3- to 7-year time frame for non-WERP efforts to achieve their objectives before WERP implementation is delayed,

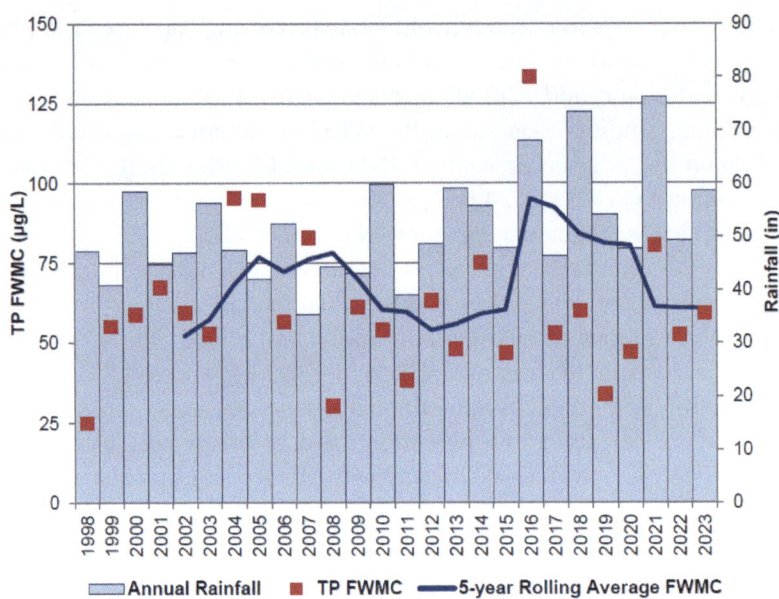

FIGURE 2-25 Annual flow-weighted mean TP concentrations for WY 1998–2023 and 5-year rolling averages at the West Wier.

SOURCE: Wang and Mahmoudi, 2024.

---

depending on the WERP feature in question. An inherent assumption of this component is that BMP implementation will be accomplished as envisioned and as scheduled. Julian and Davis (2024) show potential reductions of 25-51 percent in flow-weighted mean TP concentrations associated with BMP implementation in the region compared to up to 73 percent reduction if an STA were included. However, the effectiveness of BMPs can vary substantially based on a variety of factors and with time (Ator et al., 2020; Gitau et al., 2005). Furthermore, BMP implementation often relies on a voluntary process, although Florida has regulatory processes that could be implemented if necessary. Initial voluntary and regulatory efforts will be reviewed approximately 3 years into the project with the plan to refine existing source control projects, initiate new projects, and/or move toward mandatory source control, if needed. A mandatory state source control program would incorporate water quality monitoring, data collection and analysis protocols, and compliance inspection, in addition to BMP selection, and its development is expected to take 3 to 5 years. Because of this dependence on the BMP component, there is the need to closely monitor BMP implementation and functionality, and consider alternate solutions as needed, to ensure timely WERP implementation.

Operation of gated culvert S-223 on Kissimmee Billie Slough. As part of WERP features in Region 2 (Figure 2-24), a 180-cfs culvert (S-223) will be constructed and used to direct and control flows from the Kissimmee Billie Slough into the Seminole Big Cypress Reservation Native Area. The Seminole Tribe requested inclusion of gates in this structure, as well as to control structure operations using Indigenous Knowledge (Osceola, 2022; see also Chapter 3). Osceola (2022) explained:

> Allowing Seminole Tribe ITEK [Indigenous Traditional Ecological Knowledge] to govern the operation of the gates will ensure that WERP provides environmental benefits in the Native Area. If the Seminole Tribe determines the water is fit to introduce to the Native Area, opening the gates will provide much needed restorative hydration. Alternatively, if the Seminole Tribe determines the water is unfit for the Native Area, closing the gates will provide environmental benefits by preventing potential harm to the Native Area.

Thus, the success of this part of this part of WERP will depend on water quality in the Western Basin. The WERP adaptive management plan includes a review of gate operations if the desired conditions are not met considering prevailing gate operations.[4]

Other issues that may impact progress. There is concern that WERP-related increases in water levels will negatively impact areas occupied by endangered and threatened species, particularly the Florida panther (*Puma concolor coryi*). Although it historically occupied a vast area of the southern United States, stretching from Arkansas to the Carolinas, the panther now only exists as a single breeding/reproductive population, located south of the Caloosahatchee River. However, there has been recent documentation of panther activity north of the Caloosahatchee River, which could potentially signify population expansion north of the river (FWS, 2020). According to the draft PIR (USACE and SFWMD, 2023a), the North Feeder STA (Figure 2-24) will increase water levels in areas that are important to the Florida panther, including currently occupied areas and contiguous areas necessary for its long-term viability and persistence. The USACE Biological Assessment included in the draft PIR concluded that the project will not intersect any travel corridors and will, thus, not limit panther movement between areas south and north of the Caloosahatchee River. The 2024 Biological Opinion (FWS, 2024) outlined a plan for the USACE to use habitat units from the CERP Picayune Strand Panther Conservation Bank to offset losses due to the North Feeder STA with an anticipated 24,831 units needed. Concerns have also been raised about potential impacts on the panther's primary prey, the white-tailed deer (*Odocoileus virginianus*) (FWS, 2020), whose survival is greatly impacted by water levels in the wet prairies (Bled et al., 2022;

---

[4] A correction was made after the report was released to clarify the Seminole Tribe's approach to gate operations.

MacDonald-Beyers and Labisky, 2005). MacDonald-Beyers and Labisky (2005) recommend that water rise no higher than 0.5 m (1.6 ft) to reduce deer mortality and improve its viability. Expected water levels fall within this recommended threshold. Analyses in the Biological Assessment (USACE and SFWMD, 2023a, Annex A) show that concerns about potential water level increases in the Wingate Mill and Lard Can Canal areas due to proposed canal modifications have been alleviated, and no adverse impacts are expected to the Florida Panther Conservation Bank as a result of WERP.

Risk of increased flooding due to WERP has been identified in the Feeder Canal and L-28 Gap basins. The Savings Clause (WRDA 2000, Section 601) requires that CERP projects maintain existing levels of flood protection (relative to time of CERP enactment) or work with affected landowners to mitigate the impacts of flooding. For western areas along L-28i, upstream of canal modifications, simulated ponding depths show an up to 70 percent probability of exceedance in wet years attributable to WERP. Model results suggest increased risk of flooding in areas immediately west of L-28i, including the Looneyville community (USACE and SFWMD, 2023a). Unless the plan is modified, this increase in flooding would necessitate solutions such as the raising of structures, perpetual flowage easements, or land acquisitions for impacted privately owned properties. More localized assessments will be conducted in the pre-construction engineering and design phase to better delineate areas and structures that might be impacted.

Lake Okeechobee Watershed Restoration Project

Located north of the lake, the LOWRP was designed to capture, store, and redistribute water entering the northern part of Lake Okeechobee. Its goals are to "improve water levels in Lake Okeechobee; improve the quantity and timing of discharges to the St. Lucie and Caloosahatchee estuaries; improve water supply for existing legal users of the Lake Okeechobee Service Area (LOSA); and increase the spatial extent and functionality of wetlands" (USACE, 2023b).

LOWRP planning has been actively ongoing since 2016. In November 2020 a revised final PIR was released with a draft report of the Chief of Engineers proposing a 46,000-AF above-ground water storage feature (termed "wetland attenuation feature"), 80 ASR wells, and approximately 4,800 acres of wetland restoration.[5] Under a new administration, this plan was not approved by USACE headquarters "due in part to concerns raised by the Seminole Tribe of Florida" (USACE, 2024f) (see Chapter 3); the project was subsequently revised to remove the wetland attenuation feature and reduce the number of ASR wells

---

[5] This sentence was updated after the report was released to accurately reflect the approximate acreage of wetland restoration.

to 55 (USACE and SFWMD, 2022a), but the wetland restoration component was retained. The 2022 PIR was not approved by USACE headquarters "due to concerns with risks posed by the ASR system and the increase in estimated costs" (USACE, 2024f). The planning team was then advised to reconfigure the tentatively selected plan to consider other storage alternatives, including those previously screened out. The team was also asked to "identify and investigate the feasibility of measures to quantify and mitigate potential health and environmental risks posed by ASR wells" and "specifically focus on determining uncertainties related to (1) well recovery performance, (2) water quality, (3) construction cost, and (4) O&M [operating and maintenance] costs" in coordination with the USACE Engineering Research and Design Center (USACE, 2024f).

In response, as of January 2024, the project was progressing on three fronts: (1) advancement of a draft PIR to authorize only the wetland restoration features of the original tentatively selected plan, (2) ongoing scientific research on ASR by both the USACE and the SFWMD, and (3) independent rescoping of the North of Lake Okeechobee Storage Reservoir by the SFWMD via the Section 203 process. Progress on these three fronts is discussed below.

Wetland restoration features. In terms of progress on the first front, as of April 2024, no new revised LOWRP PIR has been publicly released that focuses only on the wetland restoration features. Presentations to the committee in January 2024 (Vega-Liriano, 2024) noted that the project team was modeling project benefits from ASR as well as the wetland restoration features for the PIR, even though the authorization would cover only the wetland restoration features. No analysis was being conducted to separate the benefits of the project component seeking authorization. As such, it may be difficult for Congress to evaluate the benefits of the wetland restoration features seeking authorization, especially if USACE headquarters has requested additional research to address ASR uncertainties before it would support full implementation of ASR. The USACE reported it was aiming for WRDA authorization in 2026 (G. Ralph, USACE, personal communication, 2024).[6]

Research to resolve concerns regarding ASR in the LOWRP. Both the USACE and the SFWMD are conducting research on ASR to resolve critical uncertainties and inform the design of this storage component north of Lake Okeechobee. The 2021 ASR Science Plan (USACE and SFWMD, 2021) outlines a program of studies intended to inform phased implementation of 55 ASR wells within the LOWRP. This plan provides a comprehensive accounting of the numerous

---

[6] The preceding paragraphs were updated after the report was released to accurately reflect the status of activities and plans at the time that the report was written.

uncertainties identified by NRC (2015) and summarizes 26 studies involving geochemical measurements, hydrogeophysical characterization, laboratory experiments, field testing of reactivated ASR wells, and clusters of new ASR wells to be located along the northern perimeter of Lake Okeechobee. The studies described in the 2021 ASR Science Plan are expected to be completed in 2030.

The 2022 ASR Science Plan (USACE and SFWMD, 2022b), developed under the guidance of an independent peer-review panel, is intended to serve as an update of the 2021 ASR Plan and was released in draft form in October 2022. The 2022 ASR Science Plan is organized around the following principal uncertainties identified in the 2015 NRC report: (1) project sequencing, (2) future construction and testing to evaluate aquifer properties, (3) understanding phosphorus reduction potential, (4) operations to maximize stored-water recovery, (5) disinfection and treatment of recharge and recovered waters, (6) ecotoxicology and ecological risk assessment, (7) water quality, and (8) ASR cost-benefit analysis (USACE and SFWMD, 2022b). Progress on the 2022 ASR Science Plan is discussed in Box 2-2.

North of Lake Okeechobee Storage Reservoir Section 203 study. In a rapid planning effort started in April 2023 and led by the SFWMD, three above-ground storage alternatives, including previous and newly developed alternatives, were analyzed. By October 2023, a tentatively selected plan was identified for the North of Lake Okeechobee Storage Reservoir (also known as the Lake Okeechobee Component A Reservoir [LOCAR]). By February 2024, the SFWMD released the final feasibility study (SFWMD, 2024b), and the USACE released the final EIS (USACE, 2024c). The plan proposes a deep reservoir (average depth = 18 ft) that would add 200,000 AF of above-ground storage north of Lake Okeechobee. The reservoir is adapted from a design that was previously screened out in the LOWRP process (for more details, see Box 3-4; USACE and SFWMD, 2020b). The reservoir would contribute much needed storage to the CERP as described by the University of Florida Water Institute (2015) and NASEM (2016) at an estimated cost of $3.5 billion (SFWMD, 2024b).

Modeling for the project suggests that LOCAR will increase the percent of time that Lake Okeechobee spends within the preferred stage envelope (11.5–15.5 ft NGVD29) and the number of events of optimal flows to the Caloosahatchee and St. Lucie Estuaries. The analysis also showed an increase in the percent of time (to a small degree) when lake water levels are below preferred levels, potentially causing harm to the lake's littoral zone. However, the comparisons against existing conditions can be difficult to interpret because the existing conditions baseline (ECB) and alternatives are modeled using the most recent LOSOM while the future without (FWO) uses the 2008 Lake Okeechobee Regulation Schedule (LORS); thus, the results are not solely demonstrating the

> **BOX 2-2**
> **Progress on the 2022 ASR Science Plan**
>
> Progress documented within the 2022 ASR Science Plan includes collection of two continuous cores and vertically profiled groundwater samples under way as of October 2022. Analyses of the cores and groundwater are supporting several objectives, including the provision of information needed to develop and parameterize local-scale groundwater flow and solute transport models. These models will be used to inform decisions on well-cluster configuration and cycle testing, as well as improve understanding of water quality responses that occur as recharge water interacts with bedrock and mixes with native groundwater. Other progress has involved evaluation of the L-63N ASR facility for reactivation and refurbishment of the Kissimmee River ASR facility, as well as construction of new exploratory ASR and monitoring wells at C-38S and C-38N. While the L-63N and Kissimmee River ASRs are single-well systems, those constructed at C-38S and C-38N were completed as multi-well clusters. Once operational, hydrological observations from these ASR clusters can be used in conjunction with flow and transport models to evaluate alternative pumping scenarios with respect to their efficacy in maximizing freshwater-recovery efficiencies.
>
> In addition to ASR well refurbishment and construction, a scope of work was developed for flow-through column experiments with portions of cores collected from C-38S and C-38N. The purpose of these experiments is to evaluate biofilm growth and potential for aquifer clogging that may arise from introduction of phosphorus-enriched surface waters. Other work during 2021-2022 focused on evaluation of alternative disinfection technologies. Disinfection is needed to comply with Florida Department of Environmental Protection and Environmental Protection Agency rules that require injected ASR waters to meet primary drinking-water standards. The report describing findings from five pilot-scale treatment trains is forthcoming and will inform selection of the treatment process employed in the initial ASR well clusters. Some attention has also been devoted to lowering uncertainties associated with ecotoxicology and ecological risk, although progress here has been limited to development of a scope of work for a quantitative ecological risk assessment, design of mobile-lab facilities to enable execution of bioaccumulation tests, and planning of ecological monitoring during cycle testing of new ASR wells along the C-38 Canal.
>
> The USACE Engineering Research and Design Center has also been engaged to conduct research on ASR mobilization of hazardous contaminants, as well as develop cost estimates for construction and long-term operation and maintenance.
>
> SOURCES: Caneja, 2024; USACE and SFWMD, 2022b.

effects of additional storage, but rather they illustrate the effects of storage addition coupled with Lake Okeechobee operational changes. Overall, however, the additional storage in the system should have positive effects, assuming reasonable operations (Julian and Reidenbach, 2024). Future studies will need to identify the system operations that work best with LOCAR as it comes online, similar to the LOSOM planning study.

Biscayne Bay and Southeastern Everglades Restoration

The freshwater and coastal wetlands and subtidal habitats of Biscayne Bay and the southeastern Everglades have been affected by over-drainage and by damaging freshwater releases from canals. Specific problems that restoration goals aim to address include salinization and invasive plant encroachment of freshwater wetlands; reduction in near-shore estuarine habitat quality for aquatic organisms; degraded habitat near canal release points due to unnatural pulsed releases; and expansion of the "White Zone," a zone of white marl soil, high soil saltiness, and low productivity between the freshwater and saline wetlands. The specific project objectives (USACE and SFWMD, 2020c) are to do the following:

1. Improve freshwater wetland water depth, ponding duration, and flow timing within the Model Lands, Southern Glades, and eastern panhandle of Everglades National Park to maintain and improve habitat;
2. Improve the quantity, timing, and distribution of freshwater to estuarine and near-shore subtidal areas, including mangrove and seagrass areas of Biscayne National Park, Card Sound, and Barnes Sound, to improve salinity regimes and to reduce damaging pulse releases;
3. Improve ecological and hydrological connectivity between Biscayne Bay coastal wetlands, the Model Lands, and Southern Glades; and
4. Increase resilience of coastal habitats in southeastern Miami-Dade County to sea-level rise.

The CERP planning constraints include not reducing flood protection (Savings Clause) while maintaining water deliveries to Everglades National Park, including Taylor Slough and Florida Bay.

The broad vision for the project included sourcing available water in the northern portion of the project footprint, conveying water through existing or new canals southeasterly through the center of the project area, and redistributing water to Biscayne Bay and southeastern Everglades wetlands (Figure 2-26). The project team has developed an array of project alternatives and as of May 2024 is in the process of comparing the outcomes with respect to nine performance measures,[7] including one that considers coastal ecosystem resilience to sea-level rise (see Chapter 4, Box 4-1). The second and third rounds of alternative evaluations include an intermediate sea-level rise scenario of 1.6 feet over 50 years (USACE, 2019).

As of May 2024, the Project Delivery Team was completing its evaluation of Round 3 Alternatives, and a tentatively selected plan is expected for summer 2024. Key features for a tentatively selected plan include water sourcing from

---

[6] The BBSEER performance measures are (1) near-shore salinity, (2) wetland porewater salinity, (3) water depth, (4) hydroperiod, (5) vegetation, (6) direct canal releases, (7) timing and distribution of flow sources to Biscayne Bay, (8) adaptive foundational resilience (see Chapter 5), and (9) connectivity between habitats (spatial extent of mangroves).

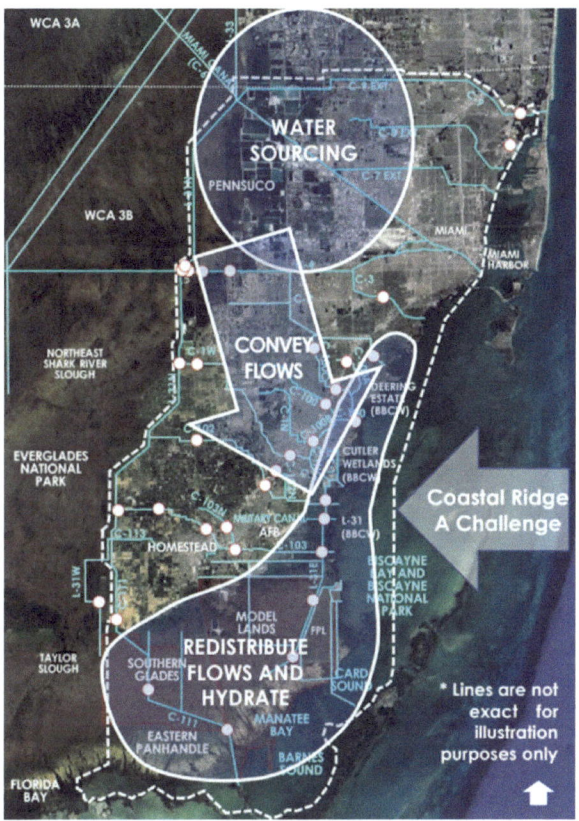

FIGURE 2-26 BBSEER hydrologic flow objectives.

SOURCE: USACE and SFWMD, 2023b.

---

the northwest area, optimization of water volumes across the coastal ridge, maximum conveyance flexibility, and storage of enough water to rehydrate the largest extent of BBSEER's target footprint. Because the analyses are ongoing, the committee did not review outcomes or the alternative selection process in detail.

NASEM (2023) noted that several of the performance measures are closely related to each other (e.g., near-shore salinity, wetland salinity, freshwater depth, hydroperiod, and timing and distribution of flow), such that the relative impact of other performance measures such as resilience and connectivity is likely diluted. The committee also recommended a nuanced approach to comparing alternatives including evaluating performance measures for each habitat type, evaluating trade-offs among habitats, and weighting performance measures based on relative importance (NASEM, 2023). In subsequent analyses, alternatives have been evaluated by indicator regions and zone, which enable a comparison of impact among geographic areas and habitat types. However, it is not evident that performance measure weighting has been incorporated into the evaluation process. At this critical stage of selecting a tentatively selected plan, consideration of these recommendations may help the evaluation of alternatives that will improve ecological outcomes. Additional discussion on the analysis of sediment accretion in the BBSEER planning process is discussed in Chapter 4.

ISSUES THAT MAY IMPACT PROGRESS: WATER QUALITY

As discussed in NASEM (2023), progress in attaining adequate water quality in the STAs has important implications for CERP progress. Because the functioning of the STAs is impacted by upstream water quality, in this section the committee includes a discussion of water quality trends in the northern Everglades and Lake Okeechobee. Additionally, potential CERP effects on phosphorus loading and the ecosystem caused by the redistribution of water and increasing flow rates are discussed.

Water Quality in the Northern Everglades and Lake Okeechobee

TP loads into Lake Okeechobee continue to be highly variable and far in excess of the total maximum daily load (TMDL) target for the lake, with a mean of 518 tons (t) over the past 10 water years (2014–2023), which is 378 t above the 140 t TMDL target (Figure 2-27; Betts et al., 2024). In addition, TP concentrations in the lake water column have continued to remain well above the 40 µg/L goal (Havens and Walker, 2002), having stabilized the past few years at approximately 150 µg/L (Figure 2-28). This very high concentration continues to stimulate algal blooms in the lake, threatening the ecological health and recreational usage of the lake as well as affecting downstream water bodies, as they can "seed" blooms in those systems (Phlips et al., 2020).

The high nutrient concentrations leaving Lake Okeechobee have direct implications for both the STAs and the northern estuaries. In WY 2023, approximately 48 percent of the lake's TP load discharges were to the south, ultimately reaching the STAs, making it more difficult to reach the stringent STA discharge targets. The Caloosahatchee and St. Lucie Rivers received 34 and 17 percent of the lake's TP discharges, respectively (Betts et al., 2024). See NASEM (2018) for in-depth discussion of Lake Okeechobee water quality.

Restoration Progress 85

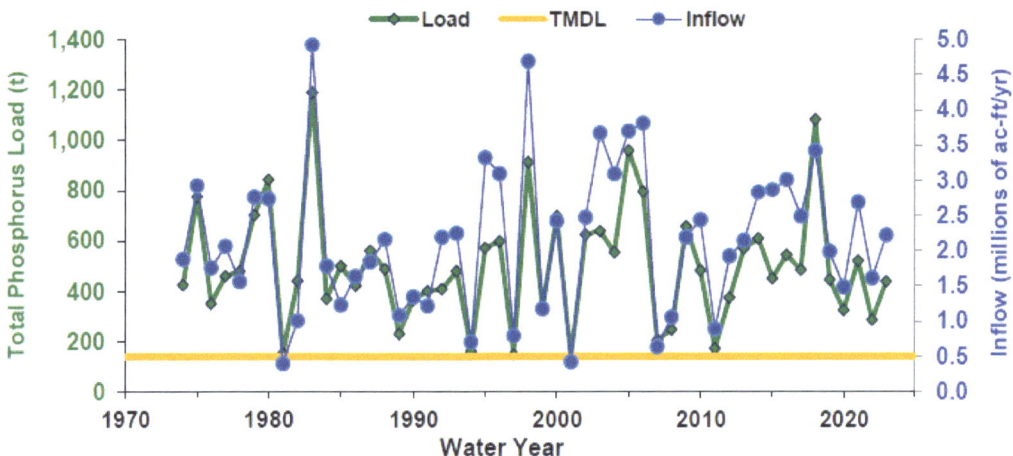

FIGURE 2-27 TP load, inflow, and TMDL for Lake Okeechobee from its tributaries calculated from the TP budget of Lake Okeechobee.

SOURCE: Betts et al., 2024.

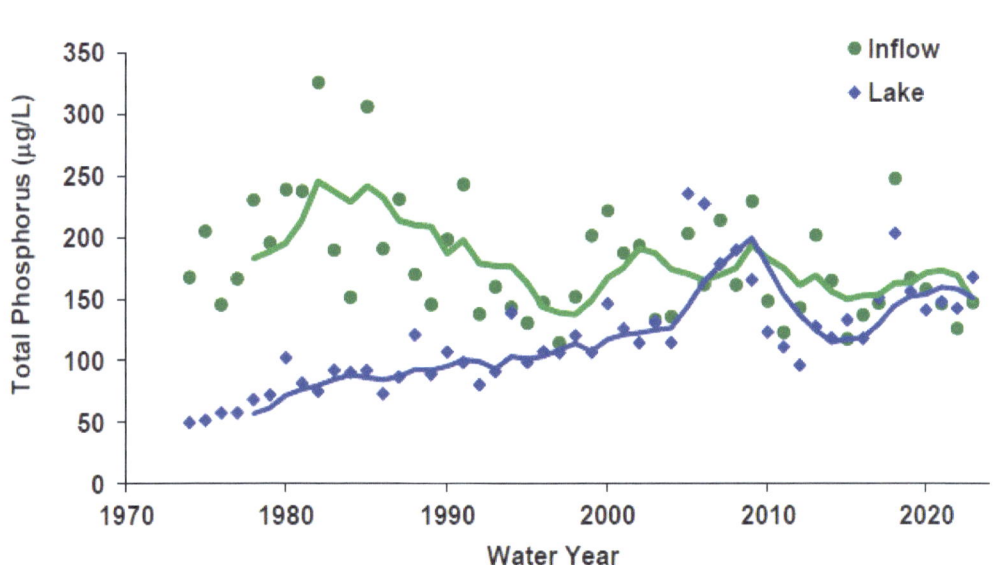

FIGURE 2-28 Mean annual TP concentrations (5-year moving average) for inflow and lake water in Lake Okeechobee.

SOURCE: Betts et al., 2024.

A positive development in the monitoring of Lake Okeechobee is the expansion of phytoplankton sampling sites over the past 3 years; the number of fixed phytoplankton monitoring stations was increased from six year-round stations to nine stations during the dry season (November 1–April 30), and to 32 stations during the wet season (May 1–October 31). This expanded coverage has helped fill data gaps in lake response, which is important given the spatial heterogeneity of the lake (Carrick and Steinman, 2001). Because this expanded monitoring period includes only 3 years of monitoring, trend analysis is not appropriate. However, as shown in Figure 2-29, WY 2023 had not only a higher percentage of samples that met bloom conditions (defined as chlorophyll *a* concentrations > 40 µg/L) but also greater chlorophyll *a* concentrations than in the prior 2 years. At present, there is no clear evidence of improving water quality conditions in Lake Okeechobee.

Stormwater Treatment Areas

The STAs are critical components of Everglades restoration managed by the SFWMD, and they play major roles in reducing TP concentrations and loads from agricultural and urban runoff and Lake Okeechobee. The Everglades STAs are essentially constructed freshwater treatment wetlands built on acquired agricultural lands located north of the Everglades Protection Area (Figure 2-30). To protect the Everglades ecosystem, the construction and operation of STAs were mandated under the 1992 Consent Decree[8] and subsequently by the Everglades Forever Act (Section 373.4592, Florida Statutes). A total of five STAs (STA-1E, -1W, -2, -3/4, and -5/6) with a combined treatment area of 62,000 acres have been built and have been operating over varying periods (Figure 2-30). Using interior levees, each STA is divided into multiple cells dominated by emergent aquatic vegetation (EAV), SAV, or a mixed marsh plant community, which includes both EAV and SAV in the same cell (Armstrong et al., 2023; Chimney, 2024). The STAs are spatially positioned along three flow paths: the Eastern Flow Path (STA-1E and STA-1W), the Central Flow Path (STA-2 and STA-3/4), and the Western Flow Path (STA-5/6) (Figure 2-30).

The Restoration Strategies Plan, launched in 2012, provides for expanding existing STA acreage and additional infrastructure improvements to meet the water quality–based effluent limit (WQBEL). The WQBEL requires that the annual flow-weighted mean outflow TP concentrations from each STA not exceed 13 µg/L in more than 3 out of 5 years (on a rolling basis) and not exceed 19 µg/L in any year (FDEP, 2012a,b). The Restoration Strategies Plan (SFWMD, 2012) included construction of approximately 6,500 acres of additional treatment

---

[7] *United States v. South Florida Water Management District*, 847 F. Supp. 1567 (S.D. Fla. 1992).

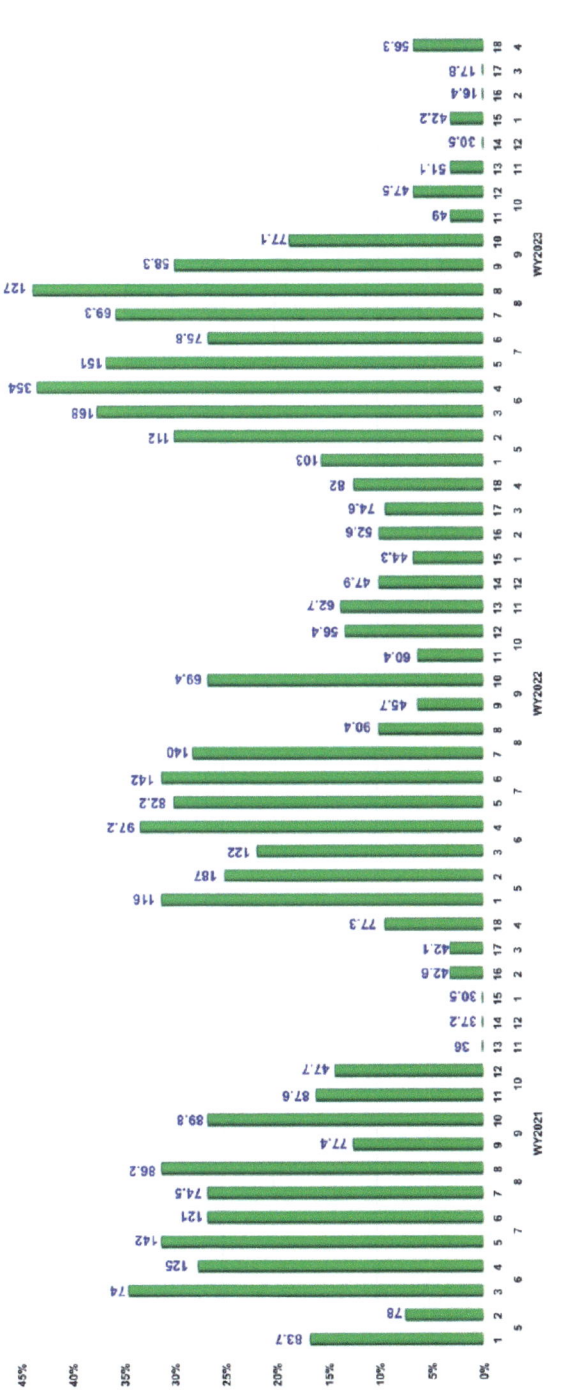

FIGURE 2-29 Percentage of collected samples during each sampling event with bloom conditions (chlorophyll a >40 μg/L) recorded for WY 2021–2023.

NOTES: Numbers in blue above the bars represent maximum chlorophyll a concentrations recorded during each sampling event. Numbers along the x-axis represent sampling events (1 through 18; top row), months (May–April; middle row), and water years (bottom row).

SOURCE: Betts et al., 2024.

FIGURE 2-30 The STAs and recent construction and rehabilitation efforts conducted through the Restoration Strategies program.

SOURCE: Shuford et al., 2024.

area, repairs to STA-1E and STA-5/6, and three flow equalization basins (FEBs) to moderate low and high flow conditions into the STAs (Figure 2-30). Restoration Strategies Plan projects are expected to be fully implemented and operational by December 31, 2025. The first year of discharge data to be incorporated into assessment of WQBEL attainment begins in WY 2027 (May 2026-April 2027). Progress on implementing the Restoration Strategies Plan is described in Table 2-6. Nearly all components of the original Restoration Strategies Plan have been completed as of early 2024, except the G-341 structure and subregional source controls in the Eastern Flow Path and internal improvements to STA-5/6 and full operation of the C-139 FEB in the Western Flow Path.

In NASEM (2023), the committee presented a detailed data analysis to determine the long-term performance of STAs during the operation period (WY 2004-2022). This analysis showed that high inflow TP concentrations coupled with high phosphorus loading rates are key external drivers of the overall performance and treatment efficiency of STAs. In this section, the committee briefly updates that review based on data focused on the recent 5 water years

TABLE 2-6 Summary Status of Major Restoration Strategies Projects

Component	Purpose	Status
Eastern Flow Path		
L-8 FEB	Attenuate flow into STA-1E and STA-1W	Construction completed in 2017, now operational
L-8 Divide Structures (G-716, G-541)	Assist movement of inflows and outflows to L-8 FEB	Construction completed in 2016, now operational
STA-1E Repairs and Modifications	Improve STA performance	Construction completed in 2022, now operational
STA-1W Expansion # 1 (Phase 1)	Increase STA-1W effective treatment area	Construction completed in 2019, now operational
STA-1W Expansion # 2 (Phase 2)	Increase STA-1W effective treatment area	Construction in progress. The initial flooding and optimization period will be complete in December 2024
G-341 Related Improvements	Divert flows (600 cfs max) to the west	Construction completion is expected in December 2024
Subregional Source Controls	Reduce inflow loads from hotspots in the basin	Pilot projects completed in 2015 and 2017; conceptual project planning ongoing
Central Flow Path		
A-1 FEB	Attenuate flow into STA-2 and STA-3/4	Construction completed in 2015, now operational
STA-2 Expansion: Compartment B	Increase STA-2 effective treatment area	Cells 7 and 8 completed in 2013, now operational
Western Flow Path		
STA-5/6 Internal Improvements	Improve the performance of STA-5/6	Construction completion is expected in December 2024. Initial flooding and optimization are expected to be complete in December 2025
C-139 FEB	Attenuate flow into STA-5/6	Construction completed in 2024. The operational monitoring and assessment period is expected to be completed by December 2024

SOURCES: Data from Chimney, 2024; Shuford et al., 2024.

(WY 2019–2023). During this period, all STAs were affected by extreme events, including hurricanes, regional droughts, and continued changes in the operation and maintenance of STAs.

Status of STAs in WY 2023

NASEM (2023) provided a detailed analysis of STA treatment performance, so this section provides an update based on new information (i.e., WY 2023 data)

since that report was released. In WY 2023, several STA flow-ways were offline for maintenance or refurbishments, reducing the effective treatment areas. For example, in WY 2023, only 59 and 60 percent of STA-1E and STA-3/4 treatment areas were online, respectively (Chimney, 2024), so STA performance should be evaluated in that context.

Table 2-7 provides an overview of STA performance in WY 2023 as well as performance over the period of record. Overall phosphorus concentration-based treatment efficiencies in the STAs were high compared to other constructed wetlands; however, within the STAs, STA-2 (74%) and STA-1E (77%) underperformed compared to the 83 to 90 percent TP treatment efficiencies attained in the other STAs. STA-3/4, with a mean TP outflow concentration of 16 µg/L, was the only STA to meet the upper WQBEL discharge concentration limit (19 µg/L) in WY 2023.

Performance of STAs in the Eastern Flow Path (STA-1E and STA-1W). Inflows to STAs in the Eastern Flow Path (STA-1E and STA-1W) are derived from C-51, L-8, and S-5A basins and during dry season regulatory releases from Lake Okeechobee (Chimney, 2024). Basin runoff can be stored in the L-8 FEB for a short period and discharged into both STAs.

In recent years, inflow TP concentrations to STA-1E have notably decreased as have TP loading rates (Figures 2-31 and 2-32), although the TP loading rate for WY 2023 (1.5 g/m^2-yr) was still 50 percent higher than that of the other STAs (Table 2-7, Figure 2-32). Construction activities in STA-1W previously diverted inflow water to STA-1E starting in WY 2021, which exacerbated TP and hydraulic loading (Figure 2-32; Chimney, 2022). Even though 40 percent of the treatment area of STA-1E was offline during WY 2023, recent reductions in both hydraulic and TP loading rates helped increase the load-based treatment efficiency from 79 percent over the period of record to 82 percent in WY 2023, with an outflow TP concentration of 26 µg/L (Chimney, 2024). Further reductions are expected once refurbishment activities are completed and 100 percent of the treatment area is available, although substantial reductions are needed to meet the 13 µg/L lower WQBEL limit.

In STA-1W, the average inflow TP concentration of 199 µg/L (WY 2023) was 86 percent higher than average inflow TP concentrations of STA-1E, -2, and -3/4 (Table 2-7 and Figure 2-31). STA-1W Expansion #1, which increased the effective treatment area by 4,266 acres (to 10,810 acres), was completed in WY 2021. Since that time, TP loading rates have decreased to below 1 g/m^2-yr, a threshold that has been judged to be important to support effective STA performance (NASEM, 2023; see Figure 2-32). Over the same time, outflow TP concentration steadily decreased from 38 µg/L (WY 2021) to 20 µg/L (WY 2023), indicating a positive trajectory toward meeting water quality goals with an impressive

TABLE 2-7 Select Water Quality Parameters for Five STAs During the Operation Period of WY 2023 (May 1, 2022 to April 30, 2023) and the Period of Record of Operation

Parameters	Eastern Flow Path		Central Flow Path		Western Flow Path
	STA-1E	STA-1W	STA-2	STA-3/4	STA-5/6
Total area, acres	4,994	10,810	15,495	16,327	14,388
(adjusted treatment area, acres)	(2,980)	(10,810)	(13,121)	(9,851)	(13,728)
Treatment cells per STA	8	10	8	6	14
% of treatment area online in WY 2023	59	100	85	60	96
Inflow, WY 2023					
Water volume (x1,000 AF)	124	152	327	327	153
Hydraulic loading rate (cm/day)	3.5	1.2	2.1	2.8	0.9
TP FWM concentration (µg/L)	115	199	113	93	288
TP loading rate (g/m^2-yr)	1.5	0.9	0.9	0.9	1.0
TP load (t)	18	37	45	38	54
Outflow, WY 2023					
Water volume (x1,000 AF)	101	158	349	307	148
TP FWM concentration (µg/L)	26	20	29	16	40
Phosphorus discharge rate (g/m^2-yr)	0.3	0.1	0.3	0.1	0.1
TP load (t)	3	4	13	6	7
Treatment Efficiency, WY 2023					
Treatment efficiency (%) (TP concentration basis)	77%	90%	74%	83%	86%
Treatment efficiency (%) (TP load basis)	82%	89%	72%	84%	87%
Period of Record					
STA start date	Sept. 2004	Oct. 1994	June 1999	Oct. 2003	Dec. 1997
Approximate years of operation	19	29	24	20	25
Inflow TP FWM concentration (µg/L)	160	181	101	101	200
Outflow TP FWM concentration (µg/L)	37	44	22	15	63
Treatment efficiency (%) (TP concentration basis)	77%	76%	78%	85%	69%
Long-term TP inflow loads (t)	491	1,108	856	1,069	842
Long-term TP outflow loads (t)	104	277	196	163	245
Long-term TP retention (t)	387	831	660	907	597
TP treatment efficiency (%) (TP load basis)	79%	75%	77%	85%	71%

NOTES: FWM = flow-weighted mean. Conversion factors are 1 acre = 0.4047 hectares (4,047 m^2); 1 AF = 1,234 m^3; 1 metric ton = 1,000 kg; 1 cm per day = 0.3937 inches per day.
SOURCE: Modified from Chimney, 2024, with additional analysis of treatment efficiency from the committee.

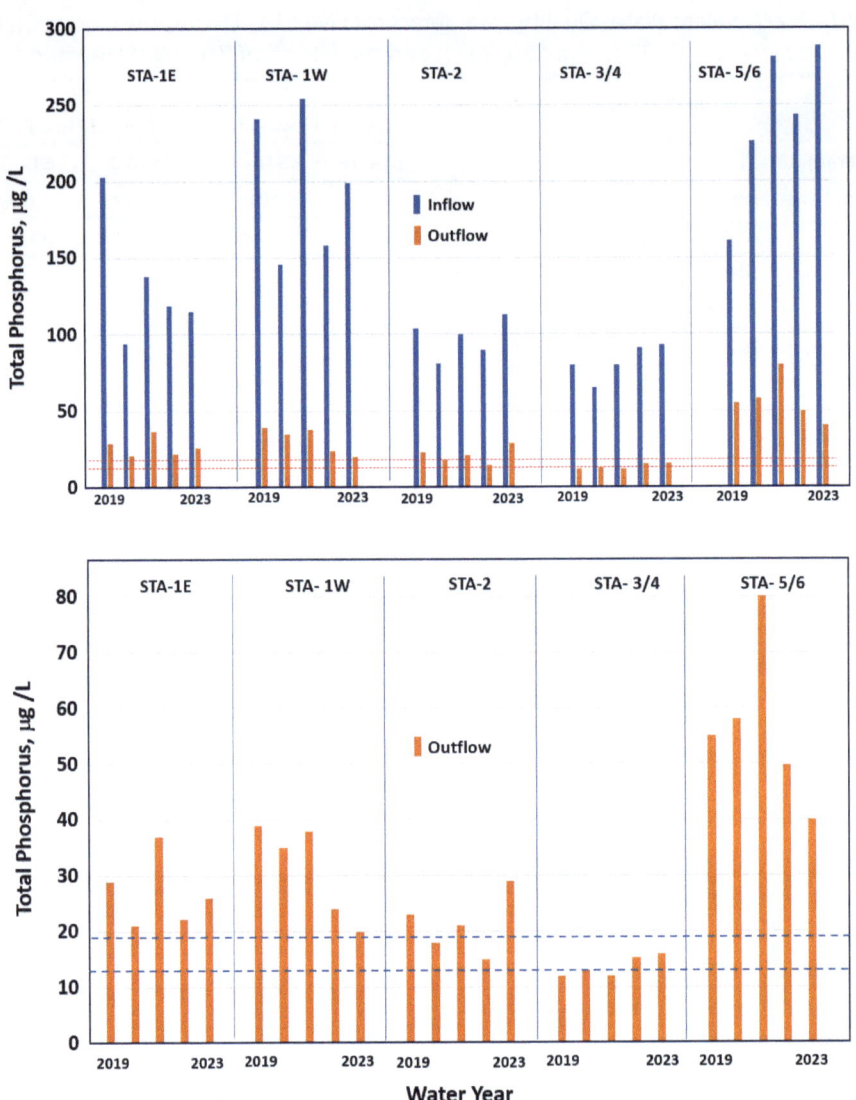

FIGURE 2-31 Top graph shows inflow and outflow TP concentrations from the five STAs during the operation period of WY 2019–2023. Bottom graph shows only the outflow concentrations relative to the WQBEL for scale.

NOTE: Red dashed lines represent lower (13 mg/L) and upper (19 mg/L) TP concentration limits set by the WQBEL.

SOURCES: Data from Chimney, 2020, 2021, 2022, 2023, 2024.

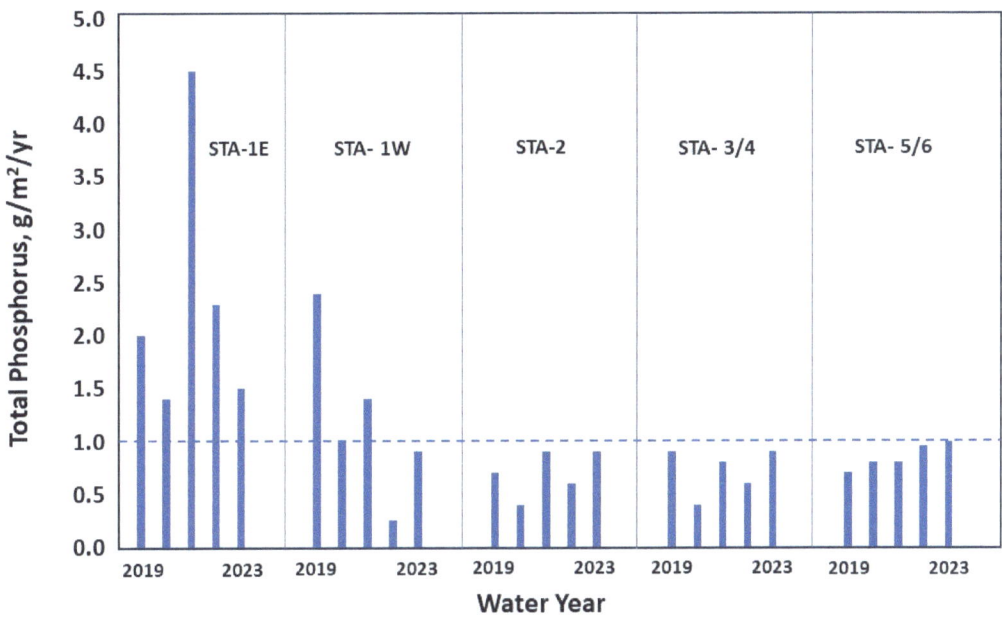

FIGURE 2-32 Inflow TP loading rate during each STA's operation period of WY 2019–2023.

SOURCES: Data from Chimney, 2020, 2021, 2022, 2023, 2024.

---

89 percent load-based treatment efficiency (Figure 2-31; Table 2-7). The addition of STA-1W Expansion #2, which will provide an additional 1,600 acres of treatment area (Shuford et al., 2024), is expected to decrease outflow TP concentrations further and bring STA-1W closer to meeting WQBEL requirements.

Performance of STAs in the Central Flow Path (STA-2 and STA-3/4). STA-2 and STA-3/4 receive agriculture runoff from EAA basins and releases from Lake Okeechobee. Inflow TP concentrations to STA-2 and STA-3/4 have been significantly lower than those to the other STAs, both recently and over the period of record (Chimney, 2024), which eases the challenge of attaining the WQBEL. The Restoration Strategies Plan in the Central Flow Path included the A-1 FEB and STA-2 Expansion (see Table 2-6; Shuford et al., 2024).

Substantial refurbishment and vegetation management was under way in STA-2 during WY 2023, with one of five flow-ways (equal to 15 percent of the treatment area) offline the entire year. Despite a low TP loading rate (0.9 g/m^2-yr), outflow TP concentrations increased from 15 μg/L to 29 μg/L from WY 2022 to

WY 2023 (Figure 2-31 and Figure 2-32), and load-based treatment efficiency decreased to 71 percent compared to its long-term average of 77 percent; this increase may be due to short-term effects of disturbance in the flow-ways, but it is important to understand the causative factors to mitigate these effects in the future if necessary.

STA-3/4 maintained ultra-low outflow TP concentrations (12 to 16 µg/L) during the recent 5 water years (WY 2019–2023) and met the WQBEL requirements during this period. During WY 2019 to WY 2021, STA-3/4 maintained outflow TP concentrations of 12 to 13 µg/L. In the most recent 2 years, outflow TP concentrations increased to 15 to 16 µg/L, likely because 40 percent of STA-3/4 was offline during WY 2022 and WY 2023 for vegetation management, but STA outflows remained within WQBEL requirements (Figure 2-31; Chimney, 2024).

The low inflow TP concentrations and phosphorus loading rates for STA-2 and STA-3/4, combined with vegetation rehabilitation in both STAs, should help improve performance and maintain low outflow TP concentrations. In addition, A-1 FEB benefits both STAs with short-term storage of runoff water and attenuation of peak inflows to improve overall treatment efficiency.

Performance of STAs in the Western Flow Path (STA-5/6). Throughout the period of record, STA-5/6 has received inflows with extremely high TP concentrations; inflow TP concentrations in WY 2023 were 170 percent higher than average inflow TP concentrations of STA-1E, -2, and -3/4 (Figure 2-31). STA-5/6 also has the lowest load-based treatment efficiency (71 percent) over the period of record (Table 2-7). During the past 5 water years, this STA has had a TP loading rate at or below 1.0 g/m^2-yr, and outflow concentrations have been declining over the latest 3 years but remain well above the target WQBEL concentrations (Figures 2-31 and 2-32). Elevated outflow TP concentrations, in spite of low phosphorus loading rates, suggest that legacy phosphorus offsets a portion of phosphorus removal occurring in the system, likely during periods of drydown and reflooding. During the past 5 years, STA-5/6 has functioned at a concentration-based treatment efficiency that ranged from 66 percent in WY 2019 to 86 percent in WY 2023.

Restoration Strategies projects for STA-5/6 include the C-139 FEB completed in 2024 with capacity to store 11,000 AF of water and internal improvements (Table 2-6; Shuford et al., 2024). The C-139 FEB, when fully operational, can provide additional water to hydrate STA-5/6 and reduce dry conditions. Additionally, as part of STA refurbishment efforts, the SFWMD is planning to develop a project that includes infrastructure to move Lake Okeechobee water to STA-5/6. This project is currently in design phase and, when completed, will move up to 300 cfs as part of regulatory releases from Lake Okeechobee to the STAs, which will help to rehydrate during dry conditions and improve treatment performance

(T. Piccone, SFWMD, personal communication, 2024). Implementing these strategies should help reduce outflow TP concentrations.

Overall Assessment and Path Forward

The SFWMD has clearly made important strides in the cumulative performance of the STAs during the recent 5-year period, with efforts to reduce the average phosphorus and hydraulic loading rates of the STAs to improve performance. The recent 5-year average outflow TP concentration of STA-3/4 is already meeting the WQBEL. All other STAs are moving in a positive trajectory through Restoration Strategies actions to reduce outflow TP concentrations. Attaining the WQBEL for outflow TP concentration depends on how well these STAs respond to Restoration Strategies, ongoing vegetation management, and soil management (as needed), to reduce TP concentrations. NASEM (2023) highlighted the importance of timely attainment of the WQBEL in order to avoid delay of restoration benefits from the CEPP. Benefits that depend on STA performance include full water deliveries from the EAA Reservoir. NASEM (2023) also noted the challenges associated with attaining and maintaining the WQBEL over time in all STAs and recommended the best available science and monitoring in an adaptive management framework to address these challenges.

In 2012 the SFWMD developed the Restoration Strategies Science Plan and updated it in 2018 in collaboration with outside consultants, governmental agencies, and universities (James et al., 2024a). Research conducted under the Restoration Strategies Science Plan was extensively published in reports and peer-reviewed journals, including an Everglades STAs special issue published in 2024 in the journal *Ecological Engineering* (Chimney, 2024; James et al., 2024b; Reddy et al., 2024). Many of these studies are useful in understanding the role of external and internal drivers in regulating hydrologic and biogeochemical processes that influence outflow TP concentrations in the STAs.

NASEM (2023) provided several recommendations on science to support STA management and decision making, related to, for example, monitoring, setting near-term milestones, assessment, modeling, research, and synthesis. It is important to extract management-specific information from the Restoration Strategies Science Plan and other new studies for possible immediate applications to support WQBEL attainment and to guide each STA in improving long-term sustainability. The committee reemphasizes the need for cell-by-cell monitoring of TP concentrations to better understand treatment efficiency. It is well known that phosphorus removed in STAs is retained in soils (long term) and in vegetation (short term). Soil and organic matter accretion and its physical and biogeochemical stability need to be determined under different biogeochemical and hydrologic conditions to understand the long-term sustainability of STAs and their capacity to provide steady performance (NASEM, 2023).

Biogeochemical Dynamics with Increased Flow in the Everglades

The structure and function of the Everglades ecosystem is predicated on oligotrophic conditions. Nutrient management in the Everglades Protection Area has been focused on meeting the phosphorus criteria established in 2003, including 10 µg/L as a long-term geometric mean.[9] With CERP restoration activities, a critical question is "How will the Everglades ecosystem and its biogeochemical processes respond to increasing flows and hydroperiods?" A suite of interrelated processes could affect ecosystem conditions: (1) increased TP loading associated with increases in flows, regardless of TP concentrations (i.e., including <10 ppb); (2) increased sheet-flow velocities leading to enhanced suspension and transport of sediment-associated phosphorus; (3) new canal-to-marsh pathways that could exacerbate phosphorus loading into unimpacted marsh areas via scour and transport of stored sediments; and (4) mobilization of legacy phosphorus in wetland soils associated with increases in reducing conditions. Characterization and quantification of changes in the transport and processing of nutrients following restoration efforts will be an important endeavor to mitigate against undesirable consequences. Several of these potential mechanisms are highlighted in the recent developments discussed in this section.

Changing Water Quality Dynamics in the Central Everglades

In general, there is a north-to-south gradient in TP concentrations in the Everglades Protection Area that results from an elevated supply of phosphorus from waters draining the EAA and urban areas followed by transport and removal by sedimentation, sorption, and assimilation within the wetlands. The spatial pattern is also characterized by higher TP concentrations along canals and lower concentrations in the interior marsh. Of the interior marsh sites, 85 percent

---

[8] Florida has a narrative surface-water quality criterion to ensure that nutrient concentrations will not be altered as to cause an imbalance in natural populations of aquatic flora or fauna. The narrative nutrient criterion was numerically interpreted for phosphorus in the Everglades in FAC §§62-302.530 (see also Rizzardi, 2001). For Class III waters in the Everglades Protection Area, a four-part test is used to assess compliance with the numeric phosphorus criterion following four provisions: (1) 5-year geometric mean is less than or equal to 10 ppb, (2) annual geometric mean averaged across all stations is less than or equal to 11 ppb, (3) annual geometric mean averaged across all stations is less than or equal to 10 ppb for 3 of 5 years, and (4) annual geometric mean at individual stations is less than or equal to 15 ppb (FAC §§62.302.540). Achievement of the criterion in Everglades National Park is governed by methods in Appendix A of the 1992 Consent Decree (see SFWMD [2009] for details).

exhibited annual geometric mean TP concentrations of 10 μg/L or less (Lopez et al., 2024). The spatial and temporal patterns of changing TP concentrations along the gradient from north to south are highlighted in Figure 2-33. The period of record has seen marked decreases in TP concentrations in inflow waters to the WCAs and within the interior marsh sites of the Everglades Protection Areas. Over the more recent interval (WY 2005-2023), TP concentrations have continued to decrease in inflows to Arthur R. Marshall Loxahatchee National Wildlife Refuge and WCA-2, but over the past 10 years, inflow concentrations appear to have increased slightly to WCA-3 and Everglades National Park. Increases in flow volumes over the past decade have also resulted in notable increases in TP loadings to Everglades National Park and slight increases in TP loadings to WCA-3A (Figure 2-33; statistical significance not available for this time frame).[10]

Recent observations indicate that processes beyond increases in the quantity of water discharge contribute to increases in TP loading and availability. The COP EIS (USACE, 2020a) presented concerns that TP concentrations increase at lower water stages (<9.2 ft), affecting concentrations and loads delivered to Everglades National Park (Figures 2-34 and 2-35). The increases in annual TP load delivered to Shark River Slough through the S-333s at lower water stages compared to past operational plans are evident in Figure 2-35. This pattern has become particularly pronounced since 2018 (Figures 2-33 and 2-35) under COP operations and the incremental testing period, which has been coincident with multiple exceedances of the Consent Decree Appendix A compliance long-term limit (Mo et al., 2022, 2024). The S-333 Working Group (2023) noted that as water levels in the marsh decrease during the dry season, water transport toward the S-333 becomes canal-dominated, "with higher TP concentrations than those observed in the marsh."

In response to concerns by the Consent Decree principals that the observed increases in TP concentrations to Shark River Slough resulted from local sources and phenomena, the S-333 Working Group was established to evaluate this condition and recommend approaches to decrease TP loading. The S-333 Working Group (2023) found that more than 10,500 cubic yards of phosphorus-laden sediments have accumulated in the canals immediately upstream of S-333, and this material could be an important source of phosphorus if entrained and transported downstream (Figure 2-36). Hydrodynamic modeling suggested that flow velocities could become high enough to entrain surface sediments in the L-29 and L-67 Canals and enhance TP transport to Everglades National Park. In response to this analysis, the Consent Decree principals proposed dredging 1,500 linear ft of L-29 and L-67 Canal sediments immediately upstream of the

---

[9] A correction was made after the report was released to clarify the data considered in this assessment of trends and to correct the magnitude of trends observed.

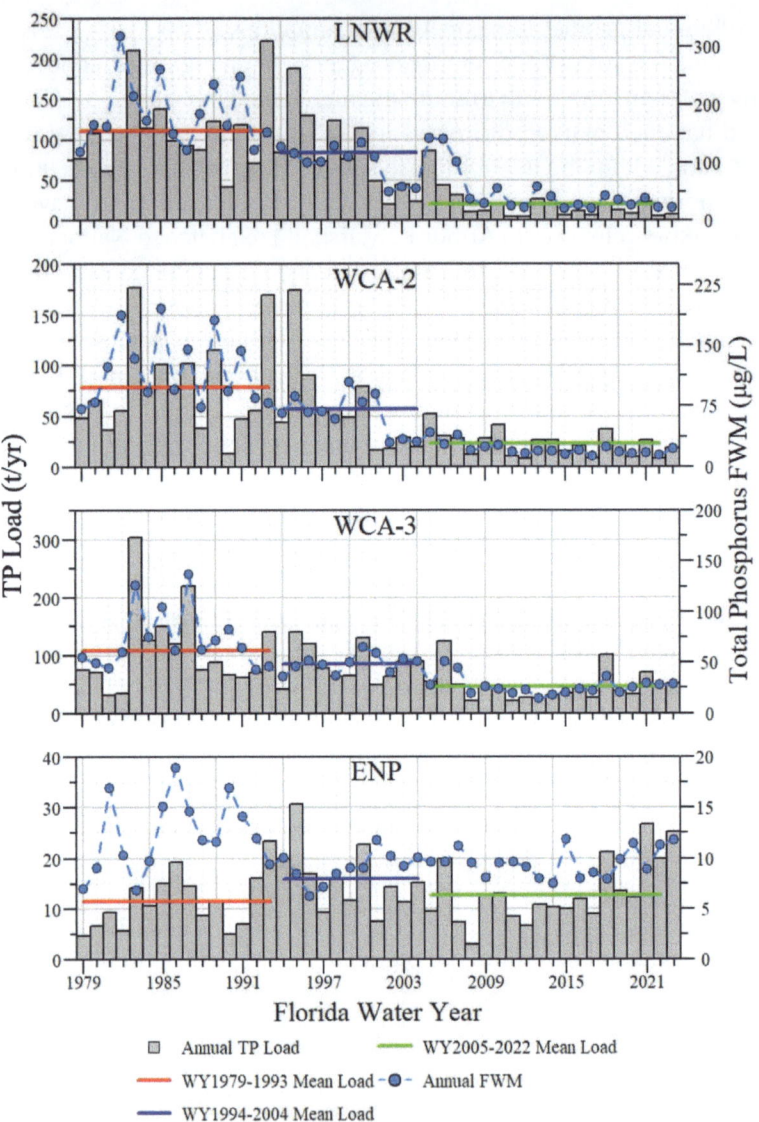

FIGURE 2-33 Annual inflow TP load in metric tons per year (t/yr) and TP flow-weighted mean concentration in μg/L for Arthur R. Marshall Loxahatchee National Wildlife Refuge, WCA-2, WCA-3, and Everglades National Park from WY 1979 through WY 2023.

NOTE: The horizontal lines indicate the mean annual loads for the Baseline (WY 1979–1993), Phase I (WY 1994–2004), and Phase II (WY 2005–2022) periods.

SOURCE: Lopez et al., 2024.

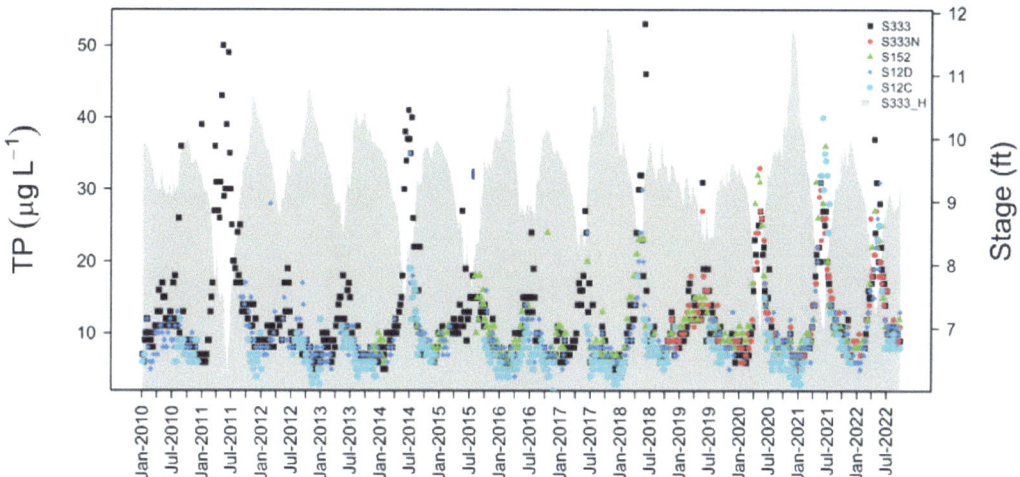

FIGURE 2-34 Water stage at S-333 headwater and TP concentrations at S-333, S-333N, S-152, S-12D, and S-12C.

NOTES: Data were secured from DBHYDRO, the environmental corporate database of the SFWMD. Concentrations of TP increase with low flow conditions.

SOURCE: S-333 Working Group, 2023.

S-333 complex and installing low-sill weirs upstream of the S-333s to reduce canal velocities and bedload transport. Completion of this project is estimated for 2026 (Hutchins, 2024). Monitoring will be conducted to evaluate the effectiveness of these actions, identify opportunities to continue to mitigate TP loading to Everglades National Park, and inform future management actions (Bartlett et al., 2023). Depending on the results of these initial actions, the S-333 Working Group (2023) noted that further work could be warranted, including a comprehensive study of an expanded portion of the canals and marsh within WCA-3A area upstream of the S-333s to determine the dominant sources (sediment, floc, and suspended in the water column) of phosphorus and the mechanisms by which these materials are contributing to TP loading to Everglades National Park.

Biogeochemical Dynamics in the Decomp Physical Model and CEPP South

The potential for enhanced transport of previously deposited phosphorus along canals or in wetland sediments has been illuminated from work in the Decomp Physical Model (DPM), with important implications for CEPP South.

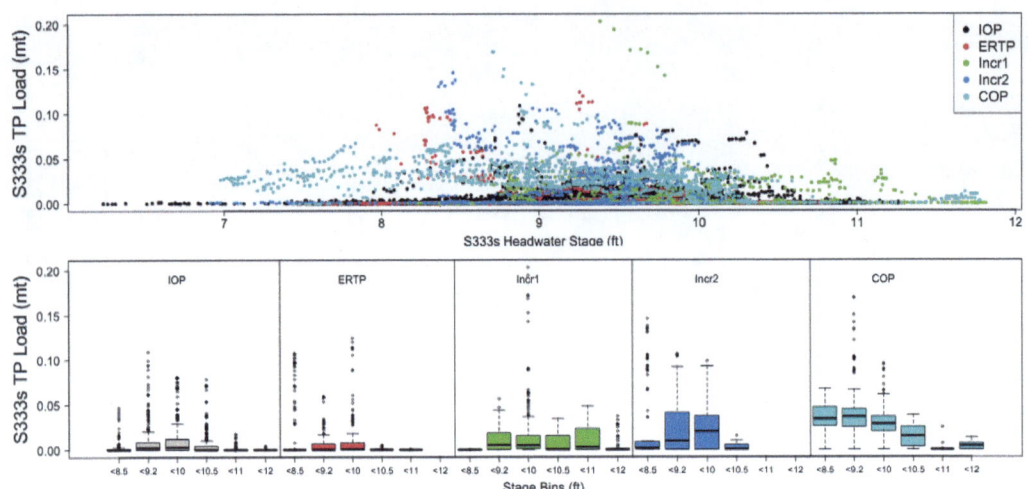

FIGURE 2-35 Daily TP load at S-333 shown by headwater stage at S-333 over five operational plans: Interim Operational Plan (IOP, 2002–2012), Everglades Restoration Transition Plan (ERTP, September 2012–September 2015), COP Increment 1 (Incr1, October 2015–February 2018), COP Increment 2 (Incr2, March 2018–August 2020), and the COP (October 2020–present).

NOTE: Top graph shows the distribution of all data, while the bottom graph shows the distribution of TP loads by stage bins.

SOURCE: D. Surratt, National Park Service, personal communication, 2024.

One of several unanticipated outcomes of DPM was that high velocity inflows with low surface-water concentrations of TP (i.e., ≤10 µg/L TP) produced phosphorus-impacted conditions downstream because of transport of phosphorus-enriched sediments and increased loads to the previously unimpacted area. This change resulted in the loss of the slough periphyton community and invasion of cattails (Saunders, 2020; Saunders et al., 2016, 2021; Sklar, 2020). Cattails continue to spread downstream within WCA-3B toward the L-29 Canal and Everglades National Park at a rate of 100 m/yr (Saunders and Newman, 2023). Ongoing work to address this issue in an adaptive management framework is discussed in more detail in Chapter 5.

Long-Term Monitoring Trends and Data Issues

Observations of TP concentrations from the Florida Coastal Everglades Long-Term Ecological Research program (FCE LTER) may yield important insight on

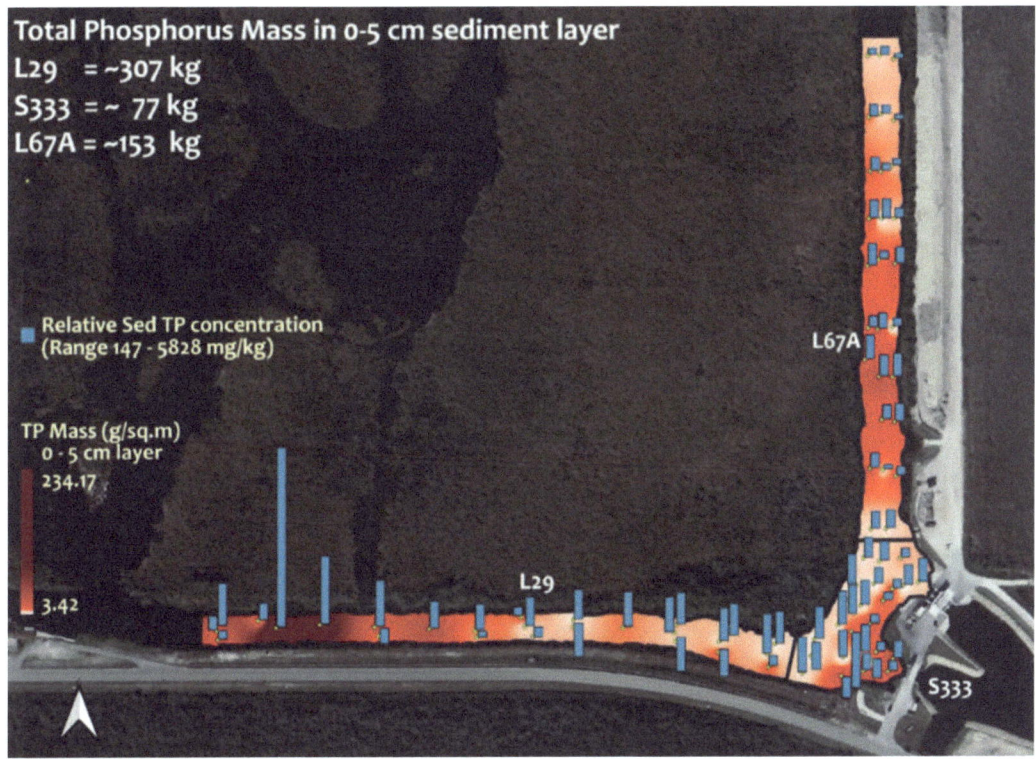

FIGURE 2-36 Concentrations of TP bar plot of sediment surface (0-5 cm) in the L-29 Canal, near the S-333 structure, and the L-67A Canal.

NOTES: The height of each bar represents relative TP concentration for the sampling location in grams per square meter (g/m). TP mass (kg) contained in sediment surface (0-5 cm) in the L-29 Canal, near the S-333 structure, and the L-67A Canal are also indicated. Note the elevated concentrations of TP in sediments, particularly in the L-29 Canal.

SOURCE: S-333 Working Group, 2023.

phosphorus dynamics in the Everglades in response to increases in discharge. The FCE LTER has established long-term monitoring stations in Shark River Slough and Taylor Slough along transects from the headwater reaches of Everglades National Park through the mangroves and into Florida Bay. The FCE LTER observed increasing trends in TP concentrations over the past decade in wetlands in Shark River Slough and Taylor Slough. However, this pattern of increasing TP concentrations in water grab samples is inconsistent with observations reported for nearby sites by the SFWMD, which do not show temporal trends (Gaiser, 2023; Figure 2-37).

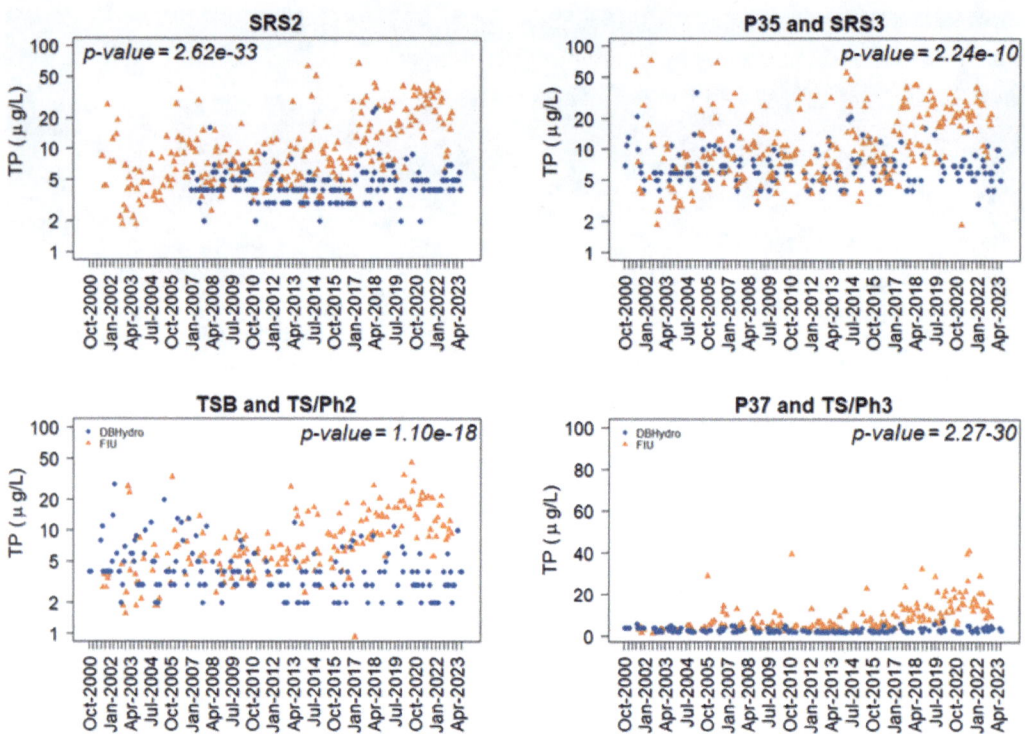

FIGURE 2-37 Comparison of time series of water measurements of TP for nearby sites in Everglades National Park that are collected and analyzed by the FCE LTER (orange) and the SFWMD (blue).

NOTE: The FCE LTER observations show an increase in TP concentrations in recent years, while the SFWMD samples do not.

SOURCE: C. Gauthier, NPS, personal communication, 2024.

The FCE LTER and Everglades National Park are collaborating to understand the reason for this discrepancy in temporal trends. They have collected and analyzed paired samples using their standard methods and compared results, which show no differences due to sample collection procedures or analytical methods. They continue to investigate whether this discrepancy is due to differences in locations or daily timing of sample collections or methods of preservation.

The FCE LTER has also been collecting two types of water samples since the early 2000s: 3- to 6-day integrated, diurnal sequential samples and daytime grab samples. Interestingly, TP concentrations in diurnal composite samples are considerably greater than values in grab samples. The reason for this discrepancy between the two sampling approaches is not clear but could potentially indicate potential night-time mobilization of phosphorus under fluctuating redox conditions. FCE LTER personnel are examining potential diurnal patterns in phosphorus in Everglades National Park to better understand the reason for different results from differences in sampling or processing approaches.

Implications for Research

Although the patterns of TP concentrations need to be fully understood and verified given the inconsistency between FCE LTER and SFWMD observations, the temporal pattern of increasing phosphorus observed in Shark River Slough and Taylor Slough coupled with the S-333 Working Group and DPM observations may suggest an emerging pattern of changes in phosphorus dynamics. Such findings necessitate additional observation and research on how a changing flow regime will influence nutrient pattern and processing in the Everglades ecosystem. The committee recommends that several of the S-333 Working Group's (2023) recommended Phase II research and modeling initiatives should proceed immediately rather than after completion of the Phase 1 activities and subsequent assessment. These proposed activities include new hydrodynamic modeling and a comprehensive study of an expanded portion of the canals and marsh within the WCA-3A area upstream of S-333 to determine the dominant sources (sediment, floc, and suspended in the water column) of TP to Everglades National Park and the mechanisms by which these materials are contributing to TP inflows. The committee also recommends a rigorous comparison of sampling methods of TP and other related analytes between the FCE LTER and the SFWMD to determine the reason for the differences in observed long-term temporal patterns and further evaluate the potential for recent TP mobilization associated with increases in discharge. Expanded monitoring efforts (e.g., all forms of TP fractions, particularly total dissolved phosphorus) would help characterize suspended and organic phosphorus fractions to better understand the mechanisms driving changes in phosphorus cycling under a shifting hydrologic regime. Such information is necessary to develop optimal strategies to mitigate against mobilization of legacy phosphorus. Ongoing work by the FCE LTER to interpret differences in TP concentrations between diurnal and grab samples, including TP, dissolved oxygen, and ferrous iron, will advance understanding of potential mechanisms of phosphorus mobilization from Everglades sediments.

CONCLUSIONS AND RECOMMENDATIONS

The pace of restoration implementation has reached historic levels, based on record state and federal investments in FY 2022 and FY 2023. Six CERP projects are under construction (Picayune Strand, C-43 Reservoir, IRL-South, Biscayne Bay Coastal Wetlands, Broward County WPAs, and the CEPP), one

CERP project (Melaleuca Eradication) and two major project components (CEPP New Water, C-44 Reservoir and STA) have been completed, and one additional project (C-111 Spreader Canal Western project) is essentially complete. The CEPP continues to progress rapidly, as befits the project that is the keystone to restoring the central heart of the Everglades. Maintaining this pace of progress requires both continued construction funding and support for other agencies responsible for facilitating restoration implementation (e.g., permitting, monitoring). With so many projects under construction, if future funding levels fall short of those used for planning, difficult decisions will need to be made as to whether to delay all projects equally or, preferably, to expedite those with the greatest near-term benefits.

Sizable restoration benefits are evident from recent progress on CEPP New Water, Picayune Strand, and the Melaleuca Eradication projects. The recent completion of CEPP New Water is a restoration milestone. It is already evident that the combined effect of the recently constructed seepage barriers—CEPP New Water and a separate project constructed by the SFWMD—will greatly reduce, and perhaps even eliminate, flood control constraints imposed by the 8.5 Square Mile Area within the CEPP footprint. Prior to the COP, these constraints completely stymied every attempt over decades to restore the historic distribution of flow between Northeastern and Western Shark River Slough. However, with the addition of these new seepage barriers, no such constraints have affected operations since WY 2022. Restoration of hydrology in Picayune Strand appears to be generating benefits to the local flora and fauna, with vegetation and macroinvertebrate communities responding favorably. Additional longitudinal monitoring will be needed to continue documentation of recovery, especially given the magnitude of seasonal and inter-annual variation. Sampling methods for some species (e.g., amphibians) need to evolve to generate a clearer picture of what species are responding to restored hydrology. CERP investments in invasive species biological control efforts have contributed to a 75 percent reduction in area dominated by Melaleuca and have largely controlled air potato reproduction to the extent that air potato is no longer a priority invasive species.

Initial monitoring results indicate that the COP has been a restoration success, generally meeting expectations in achieving hydrological and ecological restoration objectives and improving conditions in the central Everglades relative to previous water management. The rehydration of Northeast Shark River Slough in Everglades National Park represents the largest step yet toward restoring the hydrology and ecology of the central Everglades. However, achieving complex objectives that involve creating fairly precise hydrologic conditions over extended periods of time—for example, optimal recession rates and water levels during the nesting season for threatened and endangered species—has been more challenging. Like any system operations plan, the COP likely lacks the capacity to adjust sufficiently to meet the restoration targets under all rainfall regimes. Thus, the COP is proving to be what it was intended to be, not a complete solution but rather the first major step toward restoring the central heart of the Everglades. The changes wrought by the COP have revealed some issues with water quality and confirmed anticipated adverse effects on endangered Cape Sable seaside sparrows. The latter issue requires creative solutions, sooner rather than later.

Information on natural system restoration progress relative to expectations and project objectives remains difficult to find and interpret. The CERP lacks a mechanism for centralized multi-agency reporting of project-level restoration outcomes. Data, when available, are often presented in permit reports produced by a single agency or in monitoring reports produced by contractors. Increased attention to multi-agency data synthesis and interpretation is needed to support assessment and learning. The COP Biennial Report is an example of effective multi-agency analysis of extensive monitoring data on hydrologic, water quality, and ecological conditions in support of learning and adaptive management. Simplified and straightforward analyses of key metrics using geographic information system–based dashboards and easy-to-read graphics will provide increased transparency of restoration outcomes to the public and Congress.

WERP as proposed offers important benefits to the western Everglades, but implementation progress largely depends on non-CERP source control implemented by private landowners, which could lead to large delays without implementation and performance requirements. In general, WERP features, if implemented as planned, should improve hydration, hydrologic and ecological connectivity, and water quality, which have been longstanding concerns in the WERP study area. Yet, issues regarding flood risk of the Looneyville community still need to be addressed to meet Savings Clause requirements.

Downward trends in TP concentrations in STA outflow reflect extensive recent Restoration Strategies efforts, but timely attainment of the stringent WQBEL requirements will depend on how effectively the STAs respond to these efforts and the extent to which data collection (including cell-by-cell monitoring), data analysis, modeling, and synthesis are rigorously applied to inform adaptive management decisions. As noted in NASEM (2023), CERP progress and the timely delivery of restoration benefits, particularly for CEPP North and the EAA Reservoir, depend upon meeting the WQBEL in all STAs. High TP concentrations in Lake Okeechobee further the challenge of STA performance given the future plans to move more lake water south. The cumulative performance of STAs during the recent 5-year period is impressive and generally trending in the right direction, but only STA-3/4 currently meets the WQBEL. Maintaining low phosphorus and hydraulic loading rates should improve STA treatment efficiency, but high inflow concentrations in the Western and Eastern Flow Paths could pose particular challenges for WQBEL attainment.

Additional research is recommended to explore the potential biogeochemical effects from the CERP through increased flows, flow velocities, canal-to-marsh interactions, and other factors that may mobilize legacy phosphorus and impact periphyton and plant communities. The S-333 Working Group identified sediment mobilization in the L-29 Canal as a key driver of increasing concentrations and loads into Everglades National Park with recent increased flows during the dry season, and agencies are implementing strategies to reduce sediment mobilization. Research through the DPM and FCE LTER has also illuminated potential connections between increased flows and increased TP loading and/or vegetation impacts, although discrepancies between the SFWMD and FCE LTER data need to be resolved. The potential of ecological effects from increased CERP flows also merits further study. Specifically, research should examine the dominant TP sources (e.g., sediment, floc, suspended) to Everglades National Park and WCA-3B, determine the mechanisms by which these concentrations and loads may be exacerbated under the CERP through increased flows and/or canal-to-marsh interactions, and, if necessary, identify approaches to mitigate phosphorus impacts.

3

Applying Indigenous Knowledge in the Comprehensive Everglades Restoration Plan

The National Academies' Committee on Independent Scientific Review of Everglades Restoration Progress is charged with discussing scientific and engineering issues that may impact progress toward restoring the Everglades (see Chapter 1). In 2021, the Executive Office of the President released a memorandum on Indigenous Knowledge and its role in federal decision making (EOP, 2021c) and subsequently issued guidance to federal agencies on "considering, including, and applying Indigenous Knowledge in Federal research, policies, and decision making" (EOP, 2022a). Based on specific suggestions from Comprehensive Everglades Restoration Plan (CERP) agencies, the committee examines the application of Indigenous Knowledge in the CERP in this chapter and opportunities to improve its inclusion as part of the committee's charge to address scientific issues that could affect progress toward restoring the natural system and Indigenous peoples' reciprocal relationship with it (i.e., biocultural restoration).

The committee begins with a discussion of the place of the Miccosukee Tribe of Indians of Florida (hereafter the Miccosukee Tribe) and the Seminole Tribe of Florida (hereafter the Seminole Tribe) in the Everglades and the importance of the Everglades ecosystem to their cultures. Incorporation of Indigenous Knowledge into restoration planning and management requires active and effective Tribal consultation and engagement. Therefore, the current regulatory and policy context for state and federal Tribal consultation is presented, followed by an evaluation of the recent history of Tribal consultation and engagement. Indigenous Knowledge and its value to existing Everglades science and management is discussed, along with opportunities for improving the inclusion of Indigenous Knowledge in CERP processes. Finally, best practices for Tribal consultation and engagement with Indigenous Knowledge are presented based on recent practices initiated by the Miccosukee Tribe and experiences elsewhere in Indian Country.

Information for this chapter was gathered through public meetings, analysis of primary literature, and interviews with leaders and staff of the Miccosukee Tribe. The Seminole Tribe did not engage directly with the committee during this cycle, although available Tribal correspondence to CERP agencies regarding several projects was reviewed. As a consequence, this chapter draws heavily upon responses from and engagement with the Miccosukee Tribe.

PLACE OF MICCOSUKEE AND SEMINOLE TRIBES IN THE EVERGLADES AND ITS SIGNIFICANCE TO THEIR LIVELIHOODS AND CULTURES

The Florida Everglades is the unceded homeland of the Miccosukee and Seminole Tribes. Both Tribes have a deep and unique relationship with the lands, waters, biota, and ecosystem processes that make up the greater Everglades. For generations the Everglades has been an integral part of their spiritual, cultural, political, economic, and familial social fabric. The Miccosukee and Seminole Tribes have persisted in their homelands throughout brutal waves of European colonization and control of Florida by the U.S. government, three significant wars, numerous failed or broken treaties, land seizure, forcible removal in the 1800s, and displacement from their homes by increasing settlement and housing development. The Miccosukee and Seminole Tribes originated as members of the Muscogee Creek of Alabama, Georgia, and northern Florida, with ancestral ties to the Calusa people and other Native American Tribes, who moved south beginning in the early 1700s, replacing other Native people who did not survive the earlier incursions of Europeans into Florida. Both Seminoles and their Muscogee Creek ancestors were among those forcibly moved to Oklahoma on the Trail of Tears in the 1830s (Sturtevant and Cattelino, 2004).

Once occupying lands throughout southern and central Florida, the Seminole and Miccosukee Tribes now inhabit a fraction of their historical homelands. The Seminole Tribe is now scattered across six spatially disjunct small reservations in South Florida—Big Cypress, Brighton, Fort Pierce, Hollywood, Immokalee, and Tampa Reservations—totaling approximately 90,000 acres.[1] The Miccosukee Tribe has four reservations in south central Florida totaling a land area of approximately 75,000 acres—the Alligator Alley Reservation, Tamiami Trail Reservation, and two small Krome Avenue Reservations.[2] The Miccosukee Tribe also holds a perpetual lease to 189,000 acres of Water Conservation Area 3A (WCA-3A) (*Miccosukee Tribe of Indians of Florida v. U.S.*, 656 F. Supp. 2d 1375, 1378 [S.D. Fla. 2009]; Godfrey and Catton, 2011) (Figure 3-1). Both Tribes are permitted "to continue their usual and customary use and occupancy of Federal

---

[1] See https://www.semtribe.com/history/seminoles-today.
[2] See https://r4data.response.epa.gov/r4rrt/miccosukee-tribe.

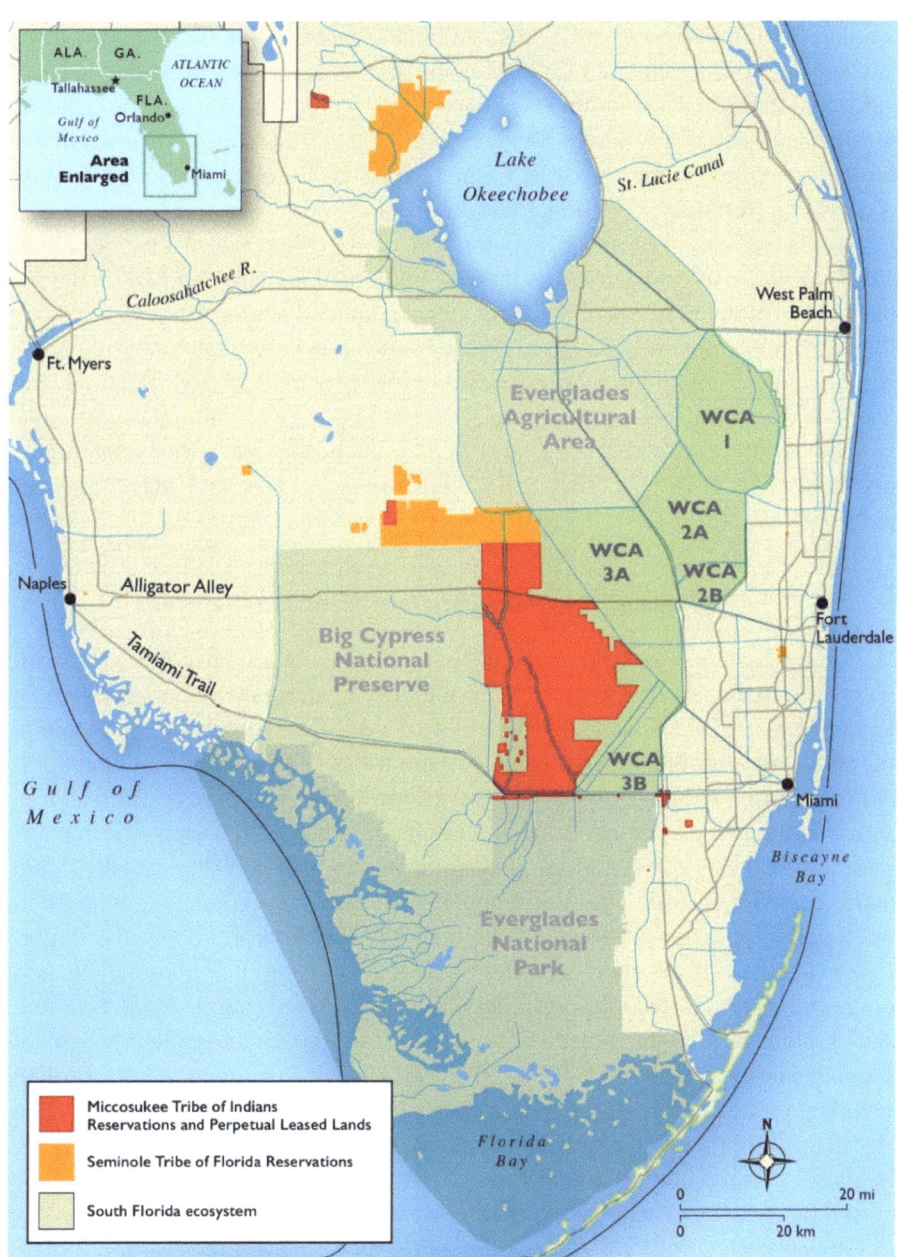

FIGURE 3-1 Map of reservation and perpetual leased lands of the Seminole Tribe of Florida and the Miccosukee Tribe of Indians.

SOURCE: Map by International Mapping.

or federally acquired lands and waters within the [Big Cypress National] preserve and the Addition,[3] including hunting, fishing, and trapping on a subsistence basis and traditional tribal ceremonials" (16 U.S.C.A. §698, West 2010; Prior, 2013). In 1957 and 1962, the Seminole and Miccosukee Tribes, respectively, gained formal recognition from the United States, thus legally establishing their status as sovereign, domestic dependent nations (Adams, 2016).

The Everglades represents a major part of the Miccosukee and Seminole Tribes' identities as a location for spiritual, traditional, kinship, and cultural connections and activities, including use of native plant and animal foods (Carr, 2002; Sturtevant and Cattelino, 2004), and this connection is passed from generation to generation. Through lived experience and intergenerational oral traditions, the Tribes of the Everglades acquire and preserve knowledge, cultural legacy, and traditions with storytelling, song, and performance—practices that continue to this day (Fixico, 2017; Jackson, 2014; LeBrasseur and Freark, 1982; Sturtevant and Cattelino, 2004).

The Significance of Everglades Tree Islands to the Tribes

Although there are many issues and aspects of Everglades restoration that are of concern to the Miccosukee and Seminole Tribes, such as water quality, habitat degradation, and the impacts of invasive species, tree islands are particularly illustrative because of their significance to the Tribes. Tree islands are topographical features in the Everglades formed by geomorphological processes. Most tree islands in the Everglades are elongated islands oriented with a long axis that follows historical water flow patterns (Sklar and van der Valk, 2002). Tree islands are characterized by woody vegetation on the upstream portion that is intolerant to prolonged inundation (Figure 3-2). In contrast, peripheral and downstream portions of tree islands are often occupied by vegetation tolerant of intermediate and longer hydroperiods (Sklar and van der Valk, 2002). These unique features of tree islands, along with their tendency to be nutrient sinks, contribute to their high plant diversity (Heisler et al., 2002; NRC, 2012) and diverse invertebrates and wildlife, including birds, mammals, reptiles, and amphibians (Meshaka et al., 2002). Tree islands are important to ecosystem processes and biodiversity, and they are among the most vulnerable landscape features of the Everglades to hydrological changes caused by decades of adverse water management practices.

For centuries, the Native peoples of the Florida Everglades—Seminoles, Miccosukee, and those that dwelled there before them—have held deep connections to tree islands. The significance of tree islands to the Miccosukee and Seminole Tribes cannot be overstated. Tree islands within the Everglades

---

[3] The "Addition" refers to the expansion of the Big Cypress National Preserve by about 146,000 acres in 1988. See https://www.nps.gov/bicy/learn/management/addition-lands-gmp.htm.

FIGURE 3-2 Tree islands in the Everglades landscape on the edge of the Big Cypress National Preserve, in WCA–3A.

NOTE: Tree islands are embedded in a matrix of sawgrass plains and ridges, emergent marshes, and deepwater sloughs.

SOURCES: Courtesy of D. Kilbane; Wetzel et al., 2005.

---

have supported every aspect of Seminole and Miccosukee peoples' lives. Tree islands have provided places to grow a variety of crops and find food, medicines, and materials for shelter and canoes, which historically provided not only a means of subsistence but also trade and self-determination in the region (Goss, 1995; Sturtevant and Cattelino, 2004). They also provided critical refuge and protection against disease, slavery, massacre, and expulsion by European colonizers and the U.S. government (and allies), and later as retreat from displacement by development of the southeast coast of Florida (Covington, 1993; Cypress, 2023; Reséndez, 2016; Sturtevant and Cattelino, 2004). It is not an exaggeration to say that the Miccosukee and Seminole Tribes would not exist in Florida today were it not for the safe harbor tree islands provided deep within the Everglades.

Tree islands are sensitive to changes in water depth, flow, quality, and hydroperiod; altered fire regimes; and invasive species, and they have been severely degraded by prolonged dry and wet hydrologic extremes (Box 3-1 and Figure 3-3). At least 70 percent of tree island land cover has been lost from the Everglades since 1940 (Sklar and van der Valk, 2012; Sklar et al., 2005), and the

BOX 3-1
Drivers of Tree Island Loss

Water management over the past 60 years caused the northern part of WCA-3A to become drier, increasing peat subsidence and fire intensity and frequency, which has led to reduced tree island elevations and tree island loss (Wetzel et al., 2005). Accordingly, much of the tree island acreage in northern WCA-3A has been lost, and because of subsidence, some remaining tree islands have experienced greater inundation during wet weather, alterations in vegetation, and reduction in biodiversity (Sklar et al., 2005). Tree islands in the northern parts of WCA-3A have increased vulnerability to fires because of their drier conditions (Cypress, 2023; NRC, 2010). In contrast, tree islands in the southern areas of WCA-3A experience higher water depths, longer hydroperiods, and ponding. These higher water depths have drowned hardwood species, shifted vegetation to more flood tolerant species, and reduced wildlife (Sklar et al., 2005). Most of these losses will require decades to centuries to recover ecological function under ideal conditions (NRC, 2012).

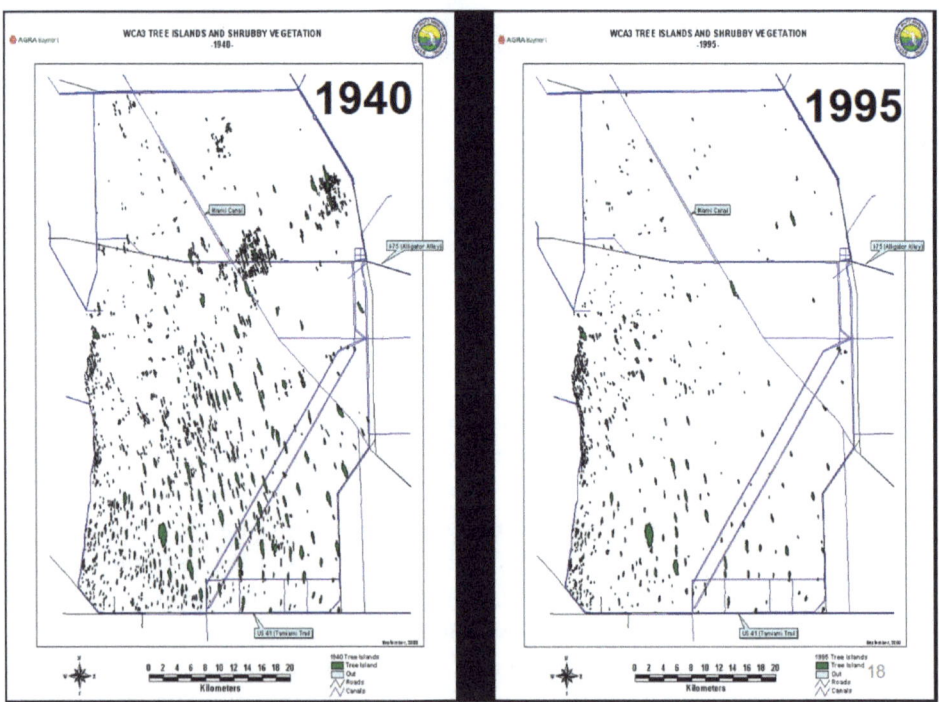

FIGURE 3-3 Maps showing the distribution of tree islands in WCA-3 in 1940 and 1995.

NOTES: Losses have been attributed to fires, subsidence, or extreme high or low water levels. Almost all remaining tree islands occur in WCA-3, Arthur R. Marshall Loxahatchee National Wildlife Refuge, and Everglades National Park.

SOURCE: https://apps.sfwmd.gov/ci/publicmeetings/viewFile/26687.

spatial extent of tree islands in WCA-3 declined by 61 percent between 1940 and 1995 (Patterson and Finck, 1999). The remaining tree island distribution overlaps significantly with the Miccosukee Tribe's reservations and current lease holdings in WCA-3A (Figures 3-1 and 3-3).

The degradation and loss of tree islands have had profound impacts on the Miccosukee Tribe. By 1960, the last family still living on the tree islands was forced to move because of uninhabitable conditions (Cypress, 2023). The Miccosukee Tribe notes that sloughs south of Tamiami Trail have become overrun with grasses and sedges due to water impoundment in WCA-3A, compounded by invasive plant species and elevated nutrients in inflows, preventing navigation by canoe and isolating Miccosukee villages in this region. Flooding has reduced opportunities for hunting as wildlife diversity decreased, and it has also interrupted the practice of ceremonies and cultural activities (Cypress, 2023).

Living in and relying on the Everglades in a deeply connected way offers unparalleled understanding of the ecosystem. The Miccosukee and Seminole Tribes' empirical knowledge and comprehensive understanding of the ecosystem is drawn from their history, current life experiences, and cultural practices. This connection to the landscape as a way of being provides the basic foundation of Indigenous Knowledge. If applied to restoration planning, monitoring, and management, Indigenous Knowledge could enhance restoration outcomes for Tribes, CERP agencies, and stakeholders alike. Careful consideration of Tribal connections to the land and their knowledge of the ecosystem will help achieve a more holistic biocultural restoration of the Everglades, in which the biophysical and sociocultural components of the ecosystem are recognized as interdependent and reciprocal (Lyver et al., 2016; Sena et al., 2022; Winter et al., 2020). Reciprocity is a cornerstone of Tribal identity and lived experience. In short, reciprocity refers to "the Earth, understood as a constantly renewing source of gifts, [and] humans having a responsibility to reciprocate for all they have been given" (Kimmerer, 2017).

In this light, an ongoing practice of meaningful engagement is necessary for effective inclusion of Indigenous Knowledge in the restoration process. In the following sections, the committee reviews the legal requirements for Tribal collaboration and consultation, reviews the recent history of collaboration, and offers context and recommendations for meaningful engagement on tree island restoration.

COLLABORATION WITH EVERGLADES TRIBES

In this section of the report, the committee describes the responsibilities for collaboration between state and federal governments with Tribes as well as the history and means of collaboration. Consultation between the lead CERP agencies—the U.S. Army Corps of Engineers (USACE) and the South Florida Water Management District (SFWMD)—and the Miccosukee and Seminole Tribes is emphasized, while noting that other federal agencies also have Tribal trust responsibilities and play a role in the CERP.

State Duties to Consult

CERP authorizing and implementing legislation obligates the State of Florida and its agencies involved in restoration to consult with the Tribes (discussed further in the next section). A unique relationship between the SFWMD and the Seminole Tribe was established by a Water Rights Compact that was approved by Congress and ratified by the Florida Legislature in 1987 as part of the Seminole Indian Land Claims Settlement Act of 1987 (Public Law 100-228). The Water Rights Compact of 1987 ensures Tribal water rights and establishes rules and authorities for managing water quality and quantity on Seminole lands (Shore and Strauss, 1990). To date, there is no similar agreement between the State of Florida and the Miccosukee Tribe.

Federal Duty to Consult and Protect

Since 1831,[4] federal law establishes that the U.S. government has a trust responsibility to Indian Tribes. Although the trust responsibility has often been ignored, the courts have clearly stated that it consists of "the highest moral obligations that the United States must meet to ensure the protection of tribal and individual Indian lands, assets, resources, and treaty and similarly recognized rights" (DOI, 2014). The CERP authorizing legislation under the Water Resources Development Act of 2000 (WRDA 2000; Public Law 106-541) identified responsibilities and consultation requirements throughout the restoration process for the state and federal agencies. WRDA 2000 noted that "tribal lands designated and managed for conservation purposes, as approved by the tribe" were part of the natural system inclusions in the CERP. The CERP authorizing legislation also stated that with respect to the restoration, "the Secretary of the Interior shall fulfill his [sic] obligation to the Indian tribes in South Florida under the Indian trust doctrine as well as other applicable legal obligations." The CERP required the Secretary of the Army to consult with both Tribes in the promulgation of programmatic regulations and stated that nothing in the authorizing legislation would impact existing water rights or existing legal sources of water for the Tribes (Box 3-2).

---

[4] *Cherokee Nation v. Georgia*, 30 U.S. 1, 16 (1831).

> **BOX 3-2**
> **Consultation Rights of Tribes of South Florida**
>
> Five Tribes have rights to consultation and coordination as a result of their ancestral links to the Florida Everglades: the Miccosukee Tribe, the Seminole Tribe, Thlopthlocco Tribal Town, the Seminole Nation of Oklahoma, and the Muscogee (Creek) Nation (C. Thomas, USACE, personal communication, 2023). The latter three were removed from the region but maintain active interest in the Florida Everglades. The Miccosukee Tribe and Seminole Tribe currently reside in a portion of their Everglades homelands and were the sole Tribes to be explicitly included in WRDA 2000 and the Programmatic Regulations. Like all federally recognized Tribes, the Miccosukee and Seminole Tribes are independent, sovereign nations that enjoy a government-to-government relationship with the U.S. federal government; they are not simply another stakeholder group. The two Tribes have distinctive language, cultures, governance, and Indigenous Knowledge.

The Programmatic Regulations (FR 66 No. 218) established the requirement for the USACE and the SFWMD to consult with the Miccosukee and Seminole Tribes (and other agencies) as they implement the CERP "to achieve and maintain the benefits to the natural system and human environment described in the Plan." This consultation encompasses changes to water reservations; the development of System and Project Operating Manuals, the adaptive management program, and a master sequencing plan; the periodic "evaluation of the Plan using new or updated modeling that includes the latest scientific, technical, and planning information," known as the Periodic CERP Update; modifications to the Plan; and reports to Congress. RECOVER, as the coordinating office for the adaptive management program, is required to follow federal and state responsibilities and consult with the Tribes on documents or work products. Consultation with Tribes is defined in the Programmatic Regulations as follows:

> In addition to any other applicable provision for consultation with Native American Tribes, including but not limited to, laws, regulations, executive orders, and policies the Corps of Engineers and non-Federal sponsors shall consult with and seek advice from the Miccosukee Tribe of Indians of Florida and the Seminole Tribe of Florida throughout the implementation process to ensure meaningful and timely input by tribal officials regarding programs and activities covered by this part. Consultation with the tribes shall be conducted on a government-to-government basis.

The implementing regulations require the USACE and the SFWMD to encourage participation of the Tribes (and other agencies) on Project Delivery Teams and on RECOVER.

Aside from CERP-specific authorities and regulations, executive orders have also established various terms for federal agencies to consult with Tribes that affect Everglades Restoration and the many federal agencies involved. Executive Order 13175 (EOP, 2000) was issued "to establish regular and meaningful consultation and collaboration with tribal officials in the development of Federal policies that have tribal implications, to strengthen the United States government-to-government relationships with Indian tribes, and to reduce the imposition of unfunded mandates upon Indian tribes." These requirements and responsibilities for consultation obligate every federal agency action that may impact Tribal resources, which can include resource management activities of the U.S. Department of the Interior agencies (National Park Service, U.S. Fish and Wildlife Service [FWS], Bureau of Indian Affairs); other federal agencies such as the U.S. Department of Agriculture, the U.S. Department of Energy, the U.S. Environmental Protection Agency, and the Federal Emergency Management Agency, and others as directed by the federal government.

Recent executive directives emphasized the value of Indigenous Knowledge as an important element of the federal Indian trust responsibility (Box 3-3). In 2022, a memorandum from the Executive Office of the President (EOP, 2022a) called upon federal agencies to "pursue and promote inclusion of Indigenous Knowledge in Federal scientific and policy decisions . . . including Tribal consultation action plans" (The White House, Office of the Press Secretary, 2022). Accompanying implementation guidance (EOP, 2022b) was provided to assist agencies "in (1) understanding Indigenous Knowledge, (2) growing and maintaining the mutually beneficial relationships with Tribal Nations and Indigenous Peoples needed to appropriately include Indigenous Knowledge, and (3) considering, including, and applying Indigenous Knowledge in Federal research, policies, and decision making" as outlined in the initial guidance (EOP, 2022a; The White House, Office of the Press Secretary, 2022). The Executive Office of the President (EOP, 2022b) notes that Indigenous Knowledge is a valid form of evidence that can be a source of accurate information, valuable insights, and effective practices. These executive directives apply to all the work of the USACE and other federal agencies working on Everglades Restoration.

ASSESSMENT OF CERP TRIBAL CONSULTATION AND COLLABORATION OVER TIME

Consultation and coordination with Tribes have been part of the Florida Everglades Restoration Project mandate from its inception. However, it has not always been honored. The volume and focus of lawsuits brought by the Tribes early in CERP planning and implementation are an indication of the adverse consequences of lack of consultation and failure to include Tribal needs, perspectives, and

> **BOX 3-3**
> **Executive Orders and Memoranda from the Executive Office of the President Relating to Indigenous Knowledge**
>
> - Executive Order 13990 (January 20, 2021) on *Protecting Public Health and the Environment and Restoring Science to Tackle the Climate Crisis,* which recognized the value of traditional knowledge and directed the establishment of a Federal Task Force and Tribal Advisory Council (EOP, 2021a).
> - Executive Order 14049 (October 11, 2021) on the *White House Initiative on Advancing Educational Equity, Excellence, and Economic Opportunity for Native Americans and Strengthening Tribal Colleges and Universities,* which committed to promoting Indigenous learning through the use of traditional ecological knowledge (EOP, 2021b).
> - Office of Science and Technology Policy (OSTP) and Council on Environmental Quality (CEQ) Memorandum (November 15, 2021) on *Indigenous Traditional Ecological Knowledge (ITEK) and Its Role in Federal Decision Making,* which acknowledges that ITEK is owned by the Indigenous people that collected it (EOP, 2021c).
> - Executive Order 14072 (April 22, 2022) on *Strengthening the Nation's Forests, Communities and Local Economies,* in which policy was enacted to support Indigenous Traditional Ecological Knowledge and cultural and subsistence practices in our national forests (EOP, 2022c).
> - Executive Office of the President Memorandum (November 30, 2022) on *Uniform Standards for Tribal Consultation,* which established baseline standards across all agencies for consulting Tribal Nations (EOP, 2022d).
> - Executive Office of the President Memorandum (December 1, 2022) on *Guidance for Federal Departments and Agencies on Indigenous Knowledge,* which assists agencies in understanding Indigenous Knowledge, developing relationships with Tribal Nations, and considering, including, and applying Indigenous Knowledge to federal research, policies, and decision making (EOP, 2022a).
> - Executive Office of the President Memorandum (December 1, 2022), on *Implementation of Guidance for Federal Departments and Agencies on Indigenous Knowledge,* which sets expectations for agencies to implement engagement with Tribal Nations, and inclusion of Indigenous Knowledge according to the Guidance (EOP, 2022b).
> - Executive Order 14096 (April 21, 2023) on *Revitalizing our Nation's Commitment to Environmental Justice for All,* which stated clearly that Indigenous people were to be recognized for having unique knowledge of their resource and were to be included in decision making to ensure equity in process (EOP, 2023a).
> - Executive Order (December 6, 2023) on *Reforming Federal Funding and Support for Tribal Nations to Better Embrace Our Trust Responsibilities and Promote the Next Era of Tribal Self-Determination,* which in addition to increasing the flexibility and utility of federal funding and support programs for Tribes, orders agency heads to "respect Tribal data sovereignty and recognize the importance of Indigenous Knowledge by, when appropriate and permitted by statute, allowing Tribal Nations to use self-certified data" (EOP, 2023b).

knowledge. Against this background, it is noteworthy that current representatives of the USACE, the SFWMD, and the Miccosukee Tribe each expressed that there is a growing culture of coordination between the agencies and the Tribes.

The USACE Jacksonville District and the SFWMD each employ a Tribal liaison who manages frequent interactions with both the Miccosukee and Seminole Tribes. The SFWMD liaison endeavors to interact with the Tribes daily "as neighbors" (A. Ramirez, SFWMD, personal communication, 2023). In addition, the liaison coordinates regular interactions between Tribal leadership and SFWMD leadership, with a goal of a minimum of two meetings per year. The USACE Tribal liaison interacts extensively with the Miccosukee and Seminole Tribes, generally logging multiple contacts per day. The USACE liaison coordinates consultation between USACE and Tribal leadership, as well as scientists and Tribal staff. The USACE Tribal liaison also participates in key events hosted by the Tribes throughout the year and ensures appropriate participation by other USACE staff and leaders (C. Thomas, USACE, personal communication, 2023). Both agency liaisons indicate that effective engagement with the Tribes requires constant commitment to cultivating and maintaining a long-term relationship.

Both Tribes have also had strong support from the Environmental Protection Agency and have successfully gone through the federal administrative process to establish water quality standards for areas in their jurisdiction (EPA, 2024). Notably, the Miccosukee Tribe was the first entity in the State of Florida to establish a numeric nutrient criterion for phosphorus—10 ppb for their Outstanding Waters (Godfrey and Catton, 2011).

In the following section, the committee discusses the evolution of CERP Tribal engagement over time through presentations of three examples: Lake Okeechobee Watershed Restoration Project (LOWRP), Western Everglades Restoration Project (WERP), and operations of the S-12A and S-12B structures. Then, the committee provides a high-level assessment of current consultation and engagement in the context of establishing a necessary foundation for engagement with and application of Indigenous Knowledge in CERP planning and management.

Examples of the Evolution of CERP Engagement

Lake Okeechobee Watershed Restoration Project

The LOWRP is an example of a project that experienced lengthy delays because of poor quality of Tribal engagement and/or lack of serious consideration of Tribal concerns. Although the time frame of this project overlapped with the more positive example from WERP (see next section), the LOWRP differed in its consideration of Tribal input. An early Lake Okeechobee Watershed planning effort began in the early 2000s but was halted in 2006. LOWRP planning was restarted in 2016, and early on in this process, the Seminole Tribe objected to a large storage reservoir near its Brighton Reservation. A draft Project Implementation Report (PIR) was released in 2018 that proposed a shallow 46,000 acre-feet (AF) above-ground water storage feature (termed "wetland attenuation feature"; Figure 3-4) after considering two alternatives with shallow reservoirs near the Brighton Reservation and one deep reservoir

FIGURE 3-4 The tentatively selected plan for the 2020 Final PIR for the LOWRP shows the location of the shallow storage feature relative to the Brighton Reservation.

SOURCE: USACE and SFWMD, 2020b.

---

at a more distant site (USACE, 2024f). The Seminole Tribe voiced numerous objections to this plan, expressing concerns about the proximity of the storage features to Brighton Reservation, potential impacts to cultural resources, potential flooding, and lack of involvement of the Seminole Tribe in project planning when the preliminary alternatives were identified and screened (USACE and SFWMD, 2020b). The Seminole Tribe also expressed concerns that alternative water storage locations were not given due consideration and evaluated equally (Osceola, 2019).

The "Final" PIR was released in October 2020, but the features were essentially unchanged despite these concerns (see Box 3-4 for an overview of major LOWRP decision points and the stated rationale). However, in 2021 under a new administration, the proposed plan was not approved by USACE headquarters "due in part to concerns raised by the Seminole Tribe of Florida" (USACE, 2024f), and efforts refocused on a previously rejected alternative. As of 2024, planning is still ongoing (Box 3-4). The LOWRP is an example of how lack of meaningful engagement with Tribes led to delays in implementation and significantly impacted planning efforts.

BOX 3-4
Evolution of the Lake Okeechobee Watershed Restoration Project

2016: Project planning was launched.

2018: A draft PIR was released that proposed a shallow 46,000-AF above-ground water storage feature (termed "wetland attenuation feature") located near the Brighton Reservation, 80 aquifer storage and recovery (ASR) wells, and approximately 4,800 acres of wetland restoration (Figure 3-4).

2020: Despite objections by the Seminole Tribe throughout the planning process and explained in a letter to the USACE (Osceola, 2019), the major features of the plan were unchanged. USACE and SFWMD (2020b) stated,

> Throughout the LOWRP planning process, the project has been modified based on Tribal and stakeholder feedback to reconfigure the surface storage footprint to avoid direct northern proximity to Brighton Reservation, avoid a known significant cultural site, reduce the depth of the surface storage pool, provide a buffer between the surface storage feature and Brighton Reservation and Tribal lands, and provide a greater buffer for future commercial development along State Road 78 not approved by USACE headquarters. . . . The project will be designed so there are no changes to flood protection caused by the project.

The 2020 PIR also stated reasons why an alternative located farther from the Brighton Reservation was not selected:

> 1) this alternative is significantly more expensive than the other alternatives, 2) this alternative proposes deep reservoir storage, which increases overall seepage concerns, 3) this location does not allow co-location of the reservoir with ASR wells, which reduces the overall operational flexibility of the reservoir, 4) entire surface storage lands for this alternative are privately owned, increasing impacts on local landowners and increasing overall real estate administrative and acquisition costs, 5) the entire reservoir footprint for this alternative contains potential habitat for critically-endangered Florida grasshopper sparrows, 6) the reservoir in this alternative impacts the largest amount of wetlands of all the alternatives, and 7) the deep reservoir storage in this alternative is less suitable for the growth of wetland vegetation within the reservoir footprint than the other two alternatives that include shallow surface storage.

2021: USACE headquarters rejected the plan, "due in part to concerns raised by the Seminole Tribe of Florida" (USACE, 2024f).

2022: The project was subsequently revised to remove the wetland attenuation feature and reduce the number of ASR wells to 55 (USACE and SFWMD, 2022a,b), but the 2022 PIR was not approved by USACE headquarters "due to concerns with risks posed by the ASR system and the increase in estimated costs" (USACE, 2024f). The planning team was then advised to reconfigure the tentatively selected plan to consider other above-ground storage alternatives, including those previously screened out (USACE, 2024f).

2024: In February 2024, the SFWMD released a Section 203 final feasibility study for the Lake Okeechobee Component A Reservoir (SFWMD, 2024b)—a 200,000-AF reservoir in the same footprint as one of the 2018 draft PIR alternatives that was not selected. The LOWRP PIR is being reconfigured into a fourth revised draft.[a]

---

[a] See https://www.saj.usace.army.mil/LOWRP.

Western Everglades Restoration Project

WERP represents an example of positive progress toward more effective Tribal consultation. The western Everglades, which covers an area of 1,200 mi^2, encompasses the Big Cypress Reservation of the Seminole Tribe of Indians and is bounded by the Miccosukee Tribe reservations to the east and south. As discussed in Chapter 2, the western Everglades has been negatively impacted by hydromodifications from the Central and South Florida Project and high nutrient inflows from upgradient agricultural land uses, resulting in extreme dryness, poor water quality, habitat destruction, and other impacts on Seminole and Miccosukee Tribal lands. In the early 2010s, the Tribes were concerned that substantial efforts were being directed to the Central Everglades Planning Project (CEPP) but CERP planning was not addressing key Tribal concerns in the western Everglades. During the December 2012 meeting of the South Florida Ecosystem Restoration Task Force (Task Force), the Seminole Tribe filed a minority report concerning unmet environmental water needs in the Big Cypress Reservation including poor water quality and insufficient water flows into the Big Cypress Reservation and Big Cypress National Preserve (SFERTF, 2012). The Miccosukee Tribe expressed similar concerns and was particularly concerned about water quality in the L-28 interceptor canal (L-28i) and its impact on WCA-3A, which is designated as Outstanding Miccosukee Waters Tribe, deemed "essential to the survival of the Miccosukee Tribe" (Miccosukee Tribe of Indians of Florida, 2021). This 2012 Task Force meeting was an important turning point for Tribal consultation and engagement that launched a concerted effort to advance restoration progress in the western Everglades. A subgroup of the Task Force was established in early 2013 to address the concerns raised by the Tribes, and planning for the WERP began in August 2016 (see Chapter 2 for discussion of WERP progress).

A common theme that emerged in early meetings of the Task Force subgroup was the lack of data and appropriate modeling tools with which to analyze alternatives for the area. Little was known about the hydrology of the western basin, particularly the interactions between surface and ground waters. Existing model boundaries needed to be expanded to analyze alternatives for the region, reflecting a prior lack of priority for restoration of those regions. The Seminole Tribe cited lack of critical data for the studies, and the Miccosukee Tribe noted the need for a data inventory, assessment of data gaps, and a plan to develop better datasets (SFERTF, 2014). The need for data and new analytical tools was noted again in 2016 (SFERTF, 2016).

Since WERP planning began in 2016, both Tribes have been involved in consultation and coordination (USACE and SFWMD, 2023a), and the Seminole Tribe has also been involved as an official cooperating agency under the National Environmental Policy Act (NEPA) (Billie, 2016). Throughout the WERP planning process, the agencies and the Tribes have actively communicated through dedicated meetings, written communications, and participation at public meetings. Additionally, the draft PIR (USACE and SFWMD, 2023a) outlines a plan to apply Indigenous Knowledge from the Seminole Tribe to determine flows through the S-223 into the Seminole Tribe of Florida Native Area, with the exact details to be outlined in the final Project Operating Manual. Both Tribes have expressed strong support for the WERP tentatively selected plan, as evidenced through their communications to the Assistant Secretary of the Army (Cypress, 2022; Osceola, 2023).

The WERP process has been lengthy and challenging—from the initial Minority Report filed in 2012, to the recently released draft PIR (USACE and SFWMD, 2023a), with the study extended twice since its inception in 2016. Considerable time and effort have been expended to build trust and a well-functioning process between the agencies and the Tribes. Because trust is of paramount importance to this process, both Tribes specifically asked for assurances that their involvement and cooperation throughout the process will not result in land condemnation or Tribal members losing any part of their lands (Cypress, 2020; Osceola, 2020).

CEPP/Combined Operational Plan (COP) Operations

Management of water in WCA-3A has been a source of contention for the Miccosukee Tribe since the 1980s because of its adverse effects on the tree islands that are so fundamental to their culture and identity (Cypress, 2023) (see above; Box 3-1). For the past 25 years contention has especially focused on seasonal closures of the S-12A and S-12B structures through which water flows from southwestern WCA-3A across the Tamiami Trail and into Western Shark River Slough in order to maintain dry conditions for endangered Cape Sable seaside sparrows (*Ammodramus maritimus mirabilis*) inhabiting the marl prairies adjacent to the slough during their nesting season (Box 3-5). These seasonal closures cause ponding of water in southwestern WCA-3A, exacerbating flooding of tree islands and affecting their native flora and fauna.

Until recently, the Tribe has had no success in affecting a change in the operation of the S-12s or the nearby S-343 structures (Box 3-5) to alleviate this problem, despite persistent attempts. The structures have been opened during the scheduled closures during emergency deviations, but these deviations have been driven primarily by flood control and levee safety (USACE et al., 2023a), and not by the needs of the Tribe for relief from high water. The needs of the sparrow—and avoiding yet another jeopardy opinion for violating the Endangered Species Act—have consistently taken precedent over the needs of the Tribe.

BOX 3-5
Water Management Challenges Affecting Flows into Western Shark River Slough

Prior to the construction of the WCAs in the 1960s, approximately two-thirds of the flow into Shark River Slough came through Northeast Shark River Slough and one-third through Western Shark River Slough (see Figure 2-21 showing distribution of flows). After construction of the WCAs, conditions in Western Shark River Slough became much wetter (90 percent of total flow) and in Northeast Shark River Slough much drier (10 percent of total flow), producing a myriad of adverse ecological effects, including effects on tree islands in WCA-3A (NASEM, 2021).

Attempts to restore the historic distribution of flow, notably the Experimental Water Deliveries Program (1983–1999), had little success (Figure 2-21), chiefly because of flood mitigation constraints protecting residences in the 8.5 Square Mile Area (Las Palmas) that affected flows into Northeast Shark River Slough. The limitations of the water management regime resulted in a crisis when large regulatory releases through the S-12A and S-12B structures into Western Shark River Slough necessitated by high water levels in 1993–1995 nearly extirpated Cape Sable seaside sparrow subpopulation A adjacent to the slough (Figure 3-5). In response to these impacts on the sparrows, FWS issued a Jeopardy Opinion on the Experimental Water Deliveries Program in 1999, effectively ending the program and necessitating new water management (NASEM, 2023).

Subsequently, various operational plans governed water management at the boundary of WCA-3A and Everglades National Park from 2000 to 2016, all of which included seasonal closures of S-12A and S-12B, as well as S-343A, -343B, and -344, through which water in southwestern WCA-3A can be released into the Big Cypress National Preserve (Figure 3-5), in order to ensure that sparrow habitat is suitably dry during their (March to mid-July) nesting season. Subject to the same constraints that plagued previous water management efforts, these plans had little success in redistributing flow in Shark River Slough (Figure 2-21) or protecting sparrows. FWS issued a Jeopardy Opinion on the impact of the last of these, the Everglades Restoration Transition Plan (2012-2016), because of its impact on the sparrows (FWS, 2016). With limited capacity to convey water into Northeast Shark River Slough, closure of the S-12s and S-343s resulted in increased water levels in southern WCA-3A during wet conditions.

The COP, which was fully implemented in 2020, has made progress moving flows from Western to Northeast Shark Slough (Figure 2-21), but seasonal closures remain in place under the COP. Under baseline COP operations, S-12A and S-12B and the S-343s are generally closed October 1 and are not re-opened until the sparrow nesting season ends in mid-July. However, under specified conditions S-12A can remain open until November 1, and S-12B until December 1 (USACE, 2020a). Deviations to this schedule have occurred in 2020 and in 2023 under the COP (USACE et al., 2023a), as they did under the various operational plans in effect during 2000–2016.

continued

BOX 3-5 Continued

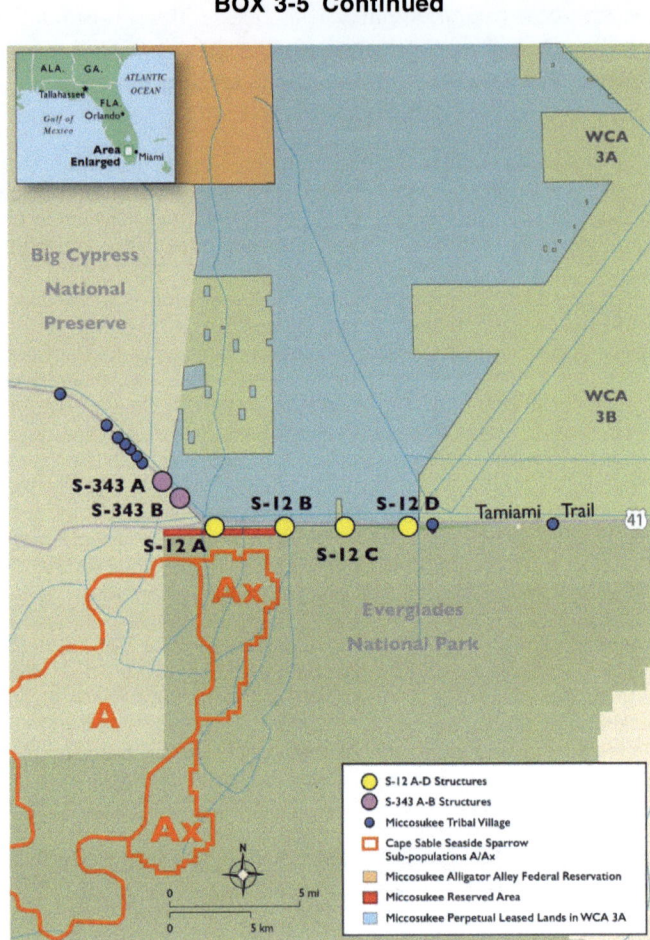

FIGURE 3-5 Area of emergency deviation, showing S-12A, -12B, -343A, and -343B structures.

SOURCE: Map by International Mapping.

However, the most recent deviation has been different. In fall 2023, heavy rains in September created high-water conditions in WCA-3A, flooding tree islands. Following up on an earlier letter that articulated the Tribe's Indigenous Knowledge relevant to the adverse impacts of the seasonal closures (Cypress, 2023), vetted through the Tribe's peer-review process (Ornstein, 2024; see below), in October 2023 the Tribe appealed to the FWS to allow the gates to be opened. Although levels did not reach those that trigger discharges under the COP, or create wildlife emergency conditions, as in past emergency deviations, forecast models indicated that these levels were likely to be reached in the near future. Additionally, no sparrows had been detected in habitat adjacent to Western Shark River Slough since 2018. The USACE proposed a planned temporary deviation, and the FWS responded to the Tribe indicating they would support it (L. Williams, 2023). A planned temporary deviation was declared and the S-12A, -12B, -343A, and -343B structures were opened in November 2023, with provisions to open them again under specified conditions through the remainder of the seasonal closure period, through July 14, 2024 (Ehlinger, 2023).

Tribal engagement and consideration of Indigenous Knowledge played an important role in the decision to proactively open the S-12 and S-343 structures to address high-water conditions (G. Ralph, USACE, personal communication, 2024). The Tribe provided a wealth of relevant information prior to the event (Cypress, 2023), and it is clear that the Tribe was highly engaged in the decision process. For the first time in decades the decision that was made about this recurring, contentious issue coincided with the Tribe's priorities.

Assessment of CERP Tribal Consultation and Coordination

The USACE and the SFWMD are fortunate to each have skilled, committed Tribal liaisons who facilitate consultation and coordination between their agencies and the Miccosukee and Seminole Tribes. The Tribal liaisons have worked to establish good relationships with Tribal staff and leadership and typically interact with them multiple times each day. They coordinate high-level government-to-government consultations, facilitate Tribal membership on project design teams, and ensure that the Tribes are invited to participate in subteams and attend agency hosted meetings. The agency Tribal liaisons also participate in Tribally led activities (C. Thomas, USACE, personal communication, 2023). The scope of their jobs and the demands on them appear to be excessive for a single individual within each agency. Furthermore, reliance on a single individual as the fulcrum for this critical function represents a risk to the continuity of good consultation and coordination should the individual abruptly be unable to fulfill the role. Progress notwithstanding, some lapses in consultation and coordination continue to occur. Given the volume of work, some measure of this may be inevitable.

However, it is worthwhile to consider and address their causes. Reasons for lapses include understaffing, simple oversight and scheduling conflicts, intensive work schedules, pressure for timely completion of agency work, and in some instances a lack of experience working with Indigenous Knowledge. Nevertheless, some CERP scientific staff consistently engage with the Tribes and consider Indigenous Knowledge in project planning (C. Thomas, USACE, personal communication, 2023), which appears to be increasingly the case. Where there has been resistance, attitudes may be changing, propelled in part by the 2022 guidance from the Executive Office of the President (EOP, 2022a) and as senior scientists learn more about Indigenous Knowledge (e.g., through careful reading of Kimmerer, 2013). The Miccosukee Tribe as well as the Tribal liaisons for the USACE and the SFWMD have stated publicly to the committee that the culture of listening and cooperation is increasing. Progress in scoping and planning for the WERP and recent deviations in operation of the S-12A and S-12B structures reflect the positive trajectory in consultation and coordination.

The Tribes may also face challenges in the consultation and coordination process because of the volume and pace of Everglades restoration work, especially given the much larger USACE and SFWMD staff sizes. It can be difficult for Tribal staff to attend to all the demands of the CERP process. Differences in the timescales on which CERP projects operate and on which Tribes deliberate may also pose challenges, emphasizing the need for early engagement and the understanding that the Tribal process may require multiple levels of Tribal concurrence and approval. In addition, although Tribes are invited to participate in the Task Force Science Coordination Team and Working Group meetings, they have not always believed that there is space for, or interest in, their concerns and knowledge (K. Cunniff, Miccosukee Tribe, personal communication, 2024). However, as relationships between the Tribes and agencies continue to develop through Tribal engagement efforts, the path toward inclusion of Indigenous Knowledge is expected to become smoother and increase in value to decision makers.

INDIGENOUS KNOWLEDGE AND THE SCIENTIFIC PROCESS

Indigenous Knowledge

Broadly speaking, Indigenous Knowledge[5] refers to the body of knowledge generated by Indigenous peoples about their environment and appropriate relationships between people and that environment. Understanding Indigenous Knowledge and its interface with western science and land management is fun-

---

[5] Over the past two to three decades, Indigenous Knowledge has been variously referred to as Indigenous Ecological Knowledge, Indigenous and Local Knowledge, and Traditional Ecological Knowledge.

damental to honoring the federal trust responsibility and the laws of the State of Florida in Everglades restoration. The Inuit Circumpolar Council[6] defines Indigenous Knowledge as

> a systematic way of thinking applied to phenomena across biological, physical, cultural, and spiritual systems. It includes insights based on evidence acquired through direct and long-term experiences and extensive and multigenerational observations, lessons and skills. It has developed over millennia and is still developing in a living process, including knowledge acquired today and in the future, and it is passed on from generation to generation.

Just as Indigenous cultures are diverse, so too is Indigenous Knowledge, its forms of expression, and its means of transmission through oral traditions (EOP, 2022a; IPBES, 2022; Robinson et al., 2021). As a result, efforts to understand and include Indigenous Knowledge in Everglades restoration require sustained commitment to understanding the specific knowledge and concerns of each Tribe.

Indigenous Knowledge and Western Science Are Different but Complementary

The fundamental differences between western science and Indigenous ways of knowing have resulted in Indigenous Knowledge being undervalued, marginalized, and often ignored in ecological restoration. Most modern restoration is based on western science, to include hypothesis testing, the scientific method, modern technology, quantitative methods, and a reliance on peer-reviewed scientific publications. In contrast, Indigenous Knowledge is rooted in an intimate holistic understanding of the environment, including the spiritual and ecological relationships between people and their environment, that is often passed down through rich traditions of oral history, rituals, and ceremony. Both pathways of knowledge bring value to ecological restoration, and both have limitations, but western institutions chronically undervalue Indigenous Knowledge because it does not follow the constructs of western knowledge systems (Zedler and Stevens, 2018).

Indigenous Knowledge is often continuously accrued and transmitted over longer timescales, sometimes for centuries to millennia, than knowledge generated from technologies and techniques of western science (Gadgil et al., 1993). It is often based on frequent (e.g., daily) observations and interactions, which is in stark contrast to periodic, seasonal, or short-term observations made in many western scientific studies. Although Indigenous understanding may not always provide the quantitative data that form the foundation of much of western science, Indigenous Knowledge is invaluable for detecting deviations from baseline

---

[6] See https://www.inuitcircumpolar.com/icc-activities/environment-sustainable-development/indigenous-knowledge.

conditions and patterns (or the "normal" range of conditions against which current conditions can be compared) that occur within ecosystems over broad timescales (Riedlinger and Berkes, 2001; also see Box 3-6). Thus, Indigenous Knowledge and western science can provide complementary insights, similar to the way that different disciplines offer unique expertise, methods, and perspectives to solving complex problems.

One shortcoming of western science and monitoring is the lack of long-term, multi-decadal knowledge of most species and ecosystems. Long-term data are especially critical to understanding gradual changes and trends in ecological processes (Hughes et al., 2017; Kuebbing et al., 2018), and are paramount to studying the demography and population dynamics of long-lived species (Clutton-Brock and Sheldon, 2010; Margalida, 2017). In recent decades, anthropogenic pressures on ecosystems have heightened the western scientific community's collective awareness that long-term research and monitoring are critical for establishing baselines and, therefore, interpretation of changes over time (Kuebbing et al., 2018). Although progress has occurred on several fronts (e.g., National Science Foundation's Long-Term Ecological Research [LTER] sites and the National Ecological Observatory Network [NEON][7]), the western scientific system still struggles to maintain studies longer than typical research funding cycles (~5 yrs) or the focus of individual researchers (approximately two to three decades in most best-case scenarios). Regardless, baselines established through western science generally are recent and developed over much shorter time periods compared to those established through Indigenous Knowledge.

There are examples of Indigenous Knowledge and western science serving as the basis for planning and implementation of successful conservation and restoration projects around the world (Box 3-7). Such collaborations may be especially effective in the development of indicators and monitoring programs (Box 3-6) (IPBES, 2022).

The lack of integration of Indigenous Knowledge in modern ecological restoration projects can manifest itself in several ways. For example, Tribal members are increasingly granted a seat at the decision-making table for restoration projects, but they are seldom placed in a leadership position of these decision-making bodies (Hernandez and Vogt, 2020). Instead, their representation is often a symbolic gesture or in fulfillment of policy requirements or administrative frameworks for best practices. This unintentional "tokenism" ultimately diminishes the importance of Indigenous Knowledge in decision making (Samuel, 2020). Such an approach is particularly problematic because Tribal members represent sovereign nations and are not merely another stakeholder in

---

[7] See https://lternet.edu and https://www.nsf.gov/news/special_reports/neon.

BOX 3-6
Case Study on Monitoring, Management, and Restoration of Natural Resources Using Indigenous Knowledge

The Cree Indians of Northern Quebec are well known for their monitoring and management of the natural resources they depend on, including moose, fish, birds, and beavers. Their cultural belief system requires that they harvest responsibly, never taking more than is given to them by the North wind, God, and the animals spirits (Feit, 1986). These cultural views, coupled with their deep multigenerational knowledge of the animals' biology and the surrounding ecosystem, have enabled the Cree to sustainably manage natural resources for centuries. But the balance they have maintained for generations has repeatedly been disrupted by non-Indigenous people, resulting in overharvest and severe population declines of species in their region. As a result, restoration of animal populations in Northern Quebec has required a combination of Indigenous Knowledge, western science, and government policy to restore the balance.

Beavers are ecosystem engineers that create and change habitats important for maintaining regional biodiversity (Wright et al., 2002) and providing ecosystem services (Thompson et al., 2021). Beavers are also among the more important animals to the Cree people (Berkes, 1998). Beavers require careful monitoring and adaptive management to maintain healthy populations, the habitats that they engineer, and the ecosystem services they provide. To accomplish this goal, the Cree rely on their local knowledge to make decisions about how many beavers can be harvested, when they can be harvested, and in what locations (Feit, 1986).

The Cree's traditional lands are divided into hunting territories that are each supervised by an individual steward, often a male elder, who maintains intimate knowledge of the status of beavers on his territory based on continuous observations of their activity and population trends, responses of local vegetation to beaver grazing, and personal knowledge of beaver harvests on their territory in prior years (Berkes, 1998; Feit, 1986). Based on the relative trends observed, they then make informed decisions about what portions of their territory, if any, can accommodate harvest. This ongoing assessment sometimes results in stewards not harvesting beavers on some or all of their territory in a particular year or until local recovery is evident, often on multi-year cycles (Berkes, 1998; Berkes et al., 2000), a practice akin to adaptive wildlife management in western science-based systems. This efficacy of Indigenous Knowledge has stood the test of time but also has been challenged by outside influences.

Since the 1700s, beaver populations in Northern Quebec have repeatedly been decimated by overharvest for the lucrative fur trade. For example, in the 1920s, non-Indigenous trappers engaged in unsustainable trapping of beaver on Cree lands, resulting in abrupt crashes in beaver populations, with important consequences for the ecosystem and Cree culture. Faced with a vanishing resource, the Cree management system broke down (Feit, 1986). In response, the Canadian government enacted new laws in the 1930s that created formal recognition and protection of Cree hunting territories, and by the 1950s beaver populations were rebounding under the Cree's effective management practices (Berkes et al., 1989). Moreover, in 1975, the Canadian government went a step further to enact legislation that gave the Cree full legal authority over beaver management in the region (Moller et al., 2004). In addition to formalizing their legal rights to manage the resource, the agreement also helped forge effective collaborations between the Cree people and western scientists (Moller et al., 2004).

> **BOX 3-7**
> **Lessons Learned: Indigenous Knowledge and Monitoring Can Provide Insights Where Western Science Falls Short**
>
> The sooty shearwater (*Ardenna grisea*, formerly *Puffinis griseus*) is one of the most common species of seabirds in the world but has experienced enigmatic population declines over the past 35+ years (Carboneras et al., 2020; Scofield and Christie, 2002; Shaffer et al., 2006). Sooty shearwaters have a broad global distribution, spending most of their life covering great distances at sea to forage and engaging in transequatorial pan-Pacific flights, but they congregate annually in exceptionally large numbers along the coast of New Zealand to breed (Carboneras et al., 2020; Shaffer et al., 2006). The Raikura Māori, Indigenous people from southern New Zealand, are permitted to harvest chicks each year in large numbers (250,000–300,000; Carboneras et al., 2020) from the surrounding Tītī islands for food, soap, oil, and trade. This practice, called muttonbirding, is an important part of the Māori's cultural identity as well as their economic well-being (Moller et al., 2004). For generations, the Māori have closely monitored their harvest in relation to their hunting effort and the body condition of the chicks harvested, and they often record their observations in multi-decadal diaries (Lyver, 2002; Lyver et al., 1999). Their observations are robust enough to document fluctuations in shearwater populations that are predictive of El Niño-Southern Oscillation patterns, as corroborated by western statistical models (Humphries and Moller, 2017; Lyver et al., 1999).
>
> Importantly, the Māori's observations also served as early evidence of long-term population declines. The Māori observed declining yields of shearwater chicks per harvest effort, with no changes to chick body condition or nesting habitat quality, leading them to conclude that shearwater populations were gradually declining and that declines were caused by factors not associated with the breeding habitat or their annual harvest. The testable hypotheses generated from the Māori led to intensive studies by western scientists to identify the mechanisms driving the population declines of shearwaters (Moller et al., 2004). Today, it is generally agreed that fisheries bycatch in nets on the open ocean, and the effects of climate change, possibly on food resources, are the primary causes of their population declines in New Zealand as well as in other portions of the world (Brooke, 2004; Department of Climate Change, Energy, the Environment and Water, 2023; Scofield, 2000; Uhlmann and Jeschke, 2011; Veit et al., 1997). Thus, the long-term knowledge of the Māori proved invaluable for detecting deviations from historical conditions as well as identifying distant factors as the most probable causes of the population declines, and it alerted western scientists to deploy technological tools and modeling to identify the underlying mechanisms (Moller et al., 2004). Both ways of knowing proved mutually beneficial for identifying and solving this conservation mystery, which could ultimately lead to management interventions (e.g., altering fishing practices).

these decision-making groups (Robinson et al., 2021). Token representation in meetings and on decision-making bodies perpetuates false western notions that Indigenous Knowledge has secondary value to western science and degrades trust between Tribal members and western practitioners.

Moreover, failure of western practitioners to effectively consider Indigenous Knowledge in ecological restoration and monitoring undermines the

> **BOX 3-8**
> **Klamath Tribes and Watershed Restoration**
>
> The Klamath River Basin is an example of where multiple Tribes have worked collectively with western science practitioners to restore land, water, and ecological resources. At the watershed level, the Karuk Tribe has focused on using traditional Indigenous Knowledge based in its cultural relationship with the Klamath watershed to establish restoration goals to enhance the social and ecological resilience to natural and anthropogenic disturbances and climate change stressors. Through a mix of methods grounded in both Indigenous and western science methods, the Karuk Tribe developed metrics for assessment centered on land use and land cover change detection and interviewed Tribal community elders and keepers of knowledge to collect information on land use history and changes over time.[a] This approach has been credited with contributing significantly to ecocultural-based restoration planning strategy.
>
> ---
> [a] See https://nature.berkeley.edu/karuk-collaborative.
> SOURCE: Eitzel et al., 2024.

rigor of their efforts and broad utility of restoration outcomes by relying on a narrow suite of perspectives and toolsets. For example, modern paradigms related to coupled socioecological systems require a clear understanding of the complex, nonlinear, and deeply interdependent relationships between human societies and the environments within which they operate and are part of (Berkes, 2017). Thus, modern ecological restoration is maximally effective when it considers similar complex relationships in pursuit of integrated ecological, social, and cultural indicators of restoration success (e.g., biocultural or ecocultural restoration; Lyver et al., 2016) (Box 3-8). Likewise, monitoring the efficacy of ecological restoration projects can be improved with the inclusion of Indigenous monitoring criteria, which often account for indicators of success that differ from those prioritized by western science (Thompson et al., 2019, 2020). Indigenous cultures generally have a richer understanding of the place and context of human communities in nature than western societies do (e.g., kincentric ecology; Salmón, 2000). Because modern restoration projects are typically societal problems, they cannot be fully resolved with technical tools and thus require a deep understanding of the bi-directional nature of the relationship between humans and their environment (Lyver et al., 2016). As a result, restoration efforts have a lot to gain by incorporating the deep understanding of socioecological systems afforded by Indigenous Knowledge, just as most modern problems benefit from the integration of multiple disciplines, perspectives, and approaches.

Challenges

There is concern among staff in some agencies about the compatibility of Indigenous Knowledge with NEPA requirements. At least two types of concern are present: (1) satisfying requirements for academic rigor, as defined by the western scientific community, for federal decision making under the Information Quality Act (2000) (Section 515 of Public Law 106-554; 67 FR 8452) and (2) safeguarding Indigenous data sovereignty and governance while meeting the federal transparency requirements. Understanding and resolving these issues is paramount to including Indigenous Knowledge in any endeavor and requires true partnership with trust.

Satisfying Agency Requirements for Western Scientific Rigor

Complicating efforts to include Indigenous Knowledge in federal decision making is a perception that such knowledge, by its very nature, falls short of satisfying western scientific scrutiny and thus federal information quality standards. For example, the Office of Management and Budget *Guidelines for Ensuring and Maximizing the Quality, Objectivity, Utility, and Integrity of Information Disseminated by Federal Agencies* (67 FR 8452) present a requirement of "objectivity" that presumes favoring peer-reviewed western academic science and research. These guidelines, in conjunction with the Information Quality Act of 2000 (Section 515 of Public Law 106-554), stipulate that "influential information . . . is required to provide sufficient transparency about data and methods to allow reproducibility of the results," constraining the type of Indigenous Knowledge that can be applied in federal decision making and potentially compromising Indigenous data sovereignty and governance. A significant obstacle to inclusion of Indigenous Knowledge in agency decision making is the enforcement of a "highly bureaucratic process of translating traditional knowledge into a format that fits federal requirements for information quality and evidence management" (Ornstein, 2024). Much of the guidance on data quality implies that knowledge-bearers possess the acumen to navigate voluminous government procedure and package Indigenous Knowledge to satisfy the constraints of western evidence, potentially altering the nature and interpretation of the information. However, as discussed in the previous section, Indigenous Knowledge can provide a deep understanding of socioecological systems necessary for successful restoration efforts (Ban et al., 2018; Jessen et al., 2021). Agencies will need to work with Tribes to gain a better understanding of the rigor underpinning Indigenous Knowledge using the recommendations for meaningful engagement highlighted in this chapter.

Indigenous Data Sovereignty and Governance

Indigenous data sovereignty is "the right of Indigenous peoples to exercise ownership and protection over Indigenous data. Ownership of data can be expressed through the creation, collection, access, analysis, interpretation, management, dissemination and reuse of Indigenous Data" (Williamson et al., 2023). Indigenous data governance is "the stewardship and the processes necessary to implement Indigenous control over Indigenous data" (Carroll et al., 2020). Indigenous data sovereignty and governance are founded on the inherent sovereignty of Indigenous peoples affirmed in the United Nations Declaration on the Rights of Indigenous Peoples (Article 31.1; United Nations, 2007):

> Indigenous peoples have the right to maintain, control, protect and develop their cultural heritage, traditional knowledge and traditional cultural expressions, as well as the manifestations of their sciences, technologies and cultures, including human and genetic resources, seeds, medicines, knowledge of the properties of fauna and flora, oral traditions, literatures, designs, sports and traditional games and visual and performing arts. They also have the right to maintain, control, protect and develop their intellectual property over such cultural heritage, traditional knowledge, and traditional cultural expressions.

Indigenous data sovereignty and governance are therefore part and parcel of Indigenous self-determination and autonomy.

In the face of climate change and environmental degradation, ecological and climate research studies increasingly include Indigenous Knowledge, often without meaningful participation or decision-making authority from the communities who are stewards of this knowledge, despite the purported value placed on Indigenous Knowledge by western scientists (David-Chavez and Gavin, 2018; Jessen et al., 2021; Williamson et al., 2023). Of particular concern to Indigenous communities is the misrepresentation and misuse of Indigenous Knowledge outside of its full cultural context in a western information ecosystem that perpetuates power imbalance (Kukutai and Taylor, 2016). For instance, Indigenous Knowledge of the medicinal properties of plants in remote biodiverse regions has been appropriated for commercial pharmaceutical breakthroughs that harm local Indigenous communities because of patenting and restricting the use of plants and animals to those same communities, a practice known as "biopiracy" (Cottrell, 2022; Shiva, 2016). The long and harmful history of appropriation and misuse of Indigenous Knowledge by westerners has led to recognition of "the right of Indigenous peoples to autonomously decide what, how and why Indigenous data are collected, accessed and used to ensure that data on or about Indigenous peoples reflects their priorities, values, cultures, worldviews and diversity" (Williamson et al., 2023).

Without recognition of Indigenous data sovereignty and multi-lateral data-sharing agreements on the terms under which it can be used, Everglades restoration cannot fully benefit from the wealth of knowledge that Tribes have accumulated over generations—knowledge that can provide important alternative perspectives and ultimately inform a richer suite of management options and outcomes. These two obstacles to applying Indigenous Knowledge to support Everglades restoration—western perceptions and requirements applied to the quality of Indigenous data and Indigenous data sovereignty—are not insurmountable; examples of and principles for overcoming these obstacles abound in the literature and in practice. In the sections below the committee highlights best practices, case studies, and recommendations for partnering with the Tribes to include Indigenous Knowledge in decision making for Everglades restoration.

BEST PRACTICES FOR INTEGRATION OF INDIGENOUS KNOWLEDGE IN FLORIDA EVERGLADES RESTORATION

Building Trust Through Meaningful Engagement

Different levels of consultation are presently practiced as part of the CERP, including informal discussions, formal information meetings, and formal agency-to-agency consultation. Effective consultation and meaningful engagement depend on the establishment of trust and good relationships between organizations and the individuals within those organizations. Frequent, open, and consistent communication between key agency leaders and staff and Tribal leaders and staff is key to establishing and maintaining good relationships. Building trust occurs across a range of activities from the ground up rather than as top-down reactions to contentious decisions or disputes, where much of government engagement with Tribes has historically occurred.

Principles of meaningful engagement should govern all interactions with the Tribes. Box 3-9 summarizes current guidance on meaningful engagement. These best practices should ideally be initiated before the start of any project with an understanding that gathering and conveying Indigenous Knowledge does not operate on the same timetable as planning requirements (e.g., see Box 3-10). However, strengthening relationships and partnerships should be viewed as a continual and ongoing process. Although Everglades restoration has been in the planning stages and under way for decades, opportunities still abound for forging strong reciprocal partnerships with the Tribes. Indeed, there is an imperative to do so.

Ultimately, the inclusion of Indigenous Knowledge in planning and restoration requires genuine partnerships based on trust, joint goals, and mutual respect.

BOX 3-9
Recommended Practices for Meaningful Engagement

Meaningful engagement is established through building strong, trusting, and reciprocal relationships. The groundwork to foster meaningful engagement takes time but ultimately strengthens relationships and outcomes over the long term. Several non-exhaustive best practices for establishing meaningful engagement are outlined below. They have been adapted primarily from Shelter, Support and Housing Administration (2019) and supplemented with recommendations from other sources (CTKW, 2014; Lukawiecki et al., 2021; Reo et al., 2017).

1. Understand the historical and current colonial context of the region and people with whom you are engaging. Understand how this context impacts Indigenous communities and your own power and privilege as it relates to Indigenous peoples. Take cultural competency/safety training and spend time reading about the history and culture of Indigenous peoples in the region.
2. Recognize Indigenous peoples' right to self-determination and autonomy. Engagement with Indigenous peoples should be viewed as a nation-to-nation interaction rather than stakeholder outreach.
3. Engagement must be mutually beneficial. Benefits must relate not only to the project's mission, values, and priorities but also to those of the Indigenous communities impacted. Ensuring the needs of the community, and not solely the needs of the project, should be the foundation of good engagement.
4. "Nothing about us without us." Indigenous partners have emphasized the importance of policy, planning, and program development being Indigenous-led or co-created in recognition of Indigenous peoples' right to self-determination and autonomy. This approach will ensure that the work is grounded in an Indigenous perspective, follows appropriate protocols, and better addresses the needs and priorities of Indigenous communities.
5. Good engagement is a process that focuses on relationship building. Engagement is an ongoing, reciprocal, and cyclical process. Consider how you will know whether you have done good consultation work and how you will obtain this feedback. How will it be determined that effective engagement has been achieved and the opinions of Indigenous communities have been heard and understood?
6. Engagement should begin early in the project and continue throughout all stages of the project (project initiation, planning, implementation, reporting back to communities, and evaluation). In some cases, multiple meetings will be necessary to ensure the community has been thoroughly informed as to the outcomes of the consultation.
7. Whenever possible, meet Tribes in the location of their choosing. Endeavor to understand and respect Tribal protocols and be attentive to opening ceremonies. In Tribal territory, know that meetings commonly begin with prayers and/or ceremonies, often conducted in the Tribe's native language. These rituals set the tone and aspiration for the event and much can be learned from them. Take them in with respect and endeavor to maintain that tone throughout the engagement.
8. Engagement is not outcome-based and will not necessarily result in Indigenous peoples agreeing with or supporting the intentions or goals of the project. Using appropriate engagement is a relational practice and not intended to sway the opinions of communities.
9. Take time to learn and understand the concerns of the community. Often the community will raise concerns that will strengthen the project and the well-being of the community. Do not assume that if a certain approach or form of engagement worked with one group/community that it will work with another. Although it is helpful to draw on previous engagements, there is no "one size fits all" approach. Seek the community's advice on appropriate engagement.
10. Develop reciprocal processes for knowledge sharing that respect Indigenous Knowledge and data sovereignty and governance. Create opportunities for co-production of knowledge and co-authorship on publications.
11. Build capacity for Indigenous communities to participate in the project and find ways to support Indigenous-led initiatives to govern and manage aspects of the project that are important to them.

BOX 3-10
Tribal Engagement in the Klamath Dam Removal Project

The Klamath River Hydroelectric Project has blocked fish passage and altered the Klamath River flows for more than 100 years. The Klamath River watershed is the traditional homelands of the Klamath Tribes, Yurok Tribe, Karuk Tribe, and Shasta Indian Nation. The Klamath River has been diverted, dammed, and impacted by changed water quality and blockage of access for salmon due to upstream agricultural, logging, and development pressures. To the Tribes, the Klamath River and the salmon that it supports are central to their culture. Socially and culturally the river itself represents the essence of life and is essential to the health and social well-being of the Indigenous people and the watershed.

After years of discussion, debate, and failed legislative attempts at resolution of Tribal and conservation concerns, an agreement was reached with the dam owners to remove four Klamath River dams. In the early 2020s consultation with the Tribes was initiated, and baseline scientific data and cultural resource work was initiated. In 2023 work on the removal of the four dams was initiated. The Tribes have been invested in the dam removal and river restoration process from the beginning. The company hired to design, implement, and oversee the restoration process consulted with the Tribes to integrate Indigenous Knowledge. The Yurok Tribe and the Shasta Indian Nation have been active participants in the restoration and revitalization of the Klamath River canyon, the watershed, and the salmon runs that historically populated the river.

Indigenous Knowledge has been applied in conjunction with western science, resulting in a dynamic approach to river restoration. Examples of how Tribal knowledge has been applied in the Klamath restoration effort include the following:

- Early and continuous conversations with the Tribes ensured their history with the river was understood and considered in the restoration planning.
- For 5 years prior to the first dam removal, the Klamath River Tribes have been gathering native seeds by hand and sending them to nurseries for storage until conditions are favorable in the exposed reservoir sediments along the restored river corridor for replanting.
- The Tribes identified areas where temporary habitat can be placed in the new river channel to aid migrating salmon and trout.

In February 2024 the U.S. Department of the Interior in cooperation with Klamath Water Users Association, the Klamath Tribes, Yurok Tribe, and the Karuk Tribe signed an agreement to restore the Klamath Basin ecosystem and improve water supply reliability for Klamath Project agriculture. A Memorandum of Understanding commits the parties to working together to identify, recommend, and support projects that advance shared Klamath Basin restoration goals, including ensuring water and irrigation stability and reliability; strengthening ecosystem resilience; protecting fish populations; and advancing drought resilience. The agreement formally recognizes the significant value of Indigenous Knowledge and commits the parties to include Indigenous Knowledge in restoration efforts throughout the Klamath River Basin.

The Bureau of Reclamation has committed to providing additional funding directly to the Klamath Tribes, Yurok Tribe, Hoopa Valley Tribe, Karuk Tribe, and the Modoc Nation for projects that restore watersheds and revitalize water infrastructure (DOI, 2024).

Achieving this goal will require long-term building of relationships and trust through meaningful engagement. When individuals in key positions change, whether at the leadership, scientific, or operational level, thorough orientation in the onboarding of new personnel with respect to consultation and coordination, as well as an introduction to relevant Tribal counterparts, is essential to sustaining good relationships. Building and maintaining trust with the Tribes requires a commitment to understand and appreciate the Tribes' historic and ongoing relationship to the Everglades and with the federal and state agencies. This effort takes time, so strategies within organizations (such as having multiple Tribal liaisons within an agency) to ensure a culture and continuity of trust are paramount to the development and maintenance of long-term vitality of relationships with the Tribes.

Indigenous Data Sovereignty and Data-Sharing Agreements

Indigenous data sovereignty and the historical context of colonial appropriation and exploitation of Indigenous data present challenges for data sharing. This era of blanket open data-sharing policies and norms conflicts with Indigenous data governance (James et al., 2014; McCartney et al., 2022; Rainie et al., 2019). These challenges have prompted the development of conceptual frameworks to inform processes for governance of Indigenous data upon which data-sharing agreements can be based (Carroll et al., 2020). Box 3-11 and Figure 3-6 outline the most prominent of these frameworks—the CARE principles—aimed at framing the issues surrounding Indigenous data sovereignty and governance that intersect with western governmental and institutional interests. Presentation of these principles here is not intended to be prescriptive for the Miccosukee and Seminole Tribes because "Indigenous Data Sovereignty can be exercised only by Indigenous Peoples as rights holders through the retention and control of their data" (Jennings et al., 2023). However, agency staff and researchers should be aware of these principles when partnering with Tribes and should endeavor to emulate these principles when developing Indigenous data-sharing agreements with the Tribes. Formalized Indigenous data-sharing agreements are necessary to protect Tribal data, confirm Tribal governance of data for and about their communities, and establish clear expectations for all parties. The public health sector provides many examples of formalized Indigenous data-sharing agreements that espouse the CARE principles, at least in part (Harding et al., 2012; for the Model Tribal Data Sharing Agreement created by the American Indian Health Commission, see AIHC [2023]). Data-sharing agreements should address research roles, responsibilities, data sharing, data housing, funding transparency, publication process, and intellectual property, and should shift power to the Tribal community

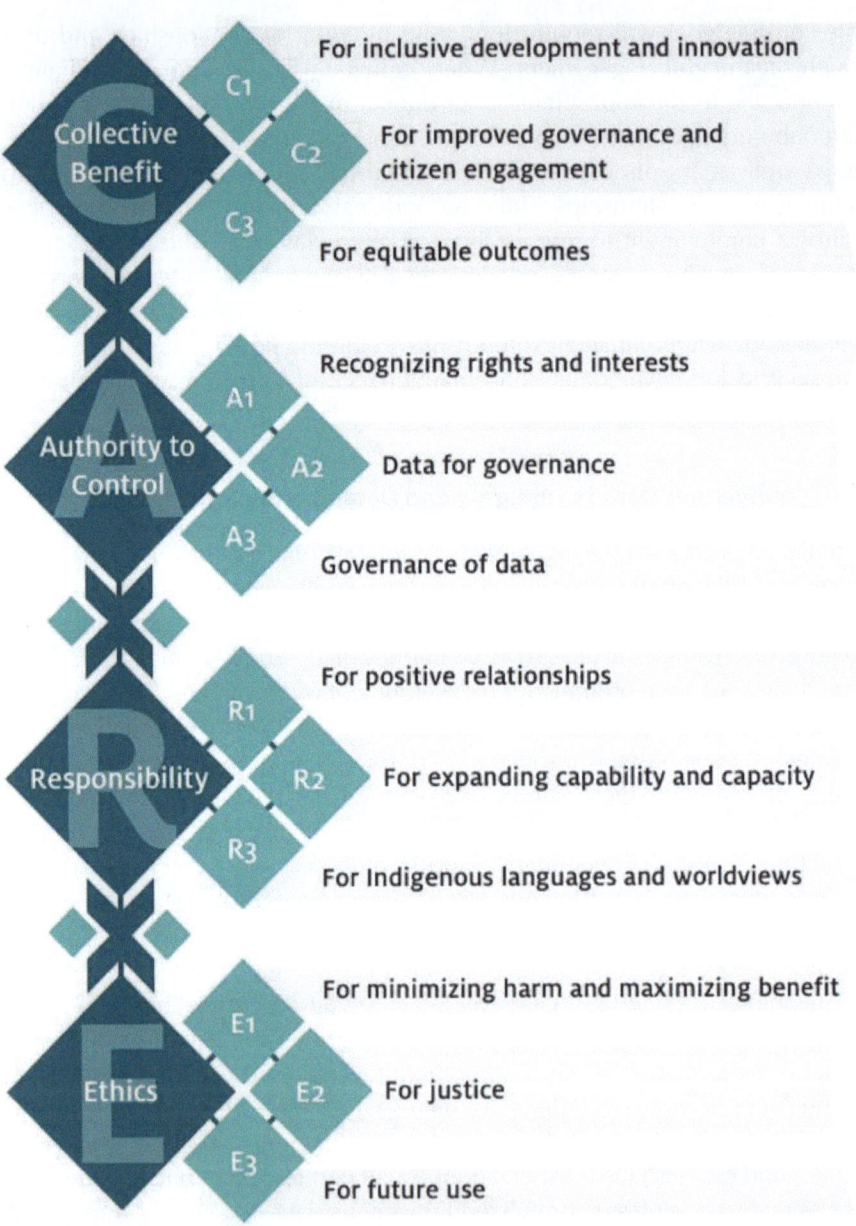

FIGURE 3-6 The CARE Principles of Indigenous Data Governance were created to advance the legal principles underlying collective and individual rights by considering power differentials and historical contexts of data in advancing Indigenous innovation self-determination.

SOURCE: Carroll et al., 2020. CC-BY 4.0.

> **BOX 3-11**
> **CARE Principles for Indigenous Data Governance as Defined by Research Data Alliance International Indigenous Data Sovereignty Interest Group**
>
> **Collective Benefit**
>
> Data ecosystems shall be designed and function in ways that enable Indigenous Peoples to derive benefit from the data.
>
> - *For inclusive development and innovation*
>   Governments and institutions must actively support the use and reuse of data by Indigenous nations and communities by facilitating the establishment of the foundations for Indigenous innovation, value generation, and the promotion of local self-determined development processes.
> - *For improved governance and citizen engagement*
>   Data enrich the planning, implementation, and evaluation processes that support the service and policy needs of Indigenous communities. Data also enable better engagement between citizens, institutions, and governments to improve decision making. Ethical use of open data has the capacity to improve transparency and decision making by providing Indigenous nations and communities with a better understanding of their peoples, territories, and resources. It similarly can provide greater insight into third-party policies and programs affecting Indigenous people.
> - *For equitable outcomes*
>   Indigenous data are grounded in community values, which extend to society at large. Any value created from Indigenous data should benefit Indigenous communities in an equitable manner and contribute to Indigenous aspirations for well-being.
>
> **Authority to Control**
>
> Indigenous Peoples' rights and interests in Indigenous data must be recognized and their authority to control such data be empowered. Indigenous data governance enables Indigenous Peoples and governing bodies to determine how Indigenous Peoples, as well as Indigenous lands, territories, resources, knowledges, and geographical indicators, are represented and identified within data.
>
> - *Recognizing rights and interests*
>   Indigenous Peoples have rights and interests in both Indigenous Knowledge and Indigenous data. Indigenous Peoples have collective and individual rights to free, prior, and informed consent in the collection and use of such data, including the development of data policies and protocols for collection.
> - *Data for governance*
>   Indigenous Peoples have the right to data that are relevant to their world views and empower self-determination and effective self-governance. Indigenous data must be made available and accessible to Indigenous nations and communities to support Indigenous governance.
> - *Governance of data*
>   Indigenous Peoples have the right to develop cultural governance protocols for Indigenous data and be active leaders in the stewardship of, and access to, Indigenous data especially in the context of Indigenous Knowledge.
>
> *continued*

BOX 3-11 Continued

Responsibility

Those working with Indigenous data have a responsibility to share how those data are used to support Indigenous Peoples' self-determination and collective benefit. Accountability requires meaningful and openly available evidence of these efforts and the benefits.

- *For positive relationships*
  Indigenous data use is unviable unless linked to relationships built on respect, reciprocity, trust, and mutual understanding, as defined by the Indigenous Peoples to whom those data relate. Those working with Indigenous data are responsible for ensuring that the creation, interpretation, and use of those data uphold, or are respectful of, the dignity of Indigenous nations and communities.
- *For expanding capability and capacity*
  Use of Indigenous data invokes a reciprocal responsibility to enhance data literacy within Indigenous communities and to support the development of an Indigenous data workforce and digital infrastructure to enable the creation, collection, management, security, governance, and application of data.
- *For Indigenous languages and worldviews*
  Resources must be provided to generate data grounded in the languages, worldviews, and lived experiences (including values and principles) of Indigenous Peoples.

Ethics

Indigenous Peoples' rights and well-being should be the primary concern at all stages of the data life cycle and across the data ecosystem.

- *For minimizing harm and maximizing benefit*
  Ethical data are data that do not stigmatize or portray Indigenous Peoples, cultures, or knowledges in terms of deficit. Ethical data are collected and applied in ways that align with Indigenous ethical frameworks and with rights affirmed in the UN Declaration on the Rights of Indigenous Peoples (UNDRIP). Assessing ethical benefits and harms should be done from the perspective of the Indigenous Peoples, nations, or communities to whom the data relate.
- *For justice*
  Ethical processes address imbalances in power, resources, and how these affect the expression of Indigenous rights and human rights. Ethical processes must include representation from relevant Indigenous communities.
- *For future use*
  Data governance should account for the potential future use and future harm based on ethical frameworks grounded in the values and principles of the relevant Indigenous community. Metadata should acknowledge the provenance and purpose and any limitations or obligations in secondary use inclusive of issues of consent.

SOURCE: Research Data Alliance International Indigenous Data Sovereignty Interest Group, 2019.

Applying Indigenous Knowledge   141

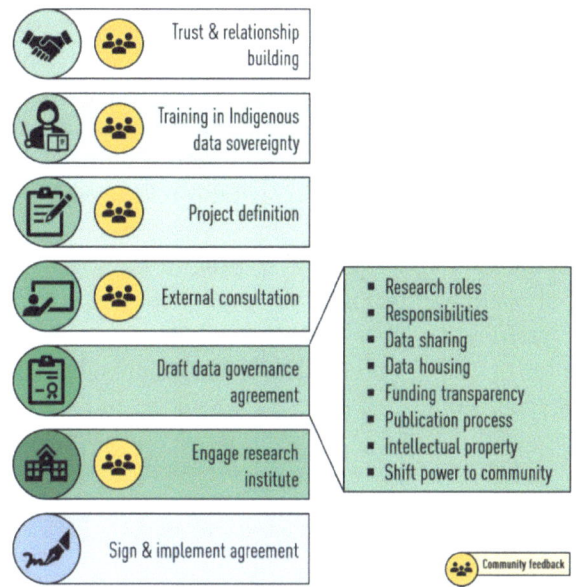

FIGURE 3-7 Process of developing data governance agreements. The development of data governance agreements proceeds through distinct stages. Community feedback and consultation (yellow) occurs throughout the process.

SOURCE: Love et al., 2022. CC-BY 4.0.

(Love et al., 2022; see Figure 3-7). *Guidelines for Considering Traditional Knowledges in Climate Change Initiatives* by the Climate and Traditional Knowledges Workgroup (CTKW, 2014) offers explicit and expansive actions for agencies and researchers when entering into data-sharing agreements to reduce risks to the Tribes of sharing confidential or sensitive Indigenous Knowledge. In part, CTKW (2014) recommends the following:

- "Determine the extent to which Indigenous Knowledge (IK) involving confidential or sensitive information can be protected from unauthorized public disclosure because of federal mandate (e.g., without express legislative authority, IKs recorded in written or electronic form provided to federal entities are subject to FOIA requests).
- Research your agency/organization's codes and policies regarding the publication or dissemination of IKs gathered for projects.

- Inform Tribes/IK holders about potential risks of disclosure. It is the obligation of agency staff and researchers to share information about what risks the project poses 'on their end.'
- Research existing intellectual property and copyright laws in your country, as they pertain to your research/project results. Will data from the project be subject to appropriation? How will this information be protected?"

Any data-sharing agreement will be specific to the focal study or project, agency, and Tribal Nation and should be developed in direct consultation and negotiation with the Tribes and their legal counsel (where possible) to create a formal Tribal resolution or formal written agreement. Additional guidance on USACE protocols for the protection of Indigenous Knowledge and applicable laws and authorities are provided in the 2023 USACE Tribal Consultation Policy (USACE, 2023a). It should be recognized that a data-sharing agreement only represents one component of meaningful engagement with Tribal Nations (Figure 3-7).

Training

The identification, administrative acceptance, and application of Indigenous Knowledge has only relatively recently been embraced as an important component for the CERP. An ongoing education effort on the value of Indigenous Knowledge is occurring with various levels of acceptance in both the academic and administrative arenas. Many U.S. organizations and agencies have identified the importance of providing Tribal affairs/engagement best practices, toolkits, and trainings to employees working on projects that include Tribal interests and connections or involve specific consultations. In general, these resources are intended to increase knowledge and understanding on building relationships, showing respect, and engaging to listen and learn about traditional cultural, spiritual, environmental, and scientific knowledge and other areas of interest. The 2022 White House Memorandum on Uniform Standards for Tribal Consultation outlines the policy direction and recommendations for Tribal consultation and engagement, including a requirement for annual training for employees who work with Tribal Nations (EOP, 2022d). At present, there is no formal training for the agency staff who engage with the Miccosukee or Seminole Tribes. Whether a function of oversight, lack of staff, agency culture, or prioritization of other goals, lapses in consultation and coordination represent failures to comply with established legislation and the federal Indian trust responsibility to Tribes. Furthermore, they highlight the need for all levels of federal and state agencies—leadership, scientists, engineers, and operational staff—to understand and be committed to observing best practices in consultation and coordination with Tribes. This necessitates training beyond the current cultural immersion course offered on an occasional, voluntary basis. Creating learning opportunities about best practices for meaningful Tribal engagement, with content tailored to and in consultation with the Miccosukee and Seminole Tribes, will help to enhance efforts to establish and sustain a robust culture around meaningful engagement (Box 3-9) by all restoration practitioners.

Building an Inclusive Agency Culture

Regular training for staff will raise collective awareness of best practices for meaningful engagement with Tribes and of effective mechanisms to apply Indigenous Knowledge in restoration, but additional steps are needed to shift agency norms and culture to become more inclusive of the Tribes. Every management agency's culture has been developed over years of administrative decisions and priorities, leadership styles, implemented policies, and the types of employees hired. The convergence of these factors influences agency culture, which can range from one that is very process-driven to one that is more open to incorporating new information and perspectives. Inclusion of Indigenous Knowledge should occur throughout the spectrum of agency cultures. How that is accomplished depends on the agency and the people who are responsible for making it happen. Deliberate and thoughtful leadership is key for establishing high expectations for engagement, as well as accountability at all levels of the agency.

Shifts in culture will take considerable time and effort to effect, but strategic investments of resources in training and staffing by each agency can support and help accelerate this process. In light of the importance of inclusion of Indigenous Knowledge in the CERP and the steep learning curve that must be confronted to effectively do so, agency investments in Tribal relations are currently inadequate to promote the necessary shift in agency culture. Increased staffing dedicated to Tribal relations by each participating agency will enable more meaningful engagement by Tribal liaisons who are currently carrying enormous individual responsibility. Because meaningful engagement requires building personal relationships and trust, additional staffing will also create institutional memory and resilience against staffing turnover. Increased resources could also be strategically allocated directly to Tribes, either to hire Tribal members as staff or to facilitate their engagement in the restoration process, decision making, and implementation (Box 3-10).

CERP-SPECIFIC EXAMPLES OF WAYS TO IMPROVE APPLICATION OF INDIGENOUS KNOWLEDGE

The above best practices can be applied broadly within the CERP or any restoration program, but the following sections highlight a few specific CERP opportunities for improving inclusion of Indigenous Knowledge.

Miccosukee Tribe Peer-Review Process

Guidance for Federal Departments and Agencies on Indigenous Knowledge (EOP, 2022a) describes numerous strategies for federal departments and agencies to engage with Tribes to include Indigenous Knowledge in decision making. However, meaningful inclusion of Indigenous Knowledge to inform decision making in Everglades restoration has been sparse, potentially due, in part, to the perceived challenges of reconciling the Foundations for Evidence-Based Policymaking Act of 2018 (Evidence Act) and the Information Quality Act with Indigenous ways of knowing and data sovereignty. The Miccosukee Tribe has taken a lead in providing solutions to what is essentially a government agency problem—navigating the constraints on data quality and satisfying federal requirements for western scientific rigor while protecting sensitive Tribal knowledge and ways of knowing (Ornstein, 2024). The Miccosukee Tribe has developed a procedure for collecting and presenting its Indigenous Knowledge that broadly satisfies many of the recommended criteria and procedures in Appendix C of the 2022 OSTP-CEQ Guidelines to "ensure that Indigenous Knowledge can be considered consistent with the IQA" (EOP, 2022a). These criteria include the following (in part):

- "Ability to substantiate that the information is part of a relationship or kinship of people across generations interconnected to social, spiritual, cultural, and natural environmental or ecological systems";
- "The inherent use and value of the information and expertise of the knowledge holders, including lived experience, is retained and respected;
- Language and names within the information, in which Indigenous Knowledge and values may be nested, are preserved;
- The Indigenous Knowledge is considered through an Indigenous lens, voice, or style and weaved together with other forms of evidence without converting or forcing the knowledge into non-Indigenous frameworks";
- "Reference in the assessment is free of any culturally sensitive information that the knowledge holders do not want made public, including consideration of how documentation may be subject to or released under the Freedom of Information Act"; and
- "Practices for ensuring quality control and validation are appropriate to the nature of the source information, as determined by the Indigenous knowledge holders from which the information comes, such as through iterative, equitable dialogue on the interpretation of findings by community members, co-researchers, or collective knowledge systems."

The procedure adopted by the Miccosukee Tribe, explained further in Ornstein (2024), conveyed Miccosukee Indigenous Knowledge to inform a change in operations of the S-12A and S-12B gates (Box 3-5). The process involved the

Tribe's Everglades Advisory Committee, which consists of Miccosukee Tribal citizens from different clans and across different age groups, representing the consensus view of the Tribe on environmental matters, and it included the following four steps:

1. Through interviews, the traditional knowledge of the Everglades Advisory Committee related to the issue of interest was documented by the Tribe's legal and environmental staff into an internal report;
2. The contents of the internal report were verified by the knowledge holders in the Everglades Advisory Committee;
3. In a letter to the FWS, the Chairman of the Miccosukee Tribe then summarized the findings of the internal report, named the members of the Everglades Advisory Committee, and described the procedure for data collection; and
4. The internal report was kept confidential within the Tribe to ensure Indigenous data sovereignty and governance.

This protocol could serve as a best practice for other Tribes to gather Indigenous Knowledge internally, subject it to verification by knowledge holders representing a broad cross-section of their communities, and disseminate a consensus view on their own terms. It also centers the primacy of Indigenous data sovereignty and governance while providing a mechanism for "verification but not reproduction by agency officials" (Ornstein, 2024).

Performance Measures for Tree Islands

Given the deep connection of the Miccosukee and Seminole Tribes to tree islands described earlier in this chapter and the wealth of knowledge the Tribes have accumulated over generations, the development of one or more CERP tree island performance measures is a prime opportunity to better include Indigenous Knowledge in the CERP process. Performance measures have been an integral component of CERP restoration planning and adaptive management for the past two decades (Doren et al., 2009; McLean et al., 2004). Throughout the CERP, performance measures have served two main purposes: (1) as a predictive tool that is used to quantitatively evalu ate project alternatives during the CERP planning process and/or (2) as means to assess the performance of restoration actions once they are implemented and inform additional adaptive management actions if restoration objectives are not met (see Chapter 5). Depending on the specific goals of a project, performance measures can be abiotic (e.g., hydroperiod, water depth, salinity) or biotic (e.g., macroinvertebrates, aquatic vegetation, fish). However, to be used in quantitative project evaluations, it should be possible to simulate the performance measure based on approved CERP modeling tools.[8] Dozens of performance measures have been developed for the greater Everglades (RECOVER, 2015b), and they continue to be developed and refined.[9]

Although tree islands are vitally important landscape features of the central Everglades for people and nature, and their continued degradation and deterioration is a certain indicator of an ecosystem in crisis, a performance measure, or suite of measures, for tree islands does not currently exist. This is predominantly due to scientific debates and uncertainties regarding tree island resilience and restoration strategies (McLean, 2010). More than 20 years ago a report from a previous National Academies' committee (NRC, 2003b) addressed CERP performance measures for tree islands, and its assessment is worth repeating here [emphasis added]:

> Work is currently under way to define the spatial characteristics of the ridge and slough, and tree island patterns thought to represent well-preserved to highly degraded patterns with respect to historical landscape patterns (Nungesser et al., 2003; Wu et al., 2003). Such characteristics include number of ridges and tree islands, area of ridge/tree islands in the landscape, length-to-width ratio of ridges/tree islands, perimeter-to-area ratios of these features, orientation of ridges and tree islands in the landscape, and average length and width of uninterrupted slough along north-to-south and west-to-east transects, respectively.
>
> These indicators, while useful, are insufficient for determining how the landscape is responding to water level, hydroperiod, and flow, because major changes in elevation may occur before degradation is reflected in such indicators.
>
> ... Regardless of the mechanism(s) responsible for creation and maintenance of the ridge and slough and tree island patterns, *performance measures must be developed so that these patterns can be monitored*. Once there is sufficient scientific evidence to establish the role of flow and the flow rates required to maintain these landscape patterns, flow-related performance measures should be developed and added to the MAP [Monitoring and Assessment Plan].

By early 2008, a performance measure had been developed for sheet flow in the Everglades ridge and slough landscape (SFWMD, 2008), which included the timing, distribution, and continuity of flows; flow volume was not included and was noted as a goal for later development. RECOVER (2015a) notes that the sheet-flow performance measure should not be used to predict optimal conditions for tree islands. Despite two decades of active research on tree islands since the recommendations of NRC (2003b), a CERP performance measure for tree islands has not been developed.

---

[8] For example, the American crocodile growth and survival performance measure (RECOVER, 2015b) is calculated using a regional hydrologic model and linear regression salinity models to compute daily average salinity values, which are then assigned a habitat suitability index score between 0 and 1. The yearly index is calculated based on the average score between August and December.

[9] See https://www.saj.usace.army.mil/Missions/Environmental/Ecosystem-Restoration/RECOVER/RECOVER-Performance- Measures.

Significant progress was made on the development of a conceptual ecological model (CEM) for tree islands in 2010 through two SFWMD/RECOVER workshops. More than 100 science-based hypotheses that related system stressors to ecological attributes of tree islands were synthesized into a draft CEM (McLean, 2010; Sklar et al., 2011). This draft was revised and refined through real-time interactive and anonymous polling of scientists to rate which stressors, attributes, and effects were most suitable for the development of tree island performance measures (Figure 3-8). The workshops established consensus on the importance of hydrology and nutrients as stressors to tree islands. The workshops also established that

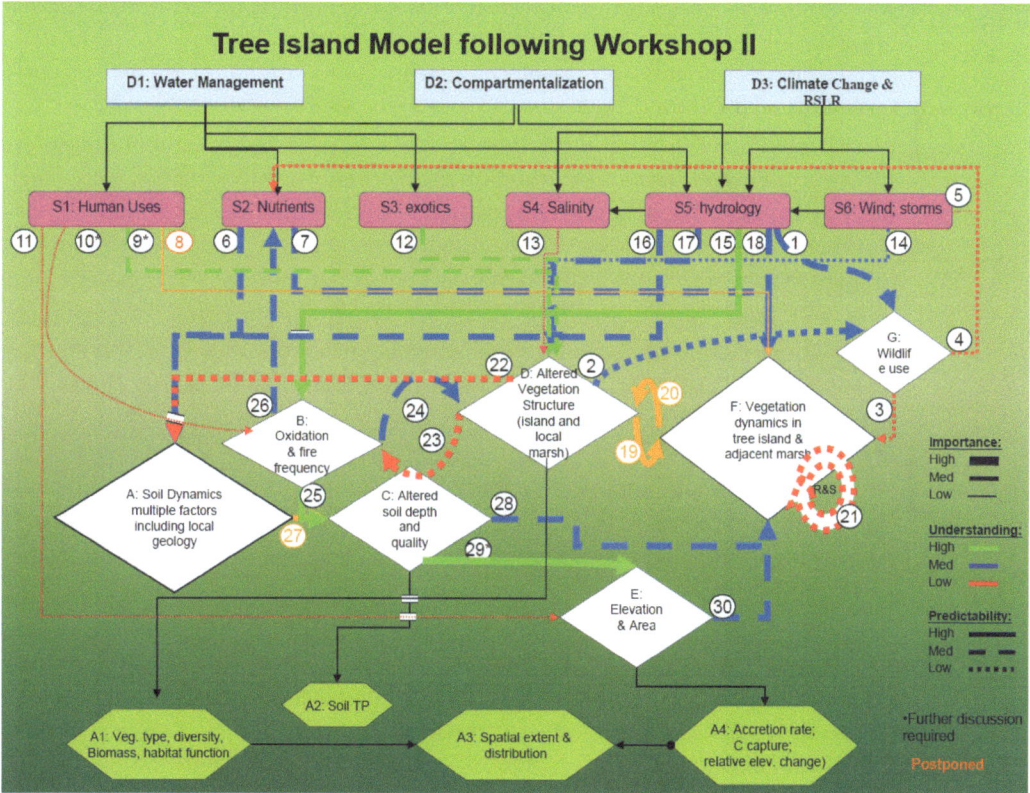

FIGURE 3-8 CEM for tree islands.

NOTES: The color, width, and style of connecting lines represent levels of understanding, importance, and predictability of linkages between components of the CEM. Pink rectangles represent stressors, white diamonds represent the effects of stressors, and green polygons represent tree island attributes.

SOURCE: McLean, 2010.

tree island vegetation structure, peat oxidation, fire frequency, soil dynamics, and wildlife use are well understood and reasonably predictable. But it was also noted that a good understanding of the processes influencing tree islands did not necessarily translate into reliable quantitative performance measures (McLean, 2010; Sklar et al., 2011). Of the 30 linkages identified in the CEM, only 3 were well understood, 12 appear to have a moderate level of understanding, 8 linkages were poorly understood, and further discussion was postponed for 4 linkages because they were too simple and did not capture important interactions (Sklar et al., 2011). No consensus was reached on which stressors, ecological responses, or tree island attributes should be focused on first in the development of performance measures, presumably indicating disagreement about the relative importance of each factor or uncertainty in the underlying science (McLean, 2010). Further progress on developing performance measures for tree islands stalled for well over a decade.

RECOVER should revisit the tree island CEM with a view to include Indigenous Knowledge through meaningful engagement with the Tribes, particularly the Miccosukee Tribe whose reservations—and livelihoods—overlap significantly with WCA-3A. However, meaningful engagement with the Tribe needs to involve thoughtful interactions that respect Tribal traditions and governance (Box 3-9). For example, real-time rapid anonymous polling from the agencies is not conducive to meaningful engagement and incorporation of Indigenous Knowledge into planning. In the Miccosukee Tribe, for example, sharing of Indigenous Knowledge outside the Tribe requires multiple rounds of interview and peer review within the Tribe, an agreement on what can and cannot be shared, and ultimate approval by the Tribal Council (Cypress, 2023; Ornstein, 2024). The polling used by agencies in these exercises has benefits in that it democratizes and anonymizes opinions, which can provide a feeling of safety in participants, but it comes at the cost of muting voices that may have a different perspective or who may have a greater stake in outcomes (Sharpe et al., 2021). This approach also ignores Tribal traditions, with a strong bias toward western methods of soliciting input from communities and stakeholders, which is not an effective means for engaging sovereign Tribes. As independent sovereign nations, Tribal Nations may not be treated as just another stakeholder among many groups. Future engagement should include meaningful protocols and methods of information gathering that avoid tokenism and support Tribal involvement. Agencies should work with the Tribes to consider cultural outcomes within performance measures where appropriate in addition to ecosystem outcomes, so that biocultural restoration can be achieved (Lyver et al., 2016; Sena et al., 2022; Winter et al., 2020).

RECOVER currently plans to develop tree island performance measures across a series of four workshops (starting March 2024) with a reduced group of participants including tree island experts from federal and state agencies, researchers, and Tribal representatives (G. Ralph, RECOVER, personal communication, 2023). The workshop series agenda and all future workshops and timelines should factor in sufficient time and mechanisms for meaningful Tribal engagement as directed by the Tribes and outlined in this chapter. This engagement should involve working with the Tribes to create agendas and timelines to ensure that the Tribes' traditions and processes for sharing Indigenous Knowledge can be fully respected.

Whatever form performance measures for tree islands ultimately take, they will need to be tied to hydrodynamical, geomorphological, and ecological processes that have characterized intact tree islands across millennia. Ideally, the performance measures should guide water operations to maintain a diversity of tree islands while supporting biodiversity including threatened and endangered species. Tree island performance measures should inform water operations not only to get the water right but also to support the livelihoods, cultures, and identities of the people who have called the Everglades home for centuries.

CONCLUSIONS AND RECOMMENDATIONS

The lands and waters of the Everglades are the geographic and spiritual home of the Miccosukee and Seminole peoples. The health and well-being of the Everglades is synonymous with that of the Miccosukee and Seminole Tribes. Therefore, the Tribes have a wealth of knowledge about the South Florida ecosystem based on their intimate reciprocal relationship with the biophysical environment that has been developed through lived experience and passed down through generations. The following conclusions and recommendations are provided in response to a request from CERP agencies for advice on how Indigenous Knowledge could be better included in CERP planning and management, which was prompted by recent Executive Office requirements to include Indigenous Knowledge in federal scientific and policy decision making.

Indigenous Knowledge, like western science, is a "body of observations, oral and written knowledge, innovations, practices, and beliefs" about the natural world that has much to offer Everglades restoration (EOP, 2022a). For example, the Miccosukee and Seminole Tribes' extensive personal and Tribal knowledge of tree islands, if applied to restoration efforts, would benefit both the ecosystem and the Tribes. Indigenous Knowledge spans much longer timeframes than western scientific studies and can therefore enhance understanding of historical ecological conditions and modern deviations from baseline conditions. WERP is an example of a project where extensive Tribal engagement has helped to improve the project through meaningful consultation and application of Indigenous Knowledge. Indigenous Knowledge should also be considered and applied in efforts to refine the RECOVER monitoring plan and conceptual models to better develop and incorporate performance measures and metrics that are relevant to biocultural restoration.

Consistent and meaningful engagement between CERP agencies and Tribal Nations is necessary to ensure a partnership in which Indigenous Knowledge is recognized, considered, and applied in restoration decision making, and notable progress has recently been made to improve the quality of Tribal engagement and cooperation in the CERP. The Everglades has offered refuge and sustenance to the Miccosukee and Seminole peoples for generations. Historically, the U.S. and state governments attempted to remove all Tribal people from the southern Florida landscape through coercive and violent means. This painful history is part of the living memory of Miccosukee and Seminole Tribal elders and remains an under-current in all consultation, engagement, and coordination with the Tribes. The lack of meaningful engagement with the Tribes historically heightens the importance today of building trustful relationships, based on integrity and with careful adherence to laws, regulations, and guidance in consultation and cooperation with the Tribes. Over the past decade, consultation has become less proforma and more meaningful, as exemplified by the application of Indigenous Knowledge to inform decision making in WERP and in a 2023 temporary deviation from the seasonal closures of the S-12A and S-12B structures. The work required to shift agency cultures to further elevate meaningful Tribal engagement will be labor- and resource-intensive for both agencies and Tribes but will reap rewards for Everglades restoration.

The recently developed Miccosukee internal peer-review process is an important step toward facilitating consideration of Indigenous Knowledge in Everglades restoration processes and provides a potential model for others throughout the nation. This process results from extensive effort on the part of the Miccosukee Tribe to speak to western scientific norms on data quality and transparency in culturally appropriate ways. The Miccosukee Tribe developed the process considering federal government requirements for data quality to ensure that the Indigenous Knowledge cannot be rejected because of quality assurance/quality control concerns, while protecting Indigenous data sovereignty and governance. In this process, the traditional knowledge of the Miccosukee Tribe related to the issue of interest was documented in an internal report that was verified by the Tribe's Everglades Advisory Committee. The Chairman of the Miccosukee Tribe then summarized the findings and described the procedure for data collection in a letter to government agencies. The onus is now upon the agencies to meet the Tribes "where they are" and develop protocols that effectively consider and apply Indigenous Knowledge even when it does not conform to western scientific norms and presentation.

RECOVER and other CERP staff should implement best practices in their efforts to engage Tribes and apply Indigenous Knowledge in Everglades restoration planning, operations, monitoring, and adaptive management. Staff should consult with the Miccosukee and Seminole Tribes to determine their desired timing and level of engagement for specific projects. Some best practices include the following:

- Recognize that Tribes are autonomous nations and interact with Indigenous partners as such, not as stakeholder groups.
- Involve Tribal members in planning, research, and monitoring efforts, with funding where necessary, to foster co-stewardship and co-production of knowledge to support restoration and priorities of value to the Tribes. This may require new contracting mechanisms that should be informed by a review of other federal-state-Tribal efforts to improve capacity building and enhance partnerships.
- Provide opportunities for the Miccosukee and Seminole Tribes to engage in CERP processes and share knowledge in meaningful and culturally sensitive ways, recognizing that independent internal discussions among Tribal members may be necessary to provide input to the decision-making process.
- Include Tribal members in meetings in thoughtful ways that avoid unintentional tokenism and encourage participation by Tribal members in planning meeting agendas.
- Create data-sharing agreements that center Indigenous data sovereignty and governance, guided by the CARE Principles (described in Box 3-11), and support the culturally sensitive integration of Indigenous Knowledge into reports and subsequent agency actions while protecting their sensitive Tribal cultural information from public disclosure.
- Build, support, and maintain an inclusive agency and CERP culture that establishes high expectations for meaningful engagement and incorporation of Indigenous Knowledge throughout agency hierarchy and operations.

To continue to improve the quality of Tribal engagement and inclusion of Indigenous Knowledge, training should be developed, in consultation with the Miccosukee and Seminole Tribes, and required on a recurring basis for all agency staff who interact with Tribal members. The integration of Indigenous Knowledge in Everglades restoration planning, operations, and adaptive management is in its infancy, and the acceptance of its application in long-established federal and state restoration processes varies within the CERP community. The CERP is not alone—restoration programs throughout the United States are grappling with how to meaningfully engage with Indigenous Knowledge at various scales. Formal training should be provided for CERP agency personnel who interact with the Tribes, including leadership, scientists, engineers, and field staff, to strengthen a culture of meaningful engagement by all restoration practitioners. This training could cover the history and governance of the Miccosukee and Seminole Tribes of Florida; Indigenous Knowledge; laws, regulations, and guidance; case studies that highlight the complementarity of Indigenous Knowledge and western science; and best practices for consultation, cooperation, and meaningful engagement. Also valuable would be education on the appropriate protection of sensitive Indigenous Knowledge to build a relationship of trust and cooperation.

Attention is needed to ensure that Tribal input and opportunities for meaningful collaboration and inclusion of Indigenous Knowledge are not lost because of staffing resource constraints. The pace of Everglades restoration planning, operations, and adaptive management and the associated requirements for effective engagement may exceed the existing staff resources of both the CERP agencies and the Tribal Nations. CERP agencies' dependence on a single staff member to coordinate Tribal relations creates vulnerabilities. CERP agencies could improve their capacity through increased agency training and staffing to ensure a breadth of expertise and uninterrupted relations in the event of personnel turnover. Other restoration programs have provided grants to Tribes, when needed, to increase Tribal staff availability for consultation and engagement.

4

Application of Tools to Evaluate the Effects of Climate Change

Several historically significant climate events occurred within the past year that bear relevance to restoration planning. Globally, the year 2023 was the hottest on record by a wide margin and heat waves were anomalously frequent, each attributed to anthropogenic climate change. Global ocean temperature also reached record highs from April through December, with parts of the Gulf of Mexico and Caribbean experiencing extreme warming. Antarctic sea ice extent reached a record low in February 2023 and remained low through the end of the year. Because of both the thermal expansion of the ocean due to warming and the melt of ice sheets and glaciers, mean global sea level reached a record in 2023. This rise in sea level is causing saltwater intrusion and loss of coastal wetlands in South Florida due to vegetation die-off and peat collapse. In a more local example, on April 12, 2023, Fort Lauderdale experienced 25.6 inches of rain in 12 hours and 25.91 inches in 24 hours. The previous recorded maximum precipitation for 24 hours in Fort Lauderdale was 14.59 inches. The extreme rainfall caused widespread damage and closures and prompted a declaration of a state of emergency. In June 2024, areas in Fort Lauderdale and North Miami received about 20 inches of rain over 2 days. Although it is premature to attribute individual extreme weather events to anthropogenic climate change, they are consistent with theoretical expectations of future climate change effects (Kirchmeier-Young and Zhang, 2020) and highlight the vulnerability of infrastructure to low probability and high impact events. In particular, these events show the need to consider climate extremes beyond those apparent in the historical record when planning long-term investments in restoration.

Such events raise the question of whether Comprehensive Everglades Restoration Plan (CERP) planners are adequately considering climate change in the restoration design and adaptive management. Adaptive management enables responses to changes in the system as key uncertainties are reduced or resolved, creating greater resilience to climate change, particularly when integrated with ecological models (Peterson et al., 1997; Williams and Brown, 2014; see also Chapter 5). Until recently, there was little formal assessment of the potential consequences of changes to climate, such as sea-level rise, warmer temperatures, and changing precipitation patterns. However, the increasing availability of climate projections for South Florida and advances in climate change planning outside of the CERP offer potential to consider climate change within the CERP. This chapter addresses four opportunities to better incorporate information about certain aspects of climate change and associated tools into the CERP:

1. use of temperature and precipitation scenarios for restoration planning,
2. use of dynamic sediment accretion modeling to consider the effects of sea-level rise,
3. improved consideration of climate in ecological models, and
4. systems operating manual updates as a form of adaptive management in light of climate change.

This chapter builds on the last committee's extensive discussions on climate change in NASEM (2023), which addressed changing sea level, precipitation, and temperature and U.S. Army Corps of Engineering (USACE) processes to incorporate this information into CERP planning, operations, and management, and a broader review of climate change and sea-level rise from that in NRC (2014). Interested readers can also consult other recent committee reports for additional in-depth discussions on saltwater intrusion and peat collapse (NASEM, 2018) and effects on estuaries and coastal systems (NASEM, 2018, 2021).

CLIMATE SCENARIOS FOR RESTORATION PLANNING AND MANAGEMENT

The success of Everglades restoration at least partially depends on the future precipitation and evaporation to be experienced in South Florida over the next decades. The extreme precipitation events of 2023 and 2024 illustrate the vulnerability of infrastructure to climate extremes that have not been previously experienced. All climate projections for South Florida indicate higher temperatures, which will in turn drive increases in potential evapotranspiration (USGCRP, 2023) and have potential to modify hydrologic processes in the Everglades. In addition, when the moisture-holding capacity of the atmosphere increases, more moisture is available to precipitate in each storm (when conditions are favorable), and precipitation is anticipated to be more extreme. Observations suggest this effect is extant in extreme precipitation events in the southeast United States (Kunkel et al., 2020; USGCRP, 2023).

For this reason, there is a clear imperative to ask whether restoration objectives will be achieved given the possible effects of a warmer climate on the hydrology of South Florida. Answering this question is difficult under any circumstances and is especially so considering the inherent uncertainties with climate change projections and the multiple objectives, constituencies, and constraints that restoration faces. However, the modeling tools developed for the CERP provide a basis for answering this question, and methodological advances provide guidance for conducting such analyses. CERP planners appear to be well positioned to begin incorporation of climate change into their analyses. The framing of these analyses will have a large influence on the insights and utility for decision making gained from the results.

Outside of the CERP, efforts in Florida aim to create climate change scenarios from the latest climate models (General Circulation Models [GCMs] used in the sixth Coupled Model Intercomparison Project [CMIP6]). In one example, the U.S. Geological Survey (USGS) and Florida Flood Hub for Applied Research and Innovation[1] developed an ensemble of changes in extreme precipitation for South Florida categorized in terms of the duration of the storm event (e.g., 1 hour to 1 day) and the frequency of occurrence (or return period) (Irizarry-Ortiz and Stamm, 2022; Irizarry-Ortiz et al., 2022). The relationships between the depth of precipitation over a given storm duration and the associated frequency are known as Depth-Duration-Frequency values and are used in stormwater management planning. The Florida Flood Hub study comprehensively addresses many of the sources of uncertainty that are pertinent in climate change studies, including unknown future greenhouse gas emissions, the particular GCMs used, the downscaling methods used, and even the observed data used as the baseline. Change factors, which are multipliers applied to historical design variables (e.g., the 1-hour design storm precipitation depth) to create "climate-impacted" future design variables, were derived from a large ensemble of CMIP5 and CMIP6 (latest generation) GCMs that were downscaled using multiple statistical and dynamical methods and bias corrected with several historical datasets. The median change factors range from a change of 5 to 20 percent in near-term projections (10 years into the future) to 10 to 60 percent in the long-term future (100 and 200 years), while some change factors suggest greater than 500 percent increases. The comprehensive approach to addressing uncertainty ensures that the true range of uncertainty in climate projections is reasonably characterized and protects against planning based on a single or arbitrary sample of climate projections, which may lead to overestimating or, worse, underestimating risk. However, because results display a wide range, translating the results to actionable information is challenging.

---

[1] See https://www.usf.edu/marine-science/research/florida-flood-hub-for-applied-research-and-innovation.

FIGURE 4-1 Future rainfall change factors for South Florida counties (25-year/1-day and 100-year/3-day duration, with median and 50% confidence interval).

SOURCE: SFWMD, 2022.

A recent example from South Florida illustrates a pragmatic attempt to translate science into practice. The South Florida Water Management District (SFWMD) produced a summary of the USGS change factor study for adoption in South Florida flood management practices (Figure 4-1; SFWMD, 2022) and maintains a Resilience Metrics Hub that provides summaries of recent trends in precipitation as well as other variables. Broward County and the South Florida Regional Climate Compact have adopted a 20 percent increase in design storm events for land development planning to manage flood risks (Irizarry-Ortiz and Stamm, 2022).[2] The approach recognizes the need to address the risks of extreme events in the face of uncertainty and provides developers and planners certainty for design. By incorporating additional stormwater capacity in their development designs, planners will be providing protection against increases in precipitation extremes at a cost that is likely to be less than the damages that could be incurred if no action were taken (Rosner et al., 2014).

---

[2] See https://www.broward.org/resilience/Planning/Pages/FutureConditions100YearFloodElevation.aspx.

Another approach that has been adopted by some water utilities, water management agencies, and the World Bank is "climate stress testing." A key distinction between climate stress testing and more typical climate change impact studies is that the objective of stress testing is to learn the climate sensitivity of the water system being studied. Thus, climate scenarios are used in a model of the water system to structurally test the response of the system to a range of plausible climate changes. The climate scenarios may be derived from GCM projections (e.g., Moody and Brown, 2013) or from stochastic climate scenario generators (stochastic models). Stochastic models are often used because they enable the creation of scenarios that reflect the particular changes in climate that are posited for the region of interest but are not necessarily represented in GCM simulations. They include pure statistical approaches (e.g., Steinschneider and Brown, 2013) and methods that link the scenarios to changes in underlying physical influences such as El Niño-Southern Oscillation (ENSO; Steinschneider et al., 2019). Alternatively, a large set of climate scenarios can be generated from GCM simulations and alternative downscaling and hydrologic modeling approaches (Gorelick et al., 2020). In either case, results of the climate stress test can be used to identify the specific changes in climate that are problematic for the water system (e.g., Ray et al., 2020) and to design adaptations based on that understanding (e.g., Herman et al., 2020).

A drawback of climate stress test approaches is the large number of simulations required, which may not be possible for the computationally intensive models used for CERP planning. In some cases, analysts have selected a small set of climate scenarios from the larger comprehensive ensemble because of computational constraints. In doing so, it is important to consider the learning goals of the analysis in order to select the scenarios that enable achieving those goals. Obeysekera et al. (2015) used two scenarios: one 10 percent wetter and one 10 percent dryer, each 1.5°C warmer with 1.5 feet of sea-level rise. The scenarios were selected to provide some baseline understanding of the sensitivity of the Everglades to plausible climate changes. The specific climate changes examined were representative of the temperature and sea-level rise signals present in observations and climate projections and addressed the uncertainty of future precipitation changes. In other cases, indices are calculated from each candidate scenario and then used to sort the scenarios in terms of the degree and kind of changes that each scenario encompasses. For example, in a study of the water supply system for San Francisco, analysts selected a small set of scenarios to represent climate variability using a drought severity metric (François et al., 2024; Whateley et al., 2016). In each case, the selection is purposeful and, in this sense, curated to obtain the specific information needed to aid decision making.

These examples illustrate how South Florida cities and agencies, including the SFWMD, are actively using climate scenarios to inform infrastructure planning, but the same cannot be said for the CERP, which has not used precipitation or temperature scenarios in its project planning to date. There are numerous ways, as highlighted above, to create carefully curated climate scenarios that can test the robustness of large infrastructure projects and inform decision making. For example, several climate scenarios could be developed from available downscaled GCM simulations or stochastic simulation and used as forcings in existing CERP modeling tools to stress test project design alternatives to identify the specific climate changes under which ecological performance metrics cannot be met at an acceptable level. In cases where computational requirements limit the number of scenarios, a small set of climate change scenarios selected to represent an acceptable level of risk could be used to test each restoration alternative in project design. Given the available methods, the magnitude of CERP investments, and the intended duration of the infrastructure, it seems appropriate to assess their robustness under changing climate using the widely available information and tools.

SEDIMENT ACCRETION MODELING FOR RESTORATION DECISION MAKING IN LIGHT OF SEA-LEVEL RISE

Sea-level rise in South Florida averaged 2.4 mm/yr over the past century (Maul and Martin, 1993) and has resulted in saltwater intrusion, especially in the southeast Everglades. Since 2006, the sea level has been rising at a rate of between 6 and 9 mm per year (Wdowinski et al., 2016). The effects of sea-level rise on urban infrastructure, water supply, and freshwater coastal habitats are already being realized. Saltwater intrusion and its impacts on coastal wetlands have been exacerbated by the historic reduction in freshwater flows in the Everglades of approximately 70 percent (Meeder et al., 2017; Perry, 2004). In the past, the southeastern Everglades was predominantly freshwater wetlands and prairie that extended to the coast where a fringe of mangroves, primarily *Rhizophora mangle*, grew along the shoreline. However, over the past century, freshwater wetlands have receded inland by 3.3 km, and a low-productivity "white zone" has shifted inland by 1.5 km (Ross et al., 2000). Mangroves have also moved landward as marine and brackish water has encroached into what were previously freshwater wetlands (Ross et al., 2000). Storm surge from hurricanes, such as 2017's Irma, push saltwater even further inland causing sawgrass die-off. In areas of thick organic peat, including much of coastal marshes of Everglades National Park, saltwater intrusion can spur peat collapse, which leads to land-surface subsidence that, in turn, amplifies sea-level rise and its adverse effects. Increasing freshwater flow to the southern and southeast Everglades can reduce the rate of saltwater intrusion caused by sea-level rise. Thus, restoration efforts have an important role to play in mitigating these effects by enhancing conditions that support sediment accretion, thereby enabling more gradual vegetation transgressions inland as sea level rises rather than sudden land loss. Ultimately, survival of coastal wetlands with sea-level rise will largely depend on their ability to accrete vertically and migrate inland. For mangroves in Biscayne Bay and the southern Everglades, accretion is almost entirely dependent on organic matter accretion from the accumulation of leaf and root biomass in the soil due to the relatively low inputs of mineral sediment (Breithaupt et al., 2017). An increase in freshwater and nutrient inputs through restoration could increase accretion through enhanced mangrove productivity in areas such as Biscayne Bay and the southern Everglades where anthropogenic impacts to hydrology have reduced freshwater flow and increased salinities. Additionally, accretion rate will also likely be affected by the rate of organic matter decomposition in the soil, which has been shown to increase with less flooding and greater soil phosphorus concentration (Poret et al., 2007). Sediment accretion models are used globally to estimate the vertical change in wetland elevation from a combination of organic matter and sediment accumulation in response to changes in flooding (including sea-level rise) and other factors such as sediment supply and wetland plant productivity.

A desired outcome of the Biscayne Bay and Southeastern Everglades Ecosystem Restoration (BBSEER) project is an enhanced accretion rate in coastal wetland habitats in response to sea-level rise with greater freshwater input. It is assumed that wetland plants within the project area will have greater productivity with a greater hydroperiod and water depth and lower salinity than under current conditions (Sklar et al., 2021). However, plant productivity will decline when optimal flooding and/or salinity conditions are exceeded. This constraint is important because the coastal environments of Biscayne Bay and the southern Everglades are relatively sediment limited and thus reliant on organic plant-based accretion (Breithaupt et al., 2017). Therefore, the survival of these important ecosystems depends on vegetation productivity for vertical accretion and/or the ability to migrate, but landward migration does not always offset marsh loss at the seaward edge (Osland et al., 2022). Sediment accretion models are being used in CERP planning for BBSEER to understand the potential outcomes of restoration alternatives. The sections that follow review the latest tool being applied in the BBSEER project and its use and limitations for informing decision making on restoration investments.

BBSEER Adaptive Foundational Resilience

Sediment accretion models are being used in the BBSEER planning process as the Adaptive Foundational Resilience (AFR) performance measure (see Box 4-1 for details), which is one of nine performance measures used to assess project

BOX 4-1
The BBSEER Adaptive Foundational Resilience
Performance Measure

The new AFR performance measure (RECOVER, 2022p, 2023) is an estimation of the sediment accretion rate based on the hydrologic outputs of the Regional Simulation Model for the Glades and Lower East Coast Service Areas (RSM-GL) and the Biscayne and Southern Everglades Coastal Transport (BISECT) models. Specifically, estimates of porewater salinity, sheet-flow volume, and depth duration at each model grid cell are translated into sediment accretion rates based on sets of empirical conversions for freshwater and saline habitats (Figure 4-2). The accretion rates are integrated over the 52-year model duration (1965–2016) and then normalized to the maximum possible accretion (based on the empirical formulations of sediment accretion), yielding a normalized performance measure score from 0 to 100. The combined scores along various transects through indicator regions of the model domain are being used to directly compare the model-estimated accretion under the different flow regimes of existing conditions, future without project, and proposed project alternatives (NASEM, 2023).

Specific details of the model include the following:

- Accretion rates respond to three key drivers—sheet-flow volume, depth-duration, and salinity—based on approximated relationships between each parameter and accretion for each habitat (e.g., mangrove, sawgrass), as shown in Figure 4-2.
- The model run is initiated with an approximated future elevation at year 2085, based on an assumption of 50 percent of the maximum accretion rate.
- The RSM-GL model runs hydrology data over the model period of record of 52 years (1965–2016) including the effects of projected intermediate and high sea-level rise scenarios and project alternatives for the 2085 time point of interest (Figure 4-3).
- In contrast, the BISECT model provides salinity projections based on the 2007–2016 period of record. The AFR assumes that the final porewater salinity in 2016 is indicative of the salinity in 2085.
- Accretion rates in the model for the year 2085 are calculated using the 52 years of hydrologic data for each of the three drivers and the assumed new elevation, accounting for accretion.
- The results are three total accretion rates in year 2085 based on sheet-flow volume, depth-duration, and salinity, respectively.
- The three final accretion rates are averaged for a single accretion rate for each project alternative.

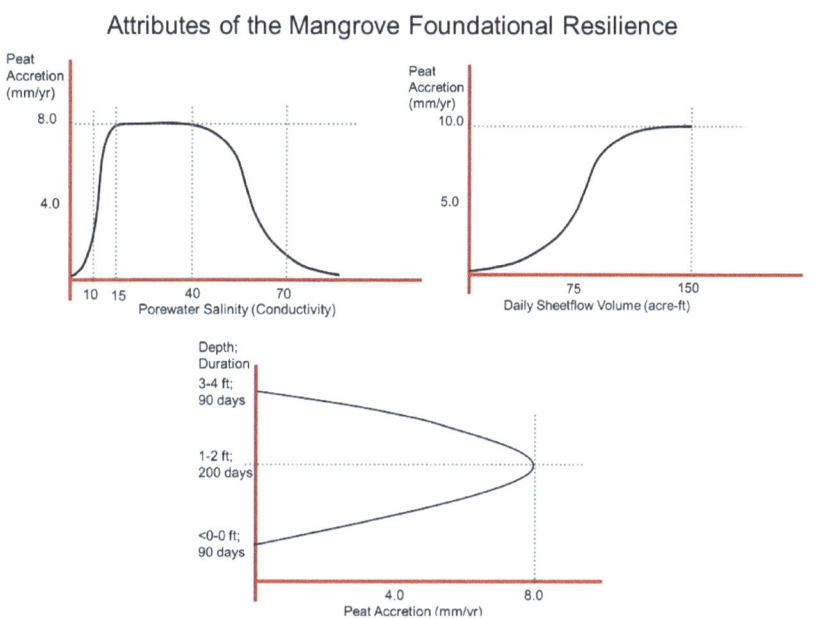

FIGURE 4-2 Hypothesized relationships between accretion and porewater salinity (mS/cm), sheet flow, and depth-duration of flooding used in the AFR performance measure for mangroves in BBSEER.

NOTES: The top left figure illustrates the prediction that as porewater salinity (conductivity) is between 0 to 40 mS/cm, accretion rates will stabilize at 8 mm/yr. When salinity increases beyond 40 mS/cm, there is a non-linear decline to 0 mm/yr above a salinity of 70 mS/cm. The top right figure shows the predicted logistic relationship between accretion and daily sheet-flow volume. With a sheet-flow volume of 100–150 acre-feet/day, sediment accretion is predicted to be 10 mm/yr in mangroves. The bottom figure shows that the rate of accretion has a parabolic relationship with the combined depth and duration of flooding with the highest accretion at 1- to 2-foot water depth and 200 days of flooding and the lowest when the system is dry for more than 90 days or with 3 to 4 feet of water for 90 days. RECOVER (2023) includes notable differences in the three curves, including a lower maximum value of accretion (6 mm/yr) relative to sheet flow. Additionally, the relationship between accretion and salinity in RECOVER (2023) increases exponentially from 0 to 8.0 mm/yr between 0 and 10 mS/cm, compared to a constant 8.0 mm/yr at low conductivities in RECOVER (2022p). These relationships may have been updated after modeling for BBSEER had begun.

SOURCE: RECOVER, 2022p.

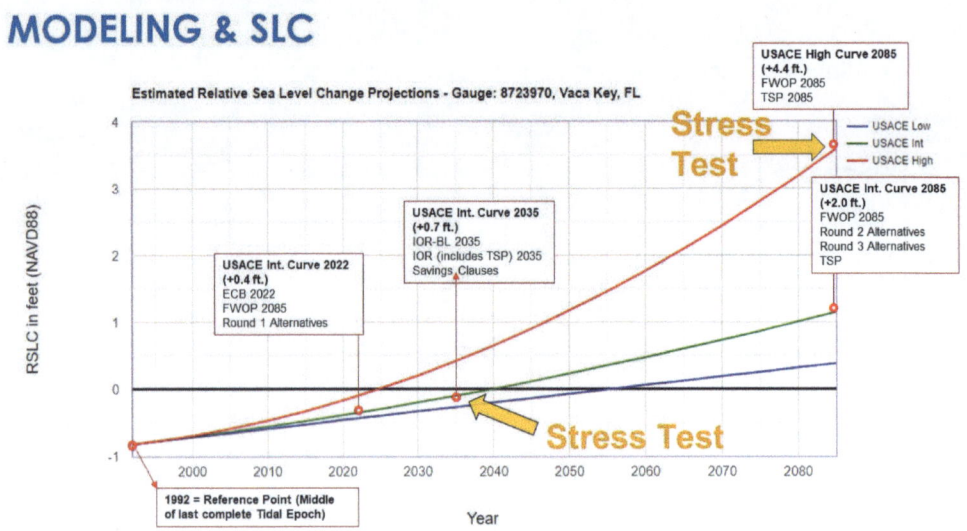

FIGURE 4-3 Relative sea-level change projections from a 1992 reference point at Vaca Key, Florida.

NOTE: Projected sea levels in 2085 are used for testing alternative project designs for the AFR performance measure.

SOURCE: W. Wilcox, SFWMD, personal communication, 2023.

---

alternatives (see also Chapter 2). As indicated in the committee's last report (NASEM, 2023), the AFR performance measure provides a *relative* comparison of *estimated* accretion under different possible future conditions, as a way to evaluate the relative benefits (e.g., higher accretion rates) of different project alternatives. It is critical to evaluate the effect of proposed restoration alternatives relative to a future without restoration on coastal wetland response to sea-level rise (i.e., resiliency), and the AFR is a useful initial step to make relative comparisons of scenarios. However, the uncertainties and assumptions incorporated into this approach limit application of the AFR performance measure to predict realistic accretion rates under future conditions. Furthermore, if the assumed relationships between accretion and hydrology or salinity are inaccurate, then the effect and relative impact of project alternatives may also not be accurate.

There are several large uncertainties in the modeled accretion rates used for the AFR performance measure. The first uncertainty is the accuracy of the accretion rate used to initialize the model for 2085. The accretion rates used to build marsh elevations from present to 2085 were arbitrarily set at 50 percent of maximum (e.g., 6–8 mm/yr for mangroves and 4 mm/yr for freshwater wetlands; Figure 4-2). This sets conditions similarly for all future comparisons but may not be at all realistic as a starting point.

Second, as currently being utilized for the AFR performance measure, the relationships between accretion rate and flow, hydroperiod, and salinity are loosely based on data but with significant smoothing based on conceptual understanding (Figure 4-2). The assumed relationships seem reasonable, but the specifics are critical (e.g., maximum and minimum rates, slopes, intercepts). For example, the maximum accretion value for mangroves (8–10 mm/yr) is slightly above the upper threshold rate of 7 mm/yr quantified from a comprehensive review of mangrove accretion associated with Holocene sea-level rise (Saintilan et al., 2020). Furthermore, accretion is highly site specific, depending on local hydrology, supply of sediment, salinity, nutrients, and other factors. Thus, the relationships used for estimating accretion may also have large associated errors not included in the modeling effort. To clarify the reliability of estimates provided by the AFR performance measure, relationships between hydrology and salinity and habitat-specific accretion rates should clearly show and reference relevant data from Everglades ecosystems or similar geomorphic settings (e.g., Breithaupt et al., 2017; Craft and Richardson, 1998; Feher et al., 2020; Lynch et al., 1989; Reddy et al., 1993; Sklar et al., 2021; Smith et al., 2009; Smoak et al., 2013). Placing existing data on curves representing relationships between peat accretion and hydrologic drivers will lend confidence to relationships and outcomes.

Third, simply averaging accretion rates derived from the three drivers (salinity, sheet flow, and depth-duration) is unrealistic; therefore, the model should incorporate some interplay among these drivers. For some habitats (e.g., sawgrass) salinity will likely be a more important constraint on productivity and accretion than flooding. But even for mangroves, which are more salt tolerant, there is little scientific support for the averaging of accretion outcomes in response to different abiotic conditions. Considering that plant productivity is limited by the most constraining abiotic factor, a limiting factors approach or a weighting approach may be a more appropriate way to combine multiple factors. Applied to the accretion response curves in the most simplistic way, this would be the minimum accretion rate of the three factors. Ultimately, sediment accretion modeling is being used in a relatively limited scope (only for the AFR performance measure) to compare project alternatives, and not as a predictive model of future conditions. There is the potential value of accretion modeling to be used in project cost-benefit analysis, planning, monitoring, and adaptive management. Modeling the effects of sea-level rise on coastal habitats is challenging because of the complexity of wetland morphodynamics; however, realistic quantitative predictive models exist and could be utilized (see Fagherazzi et al., 2020, for a comprehensive review). For example, multiple predictive models of marsh evolution, such as the marsh equilibrium model (MEM) and the WARMER model, which are point-based models of elevation and sediment accumulation related to sea-level rise, could be applied to BBSEER habitats (e.g., Fagherazzi et al., 2020; Morris et al., 2002; Schile et al., 2014). A recent steady-state model of soil accretion rates in the Everglades with sea-level rise has been developed (Chambers et al., 2021), which may be a good example or starting point. Dynamic modeling could be conducted as a parallel effort to habitat unit evaluation and can serve as the basis for comparing future monitoring efforts.

In evaluating the value of the AFR accretion modeling effort to gauging resiliency of coastal wetlands to sea-level rise within the CERP, the question is "What are the consequences of unrealistic predictions of sea-level rise effects on coastal habitats for BBSEER restoration planning and management?" Without more confidence in the assumed accretion rates under all scenarios, it is difficult to assess whether restoration investments are likely to deliver anticipated returns. To increase confidence in predicted future accretion rates, more accurate models should be incorporated, which would also have the added benefit of enhancing the outputs of ecological modeling through better predictions of the productivity and distribution of vegetation communities. Uncertainties in the distribution and health of the foundational ecosystems will compound the uncertainties of ecological models at higher trophic levels. Lastly, the AFR is one of nine performance measures (see Chapter 2) being used in the evaluation of project alternatives. All performance measures are weighted equally in the evaluation of alternatives, and with sea-level rise resiliency just one performance measure out of nine, the relative importance of this metric in the alternative evaluation exercise is very low despite the fact that it will largely determine future submergence and distributions of coastal ecosystems. Overall, while the AFR performance measure is a relative measure of project performance of alternative actions, there remains large uncertainty about the ability of the ecosystems to adjust to future rates of sea-level rise, and these uncertainties merit further use of sediment accretion data and modeling tools to inform future investments.

Restoration Planning Needs

More accurate accretion estimates necessitate a model that begins with current measured accretion rates in representative habitats under present conditions of sea-level rise, hydrology, and salinity. Such a modeling approach would use the current digital elevation model and integrated model inputs of hydrology and salinity from project alternatives and projected rates of sea-level rise to estimate annual accretion rates. By considering feedback between processes and/or limiting factors in annual accretion rate calculations, more realistic estimates of changes in wetland elevation from present day to 2085 can be determined (e.g., Schile et al., 2014). Additionally, the relationships used for model-based accretion predictions should undergo rigorous calibration to the extent possible using field data and existing literature values (see citations provided above), moving beyond the conceptual relationships that are currently used. Improved dynamic modeling approaches would build on the current utility of the AFR modeling for comparing potential restoration alternatives to provide more realistic evaluation of wetland responses and project outcomes under a range of future climate scenarios.

MODELS FOR ASSESSING ECOLOGICAL RESPONSES TO CLIMATE CHANGE

Evaluating responses of biota and ecosystems to changes in hydrology in the Everglades is a key component of measuring restoration progress and success under the CERP. The USGS-led Joint Ecosystem Modeling (JEM) collaboration updates and applies 20 ecological models spanning a number of key taxa (e.g., birds, fishes, crayfish, herpetofauna) and their habitats to a range of restoration alternatives and environmental scenarios including sea-level rise and increased salinity due to climate change. These models use as input a variety of hydrological variables from the Everglades Depth Estimation Network (EDEN), the Regional Simulation Model (RSM), and more recently BISECT, as well as data and information from monitoring of species, their habitats, and other ecosystem variables. A few models feed output into other ecological models; for example, the Snail Kite model (EverKite) requires as input population densities estimated from the Florida apple snail model (EverSnail), and the wading bird models use fish density and biomass as input (Beerens et al., 2015; D'Acunto et al., 2021; Shinde et al., 2014). Although most of the ecological models output spatially explicit habitat suitability indices that integrate environmental predictors of species presence, a small number of them estimate or project spatially explicit population abundances, densities, or biomass founded on life history characteristics of the target species (e.g., Small Fish Density Model, Prey Fish Biomass Model, EverKite, EverSnail, and Wader Distribution Evaluation Modeling), and one model estimates amphibian community richness.[3]

The models under the JEM collaboration have the capacity to incorporate the climate change impacts of altered hydrology, sea-level rise, and increased salinity where relevant and thus can be (and in some cases have been) used to project the ecological effects of hydrologic change from restoration and climate change when data are available (Perez et al., 2017). For example, Catano et al. (2015) and Romañach et al. (2023) use climate change scenarios and projected hydrological responses, including sea-level rise, as input into ecological models to predict changes in fish density, alligator habitat, wading bird distribution, apple

---

[3] See https://www.jem.gov/Modeling.

snail density, amphibian occurrence, and Cape Sable seaside sparrow probability of presence. Results from these studies indicated a decrease in suitable habitat and fish and apple snail density under climate change scenarios that project lower precipitation and higher evapotranspiration, while an increased rainfall scenario benefited fish densities and alligator habitat. Both studies emphasize the importance of freshwater flow into the southern Everglades to counteract reduced rainfall, protect habitat, and forestall the impacts of sea-level rise under climate change. These studies show that ecological models can be powerful tools to explore the effects of different climate change scenarios and restoration strategies on the wildlife species that are indicators of Everglades restoration success. When they integrate both habitat and population responses to environmental change, these tools can clarify the aspects of climate change that are likely to be the most destructive to ecosystems and the restoration activities that can have the greatest potential of mitigation. They are one of the few means available for forecasting the effects of climate change on individual species distributions and populations across large spatial scales and time periods (Wilsey et al., 2013). Despite the sophistication, longevity, and ground truthing of these models, they suffer from limitations when applied to the ecological impacts of climate change, in part because almost all of them do not explicitly include relevant effects of temperature increases on the target species. Only the EverSnail model, which projects the abundance and distribution of Florida apple snail using demographic data and environmental variables, explicitly includes the effects of temperature on reproductive rates. EverSnail output is input as prey into the EverKite model to project the movement, reproductive success, and mortality of the endangered Everglades snail kite (Darby et al., 2015).

Both hydrological and temperature changes are expected to have profound effects on biodiversity in myriad ways (Bellard et al., 2012; Mantyka-Pringle et al., 2012; McLaughlin et al., 2002). Hydrology has been the central focus of Everglades restoration; hence, much is known about the ecosystem effects of altered flow, water depth, fluctuations in precipitation, and sea-level rise. With perhaps the exception of the influence of temperature on algal blooms in Lake Okeechobee and Florida Bay (Havens et al., 1994; Koch et al., 2007), very little consideration has been given to the effects of increased temperature trends on the Everglades. The extreme surface ocean temperatures in coastal southern Florida in 2023 underscore the urgency of better understanding the effects of increased temperature on the success of ecological restoration in the Everglades. Moreover, under all emissions scenarios, projected increases in temperature are more certain than precipitation projections, which predict high variability in rainfall (Kunkel et al., 2020). Therefore, a focus on the effects of increased temperature on Everglades biota could better illuminate the effects of climate change on ecological restoration success than hydrology alone (Grieger et al., 2020).

The biodiversity of the Everglades is particularly vulnerable to increases in ambient air and water temperatures because it is already a highly impacted system, and hydrological flow rates and volumes are greatly reduced from the predrainage system (NASEM, 2016). Some of the more significant direct effects of increased temperature that have relevance to the species and ecosystems covered in the JEM collaboration include (but are not limited to)

- shifts in sex ratios from temperature-dependent sex determination in crocodilians, reducing reproductive output (Bock et al., 2020);
- shifts in breeding phenology of fish, amphibians, and crocodilians, altering community composition (Blaustein et al., 2001, 2010; Cherkiss et al., 2020; Joanen and McNease, 1989; Kushlan and Jacobsen, 1990; Todd et al., 2011; Walther et al., 2002);
- disruption of aquatic communities due to differences in thermal tolerance extremes, which can decompose food webs (Daufresne et al., 2009; Lurgi et al., 2012);
- heat stress in fish, reducing biomass (Alfonso et al., 2021);
- enhanced virulence of disease, increasing mortality rates of target species (Harvell et al., 2002);
- increased mortality in aquatic species due to temperature-induced declines in dissolved oxygen in wetlands (Bulbul Ali and Mishra, 2022; Ficke et al., 2007);
- shifts in the reproductive and feeding phenology of animals (Cohen et al., 2018; Dunn and Winkler, 2010); and
- migration phenology of birds (Cotton, 2003).

The thermal environment also plays a critical role in the development of many species, because it can influence body size, metabolism, endocrinology, and behavior (Atkinson and Sibly, 1997; Boltaña et al., 2017; Elmore et al., 2017; Ruuskanen et al., 2021). All of these effects have the potential to reduce or extirpate populations of the restoration target species and disrupt (and in some cases enhance) the aquatic food web of the Everglades, and they are exacerbated by reduced hydrological flow and increased nutrients (Lorenz, 2014; Statham, 2012; Stys et al., 2017). Omission of the potential effects of increased temperature in the JEM models, separately and in combination with precipitation effects, overlooks a suite of important physiological characteristics of these species, rendering the models insufficient for fully measuring the response of these species and ecosystems to proposed restoration actions.

With the exception of EverSnail, the current JEM models do not incorporate explicit effects of temperature, in large part because of lack of data or the omission of climate change from the original conceptual frameworks on which the ecological models are based. Some of these models could readily incorporate explicit effects of increased temperature in addition to changes in hydrology (e.g., breeding phenology [Shinde et al., 2014]). Indeed, Shinde et al. (2014) acknowledge, "Alligators nest earlier following warmer springs and delay nesting following colder springs (Joanen and McNease, 1989; Kushlan and Jacobsen, 1990). Effect of temperature on nesting will be an option to include as more data become available." Other endeavors to explicitly incorporate the effects of temperature changes would require adaptations of the existing models. For instance, a stage-structured population model of American alligators (*Alligator mississippiensis*) linked to the Alligator Production Suitability Index model may be necessary (and feasible) to investigate the effects of male-skewed sex ratios due to climate change (Brook et al., 2000; Gerber and White, 2014; Visintin et al., 2020).

Additionally, consideration should be given to the development of mechanistic niche models that link physiology with spatial data to project species ranges under current conditions and climate change scenarios (Kearney and Porter, 2009) to supplement the JEM models (see Fordham et al., 2012; Franklin et al., 2014). Mechanistic niche models can have greater relevance to climate change applications than purely correlative niche models and the types of hydrology-based mechanistic habitat suitability models used in JEM models. Mechanistic niche models can link thermal tolerances (among other physiological responses) with environmental variables and spatial occurrences to constrain species distributions to more realistic ranges that track physiological responses to the environment (Evans et al., 2015). This should currently be feasible for crocodilians, for which much is known about thermal tolerances, behavior, and responses to environmental conditions (e.g., Fujisaki et al., 2014; Lawson et al., 2018; Mazzotti, 1989; Smith, 1979). Moreover, there exists a foundation of species distribution models developed in the region on which to base adaptations to include physiological tolerances to climatic variables (Bucklin et al., 2015; D'Acunto et al., 2021). Like all models, mechanistic niche models are subject to uncertainties in input parameters and model uncertainty (Elith et al., 2002; Regan et al., 2002), which compounds under climate change scenarios. However, a greater suite of tools in the climate change modeling toolbox that can shed light on projected biotic responses to climate change, and the inherent uncertainties, can better equip managers to plan for vulnerabilities, responses, and activities to reduce uncertainty.

Any effort to explicitly incorporate the effects of temperature changes into species- and ecosystem- level ecological models will require long-term monitoring data and field experiments, to measure the response of organisms to novel environmental conditions, and spatially explicit forecasts of air, water, and soil temperatures under climate change and restoration. Such an endeavor can and should be tied to adaptive management and monitoring—the ecological models can be used to support project management decisions by comparing expectations to outcomes (see Chapter 5) and to highlight monitoring needs. Monitoring and adaptive management will provide data and information on how models could be updated and applied (NASEM, 2021), although care needs to be taken to account for non-stationary processes associated with climate change (Austin, 2007; Koons et al., 2016; Wolkovich et al., 2014). Such an undertaking requires long-term commitment and careful coordination and communication among teams overseeing models, management, monitoring, and data collection, although databases on species' thermal traits (e.g., GlobTherm, TRAD) have been compiled that can support some data needs (Bennett et al., 2018; DuBose et al., 2024; Lancaster and Humphreys, 2020; Pottier et al., 2022). Machine learning methods also offer opportunities to gain insights into complex ecological systems under uncertainty and non-stationarity (Jones et al., 2023; Peters et al., 2014; Stupariu et al., 2022). These methods span hypothesis-data-driven, data-intensive, and deep learning approaches and include random forests, neural networks, support vector machines, regression trees, and Bayesian networks, among many other methods that continue to be developed and refined as the capacity of computing and artificial intelligence expands (Stupariu et al., 2022).

INTEGRATION

Understanding and managing for the effects of climate change on ecosystems requires an overarching framework to integrate ecological and physical models with monitoring data and climate projections, a consistent set of climate scenarios and assumptions across models, and mechanisms to adapt and update each of these models according to the information that each provides. Such an endeavor is necessary to understand, plan, and execute ecological restoration under a more comprehensive suite of the likely effects of climate change and restoration. However, when predicting potential effects of climate change, it is vital to understand the many factors, and their uncertainties, that might explain or impede restoration success.

The Everglades Vulnerability Analysis (EVA) model shows promise for integration across ecological and physical models, climate projections, and monitoring data that could provide feedback to adaptive management. This recently developed framework aims to integrate physical and ecological models and, with further development, could serve as a basis for a larger enterprise that links models to monitoring and adaptive management efforts. EVA evaluates and synthesizes the impacts of hydrology and salinity on indicators of Everglades

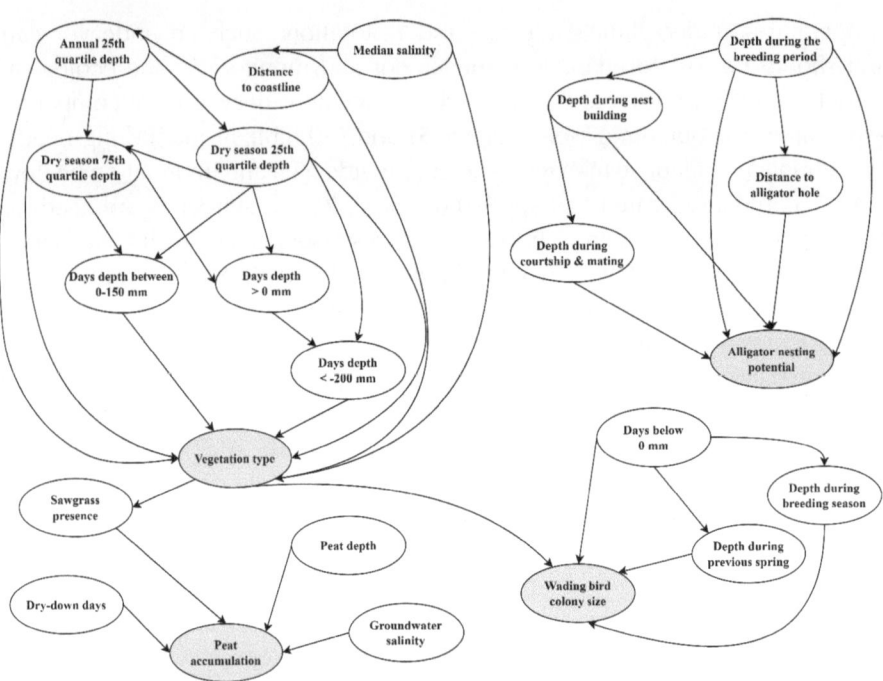

FIGURE 4-4 Structure of the Bayesian network modules currently used in the EVA. Green shaded nodes are output nodes, while white non-shaded nodes are variable input nodes.

SOURCE: D'Acunto et al., 2023. CC-BY 4.0.

---

ecosystem health through a Bayesian network (D'Acunto et al., 2023). Currently the network consists of four modules for indicators of Everglades ecosystem health that are used in RECOVER assessments of restoration activities (Doren et al., 2009; RECOVER, 2020a): vegetation type, American alligator nesting potential, wading bird colony size, and sawgrass peat accretion (Figure 4-4). The framework patches together hydrological inputs from RSM, EDEN, and BISECT to characterize hydrology and salinity across the landscape. These three hydrological system models have mismatches in spatial resolution, location, and extent, and in their capacities to produce information on salinity, long-term future predictions, effects of climate change and sea-level rise on hydrology, or hydrological responses to water management operations. The BISECT and RSM models were each converted to a 400 m grid to align with EDEN.

Each module in the EVA framework consists of variables informed by existing conceptual frameworks and known processes for the Everglades (Figure 4-4) and parameterized by output from the hydrological models, existing Everglades

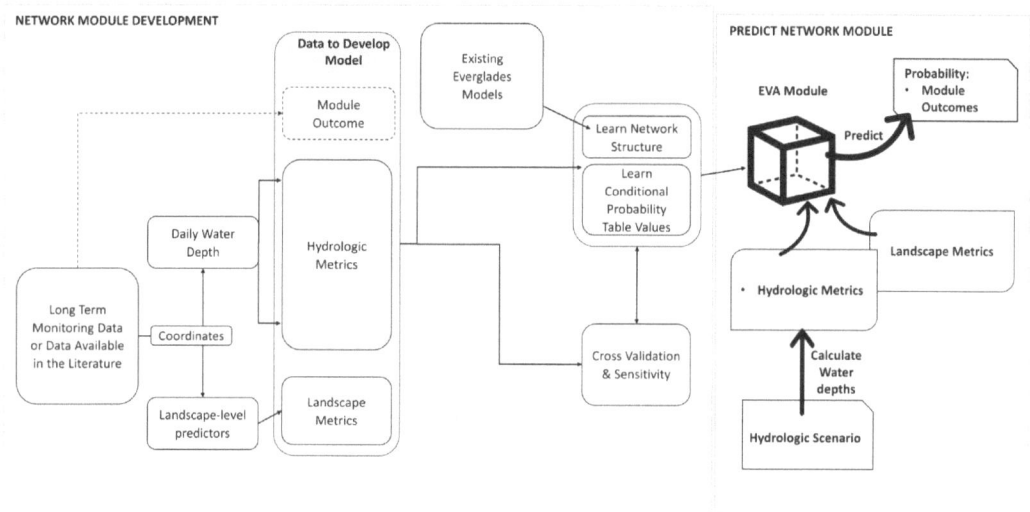

FIGURE 4-5 Conceptual framework detailing how each Bayesian network module was built and then predicted across the landscape.

NOTES: The left-hand box describes the process used to develop the module, including assessing module accuracy and sensitivity. The right-hand box describes the process used to predict the module outcomes across the landscape.

SOURCE: D'Acunto et al., 2023. CC-BY 4.0.

---

models,[4] available data, and expert knowledge (Figure 4-5). Vegetation types are classified into six freshwater marsh and coastal communities—freshwater prairie, mangrove, mangrove scrub, open water, sawgrass, and upland—the results of which feed into peat accumulation and wading bird modules. The Bayesian network outputs probabilities of outcomes generated by influence diagrams and conditional probability tables, which are used to calculate a "vulnerability score" spatial surface defined as the distance from a predefined target site outcome. As such, the EVA does not assess vulnerability per se; rather, it characterizes the expected state of the system as it compares to a set of desirable outcomes (in a complementary way to the BBSEER performance measures), thus highlighting areas of the landscape that are furthest away from ecological restoration goals. The output appears to be limited to separate spatial visualizations of model output (e.g., maps of vegetation types; wading bird colony size; alligator nesting

---

[4] See available models at https://jem.gov/Modeling.

sites in wet, dry, and median years), which can be useful information for targeted management of individual species or vegetation types but has limitations for understanding trade-offs across multiple management objectives.

A major advantage of the EVA is that it is based on a Bayesian network, which is equipped with a set of tools to evaluate model accuracy, sensitivity, and uncertainty that can be readily updated to incorporate additional modules and new data. This feature is particularly pertinent for adaptive management applications in which the learning process is a central feature (Landis et al., 2017). Management can be informed by the EVA, which in turn can be updated (with new input parameter values or modules) based on new information to further inform management and data collection. In this way, ecological modeling, management, and monitoring are integrated into an ongoing adaptive management cycle (Howes et al., 2010; Rumpff et al., 2011).

The EVA framework's decision support utility could be enhanced by the adoption of a post-hoc tool that aggregates or optimizes across the ecological indicators and landscape features. This tool might involve multi-objective optimization approaches (Brias and Munch, 2021; Chadès et al., 2017; Williams and Kendall, 2017) or an approach such as EverForecast that examines trade-offs between all pairs of species (Romañach et al., 2022). Trade-offs exist in nearly all Everglades restoration decisions (e.g., between suitable hydrology for tree island restoration and Cape Sable seaside sparrow breeding habitat [see Chapter 3, Box 3-5]). In light of these trade-offs, decision making should be supported by an integrated framework based on ecological and physical models and informed by monitoring and continuous learning (NRC, 2010). Such a framework should enable weighing the effects of restoration projects on multiple ecosystem components in the context of climate change.

APPLICATION OF CLIMATE TOOLS FOR ADAPTATION OF OPERATIONS

Climate change and associated changes in the frequency and magnitude of extreme events complicate our ability to precisely predict the future performance of CERP infrastructure. Although infrastructure often is considered static in nature, the rules and guidelines by which the infrastructure is operated add a degree of flexibility (see also Chapter 5). Operation of control structures and pump use can be adjusted in response to changing conditions and improved understanding of the system performance. System Operating Manuals (SOMs), regional water control plans such as the Combined Operational Plan (COP), and Project Operating Manuals (POMs) are the policies by which CERP infrastructure systems and components, respectively, are operated. In the context of climate change, these operating plans offer the means to adapt the operations of CERP infrastructure to the specific changes and surprises that directly affect CERP performance.

Recognizing and utilizing this potential can improve the performance of the CERP in the face of a changing and variable climate.

As CERP projects come online, they require operating plans, and those that cause downstream effects or depend on upstream projects require updates of the relevant SOM. These updates occur every 3 to 5 years, providing a means to incrementally adjust operations in accordance with prevailing and anticipated conditions. During that time increment, more information is revealed about external conditions and the performance of existing CERP projects, and improved tools are available to estimate expected future conditions; thus, uncertainty is reduced. To benefit from this opportunity for learning in the face of climate change, the system's performance should be evaluated in light of these external factors (e.g., temperature, precipitation, sea-level rise trends) and factored into the updates of the POMs, regional water control plans, and SOMs, as well as the project design itself. As the Central Everglades Planning Project (CEPP) 1.0 comes online to replace the COP (see Chapter 2), there is an opportunity to standardize incorporation of the latest climate change considerations in operating plan revisions. Effectively doing so requires communication between modelers, monitoring staff, and climate experts as well as a process for learning to inform future CERP operations. It also requires updating of the historical record of weather and sea level that is used to develop operating manuals, ideally on the same frequency as the updating of the SOMs and POMs.

CONCLUSIONS AND RECOMMENDATIONS

Attempting to include climate change information in CERP planning could overwhelm even the most intrepid project planner, and CERP planners have been cautious in their attempts to do so. The amount of climate change information, the unknown credibility of the information, the number and complexity of models used in CERP planning, and the complications of incorporating climate information into models not designed for that purpose add up to a very challenging endeavor. Although there is danger in using climate projections without a carefully considered plan that is consistent across the diverse set of analysis and models involved in the CERP, the greatest danger is making no attempt to plan for climate change. The committee offers the following recommendations to advance the use of climate change information in CERP planning and operations.

A strategy to understand the impacts of climate change should be developed with a curated set of scenarios that are used consistently across all components of planning and restoration implementation. This set of climate change scenarios should represent the range of plausible changes based on review of the scientific literature and available climate projections and be used to assess project and system vulnerability to changes in temperature, precipitation, and sea-level rise. Validated methods for the use of climate scenarios, of which there are several in the scientific literature, should be used to stress test restoration plans. These planning scenarios should be applied through existing hydrologic and ecological models to provide insights on the potential vulnerability of flora, fauna, and infrastructure to these plausible changes. This strategy can be implemented now, and the results can lead to better estimation of benefits of planning and ultimately better outcomes in the long run.

A dynamic model that predicts coastal wetland elevations through time informed by empirical data is needed to provide more accurate predictions of coastal restoration outcomes and guide investment decisions. Sediment accretion models are currently being used to compare alternative restoration plans for BBSEER in the context of sea-level rise. Because the models are based on approximated relationships between flooding and accretion and salinity and accretion, the predicted and even relative outcomes for each alternative may not reflect reality. Recommendations to improve the accuracy of accretion models, particularly if they are to be predictive of future conditions, are to (1) start models with elevation and accretion rates at present day with annual time steps that include projected sea-level rise and project alternative effects; and (2) ground truth relationships between flooding, salinity, and accretion using existing data from comparable environmental settings. A dynamic predictive model of accretion and wetland elevations over time will be more accurate and can serve as a guidepost for monitoring and adaptive management. Development of a dynamic model could provide helpful confirmation of the potential ecological return on the large, expected infrastructure investments in BBSEER and inform the pending planning of the Southern Everglades project.

Existing ecological models should be used to a greater extent and further developed to anticipate the effects of climate change, including temperature, on the wildlife indicators of Everglades restoration success. Because wildlife species and habitats are the ecological endpoints of Everglades restoration, the output of ecological models should be considered early in the process (when evaluating restoration plans) rather than late in the process (when preparing National Environmental Policy Act reports). Ecological models should be developed and applied to evaluate the effects of projected changes in precipitation and temperature on biotic indicators of restoration success. Because confidence in temperature projections is greater than that for precipitation, attention should be paid to the effects of increased temperature on life history, phenology, and physiology of wildlife species using tools such as mechanistic niche models. Furthermore, a more thorough accounting of uncertainty in ecological models with respect to climate change impacts should be undertaken. How changing climate is incorporated into models based on historical data that are unlikely to hold in the future as well as the reliability of tools under changing, or nonstationary, conditions should be evaluated. Spatially explicit estimates of air, soil, and water temperatures under climate change should be developed, and models should be linked to ongoing monitoring and adaptive management efforts and updated accordingly. Failure to incorporate ecological models early in planning and monitoring efforts is a missed opportunity to gain greater insight into the effects of different climate change scenarios on Everglades restoration.

A more cohesive integration of ecological and physical modeling and monitoring that draws together existing data, models, and efforts should be pursued to understand and mitigate the effects of climate change on Everglades restoration to better support restoration decisions. Such integration can and should be tied to applications of adaptive management in which the learning process is a central feature. In an integrated framework, management decisions can be informed by physical and ecological models, which in turn can be updated with monitoring and other data to further inform management, monitoring, and model refinement in an ongoing cycle of learning that can reduce uncertainty in projections. Such an integrated framework that enables updating based on new information, such as in the EVA modeling framework, can better support decision makers as they weigh risks and benefits of alternatives, manage trade-offs, and prepare for the effects of climate change on Everglades restoration. A long-term commitment and careful coordination and communication among teams overseeing models, monitoring, management, and decision making will be necessary to achieve this objective.

Regular revisions to the SOMs and other operational plans should incorporate the evolving understanding of climate variability and change, including extreme events, to ensure anticipation of and planning for a wide range of conditions. SOMs represent the flexibility inherent in infrastructure operations and can be leveraged for that purpose if monitoring and periodic updating are systemized and linked to operations. The evolution of the COP to CEPP 1.0 represents a prime opportunity to apply learning from several years of COP operations and consider a subset of future climate scenarios to test the response of operations to changing climate conditions.

5

Adaptive Management and Use of New Information in Decision Making

When the U.S. Congress approved the Comprehensive Everglades Restoration Plan (CERP) in the Water Resources Development Act of 2000 (WRDA 2000), the Central and South Florida Comprehensive Review Study (or Yellow Book; USACE and SFWMD, 1999) was recognized as providing a general outline for restoration of the Everglades rather than a detailed restoration plan. In consideration of Congress's direction to move forward in the face of uncertainties, adaptive management was identified and authorized in the legislation as a CERP program element and mechanism to incorporate emerging scientific information into the plan and to address unforeseen consequences of restoration projects. Congress subsequently approved funding for an Adaptive Management and Monitoring program in WRDA 2000, and the 2003 Programmatic Regulations (33 CFR §385.31) directed the U.S. Army Corps of Engineers (USACE) to adopt an adaptive management approach.

The concept of adaptive management emerged in the 1970s as an approach to address increasingly complex resource management issues. Adaptive management offers a means to proceed amid uncertainty through the iterative refinement of management actions, ideally based on experimentation and a rigorous monitoring and science program (Lee, 1999; Walters and Holling, 1990). Of the many applications of adaptive management, the most effective ones have well-structured processes that include the following:

- management objectives that are regularly revisited and accordingly revised,
- a model or models of the managed system,
- a range of management choices,
- the monitoring and evaluation of outcomes,
- mechanisms for incorporating learning into future decisions, and
- a collaborative process for stakeholder participation and learning (NRC, 2004b).

Through effective application of such processes, decision making moves from trial and error to *learning by doing* based on modeling, monitoring, assessment, reevaluation, research, and field-based pilot projects, thus improving management by minimizing costly mistakes and reducing operational and ecosystem risk associated with unintended outcomes. The literature is replete with diagrams outlining the iterative adaptive management process, and one of the more detailed examples is shown in Figure 5-1.

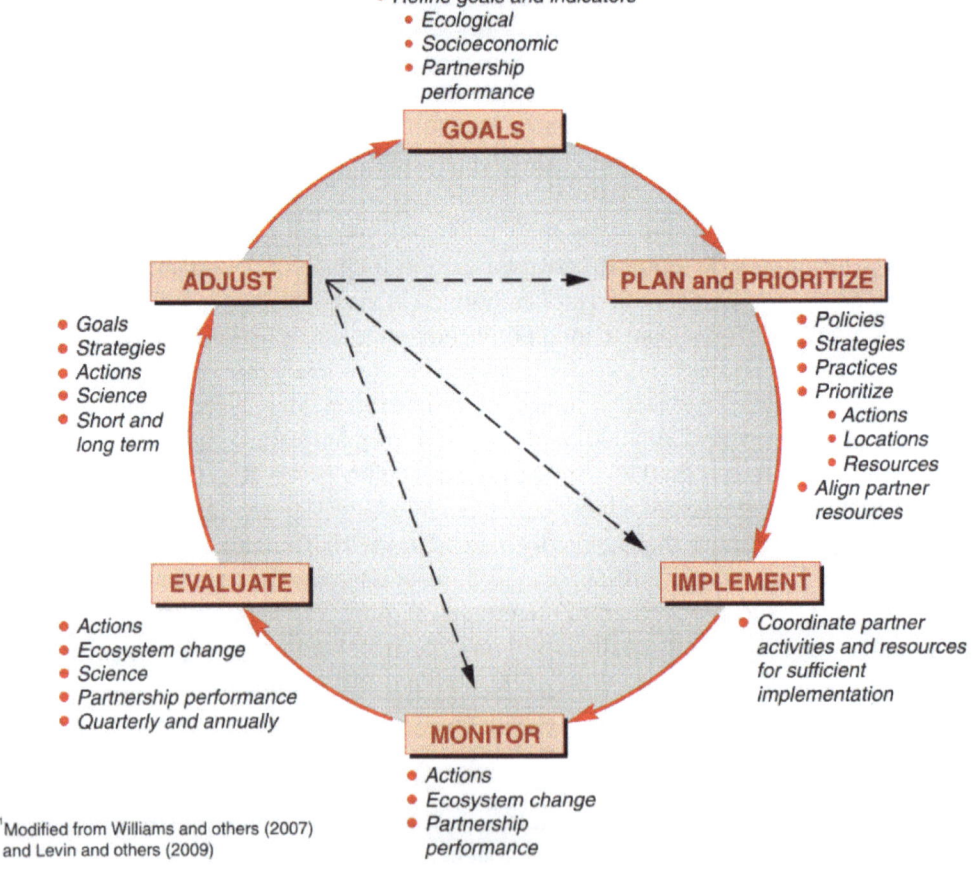

FIGURE 5-1 An ecosystem-based adaptive management framework with supporting science elements for the Chesapeake Bay Restoration partnership.

SOURCE: DOI and DOC, 2009.

Effective adaptive management programs require governance authority and policy-level support and direction as well as clear and agreed-upon adaptive management processes and objectives. These processes are critical to facilitate the timely incorporation of new information and to expeditiously adjust and improve management to better meet program goals. Importantly, effective programs and processes require general agreement among agencies and stakeholders on why adaptive management is being implemented. For example, is the objective of adaptive management to better achieve the outcomes expected at the outset of the restoration or modified outcomes based on new information, or to maximize the benefits of management actions? Without clear, agreed-upon objectives and processes, adaptive management programs will face delays and inaction.

Twenty-five years have passed since the Yellow Book was published, and although much has been learned about the Everglades, adaptive management remains central to enhancing restoration outcomes. As multiple CERP and non-CERP projects are being implemented, opportunities exist to better understand the interconnected hydrologic, geomorphic, geochemical, and biologic processes necessary to sustain the Everglades and inform future decisions on restoration design and operations. Adaptive management built on strategically designed monitoring, research, and predictive models can identify feedback mechanisms and thereby improve understanding of system responses under different operational actions and climate regimes—supporting progress toward CERP goals while building a culture of cooperation, communication, and long-term support for restoration among scientists, restoration practitioners, decision makers, Tribes, and stakeholders.

In this chapter, the committee evaluates the status and effectiveness of adaptive management and incorporation of new information in the CERP to support learning and improve decision making after an initial project plan (i.e., a Project Implementation Report [PIR]) has been developed. The committee reviewed the CERP agencies' use of new information from monitoring, modeling, and research to inform decision making at four different scales, some of which may not be considered adaptive management by all agencies but nevertheless play an important role in achieving restoration outcomes.

1. **Project-level adaptation during design and construction.** After the initial project planning and authorization, engineers develop specific designs that incorporate additional detail that was not available or included in the original PIR. New information gained and additional detailed analysis between the development of the PIR and the final design can lead to project refinements. All changes must be approved according to USACE policy.
2. **Project-level adaptive management after operations begin.** After construction of a project (or project component) is completed and operation ensues, the USACE project-level adaptive management process can begin,

including monitoring, assessment, and modifications to operations or the design if warranted. Extensive guidance documents (e.g., RECOVER, 2011, 2022l, 2022m, 2022n; USACE and SFWMD, 2018) have been developed, and all CERP projects authorized after the guidance documents were released must have an adaptive management plan.
3. **Operational adaptation at regional scales.** Regional systems have operational plans to manage the delivery of water under changing conditions. While operational plans are designed to cover a wide range of conditions, new information about species or habitat conditions or extreme hydrologic events can create a need to deviate from the existing plan over the short term and/or adapt the operational plan over the long term through periodic revisions. "Managing adaptively" is a term used to describe leveraging new science and information to make flexibility-based decisions and improve operations.
4. **Program-level adaptive management.** The CERP originally included 68 projects that were designed to function collectively to support restoration of the remnant Everglades ecosystem while balancing requirements for flood control and water supply needs. Program-level adaptive management represents system-level adjustments to new information based on an assessment of expected progress of the CERP relative to the original goals.

In this chapter the committee discusses the evolution of adaptive management in the CERP and then evaluates the effectiveness of adapting to new information at these four scales. Specifically, the committee's review focuses on assessing how CERP projects and the overall program are using new information to improve their design, development, construction, and operation. Finally, the committee discusses the key elements of the science enterprise (see also NASEM, 2023) that are essential to support adaptive management and recommends approaches to strengthening them.

ADAPTIVE MANAGEMENT PROCESS GUIDANCE FROM RECOVER

RECOVER (see Appendix A) has assumed much of the responsibility for fulfilling the requirement that the CERP employ an adaptive management approach. In 2006, RECOVER produced an early vision for adaptive management that emphasized systemwide (i.e., program-level) adaptive management and included elements of project-level adaptive management (RECOVER, 2006a). The CERP Adaptive Management Strategy was visualized in four stages (see also Figure 5-2):

1. Planning, during which uncertainties are identified;
2. Performance assessment, during which monitoring is conducted and the outcomes of restoration investments are assessed using either project-level

Adaptive Management and Use of New Information in Decision Making 181

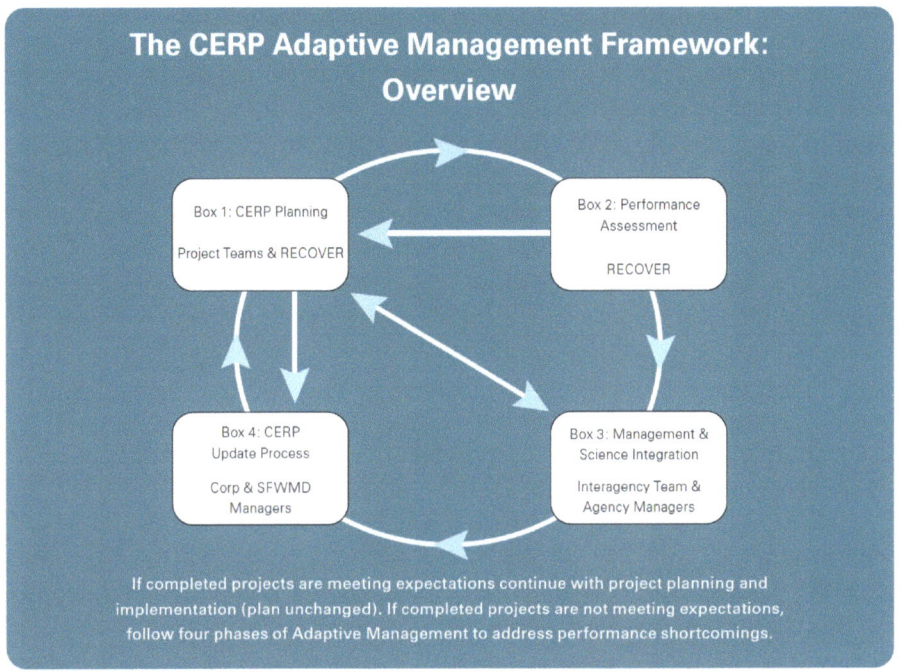

FIGURE 5-2 An early conceptual framework for CERP adaptive management, with four steps: (1) planning, (2) performance assessment, (3) integration of science and management, and (4) modifications to the CERP.

NOTES: Box 4 in the diagram differs from the Periodic CERP Update, which is defined in the Programmatic Regulations (33 CFR Part 385) as "the evaluation of the Plan that is conducted periodically with new or updated modeling that includes the latest available scientific, technical, and planning information." Under the Programmatic Regulations, a separate Comprehensive Plan Modification Report is required to make changes to the overall CERP.

SOURCE: RECOVER, 2006a.

   monitoring or the CERP Monitoring and Assessment Plan (RECOVER, 2004, 2006b, 2009);
3. Management and science integration, which involves communication of new information and its implications to management; and
4. Decision making, during which modifications to the CERP may be made.

However, this early broad vision provided little detail on the specifics of each stage or their interconnections.

Detailed Project-Level Implementation Guidance

Although the CERP adaptive management strategy articulated the importance of adaptive management at the project level, the early CERP projects (e.g., Picayune Strand, Indian River Lagoon) did not include formal adaptive management plans in their PIRs. In 2009, the USACE required adaptive management plans in all CERP projects (USACE, 2009), and the Adaptive Management Integration Guide (RECOVER, 2011) was developed to provide specific process guidance on how to integrate adaptive management within the USACE project planning process. This includes guidance for developing project-level adaptive management plans to facilitate future decision making. These project-level plans identify critical uncertainties, and for each uncertainty the plans document indicators to assess performance, thresholds for when corrective action should be taken, and management options matrices describing corrective actions if projects are not meeting performance targets.

CERP Guidance Memorandum 66 (USACE and SFWMD, 2018) was developed to provide additional direction and detailed process guidance for how RECOVER helps provide integration of new science and information into projects during post-authorization design, construction, and operations. A key element of this guidance is that each project is assigned a Point of Contact from RECOVER who monitors and supports the interactions of RECOVER with the Project Delivery Team. Additional recent detailed guidance (RECOVER, 2022a-n) describes Standard Operating Procedures (SOPs) for points of interaction of RECOVER with Project Delivery Teams during project planning (i.e., prior to project authorization; Figure 5-3) and during project implementation (i.e., post-authorization; Figure 5-4).

The detailed adaptive management guidance is conceptually consistent with the original vision for CERP adaptive management, describing in detail the roles of the various entities involved (e.g., project teams, RECOVER, decision makers) and their interactions depicted in Figure 5-2. The guidance focuses on the interactions of RECOVER with project teams and its role and responsibilities as the assessor and conduit of new information. In practice, modifying projects in response to new information is a flexible, collaborative effort rather than a rigid process with many required boxes to check. Project teams can involve RECOVER (or not) in their work to incorporate new information into approved plans, and RECOVER can choose to become involved if deemed appropriate. Many decisions to modify an authorized project can be made by CERP leadership at the USACE Jacksonville District and the South Florida Water Management District (SFWMD). However, depending on the level of change necessary, additional analysis, review, reporting, and approvals by the USACE Division or Headquarters may be required by USACE policy.

Adaptive Management and Use of New Information in Decision Making 183

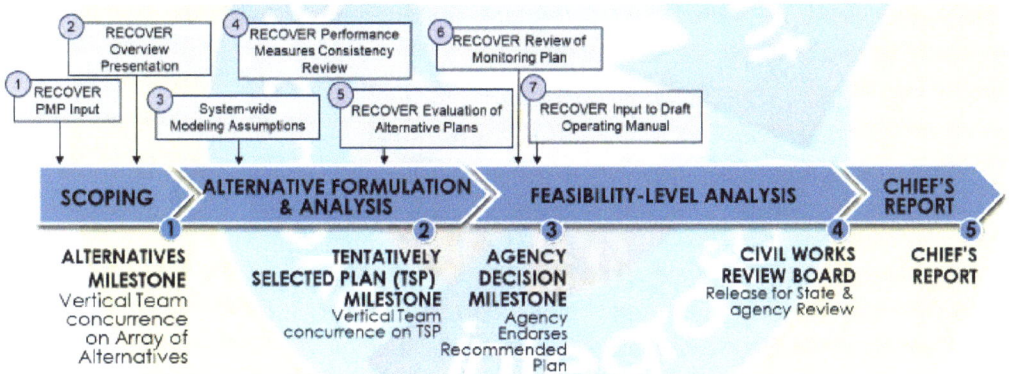

FIGURE 5-3 Seven points of interaction between RECOVER and project teams during project planning.

NOTE: Some of these items are critical steps for the development of an adaptive management plan (i.e., #4, #6), but others address the evaluation of project alternatives.

SOURCE: RECOVER, 2022a.

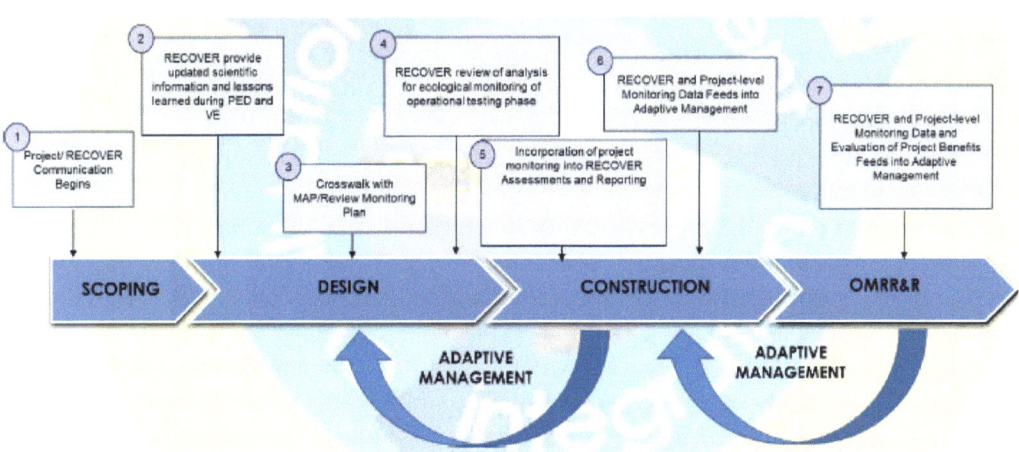

FIGURE 5-4 Seven points of interaction between RECOVER and project teams during project implementation (i.e., post-authorization).

SOURCE: RECOVER, 2022a.

Program-Level Adaptive Management Guidance

Although the Adaptive Management Integration Guide (RECOVER, 2011) encompasses program-level as well as project-level adaptive management, it provides more extensive guidance at the project level, especially for planning, which was the primary CERP activity when it was developed. The detailed description of how to implement the vision for program-level adaptive management (Figure 5-2) is provided in a Program-Level Adaptive Management Plan (RECOVER, 2015a), which follows the same framework as the Adaptive Management Integration Guide (RECOVER, 2011), identifying programmatic uncertainties facing multiple projects and strategies to address them, as well as the responsibilities of various parties. The Program-Level Adaptive Management Plan was reviewed favorably by the committee in NASEM (2016), although the committee noted that it was ambitious and lacked an implementation plan to ensure that appropriate actions were taken to support program-level adaptive management. (See below for a detailed discussion of the Program-Level Adaptive Management Plan.)

EVALUATION OF CERP ADAPTIVE MANAGEMENT AND INCORPORATING NEW INFORMATION INTO CERP DECISION MAKING

New information relevant to the CERP can come at any time, from many sources, and in many forms—from project-level monitoring, system-level monitoring, new science, new modeling, Indigenous Knowledge, and endangered species consultations to name a few. Although the term adaptive management is often used to apply to projects only when they are fully operational, the CERP adaptive management guidance and USACE processes provide the means to accomplish the broader goal of incorporating new information at all stages of program and project activities. During the planning process, new information is actively sought in the development of project plans, and RECOVER has been effective in supporting the creation of sound project-level adaptive management plans. In the process, RECOVER develops collaborative relationships with the Project Delivery Teams, which serve the CERP program well by providing a foundation for RECOVER's later involvement if project teams encounter challenges in design, construction, and operation related to new information.

For post-authorization project design, construction, and operations, the overall adaptive management strategy (Figure 5-2) and the guidance built upon it (RECOVER, 2011, 2015a, 2022a-n; USACE and SFWMD, 2018) are sufficiently sound conceptually and detailed to accomplish effective adaptive management. The guidance exhibits most of the features of effective adaptive management described previously in this chapter—modeling of the managed system, a range of management choices, monitoring and evaluation of outcomes, mechanisms for incorporating learning into future decisions, and a collaborative process for stakeholder participation and learning. Adaptive management, as implemented by the USACE, however, is not by nature a nimble process. Existing procedure requires that once a PIR is approved by USACE headquarters and authorized by Congress, it becomes subject to additional administrative requirements, which can make it burdensome and difficult to incorporate new information necessary to improve projects. Refinement of operations or project design may require new National Environmental Policy Act (NEPA) evaluation and USACE reports with approval levels from District commander to headquarters to Congress, depending on the level of project change, its potential impacts, and effects on overall project costs (Table 5-1). These requirements are put in place to provide oversight of changes to a congressionally approved project, but they challenge timely and effective implementation of adaptive management and incorporation of new information. If "minor changes in design and costs from authorizing reports" are involved, changes can be authorized via an Engineering Documentation Report (EDR; USACE, 2000). If "reformulation of plans or other sufficient major revisions is required," a Limited Reevaluation Report (LRR) or General Reevaluation Report (GRR) is required for approval (USACE, 1999). The LRR, employed when a reformulation of a specific portion of the project plan is proposed, can be approved at the USACE division level (i.e., South Atlantic Division), whereas the GRR, employed when a major reformulation of the plan is proposed, requires USACE headquarters approval. Both may require a new NEPA evaluation and an Endangered Species Act review. If project changes result in a 20 percent (or greater) increase in total authorized project costs (adjusted for inflation), a Post-Authorization Change Report (PACR) is required, which must be authorized by Congress, typically through WRDA legislation (USACE, 1999, 2000). Modifying projects during these stages usually requires negotiating changes to the Project Partnership Agreement,[1] which is the legal agreement between the state and the USACE that outlines the terms and conditions for project construction and cost sharing.

These required processes make it challenging to modify projects in a timely manner once they are authorized. The major issue with CERP adaptive management is not the plan but rather how effectively the plan can be implemented, especially when applied across many projects at once. The following sections review and evaluate how effective the adaptive management process and the incorporation of new information into decision making has been in the CERP during four stages: (1) project-level adaptation during design and construction, (2) project-level adaptive management (post-construction), (3) adaptation of operations, and (4) program-level adaptive management.

---

[1] See https://www.usace.army.mil/Missions/Civil-Works/Project-Partnership-Agreements.

TABLE 5-1 USACE Mechanism for Incorporating Post-Authorization Changes into a Project

Report	Description	Authority
Engineering Documentation Reports (EDR)	Prepared when there are minor changes in design and costs from the authorizing reports. The EDR may also be used in lieu of a GRR to document other information not included in a decision document when project reformulation is not required, and the changes are only technical changes.	District commander[a]
Limited Reevaluation Report (LRR)	If reformulation of plans or other sufficient major revisions is required during Preconstruction Engineering and Design (PED), then districts shall prepare a GRR or LRR. This study provides an evaluation of a specific portion of a plan under current policies, criteria, and guidelines, and may be limited to economics, environmental effects, or, in rare cases, project formulation.	Division
	Limited Reevaluation Studies ordinarily should require only modest resources and documentation. If any part of the reevaluation will be complex or will require substantial resources, or if the recommended plan will change in any way, a General Reevaluation is required. When a project must be reformulated or estimated costs updated, an Engineering Appendix shall be prepared.	
General Reevaluation Report (GRR)	If reformulation of plans is required during PED, then districts shall prepare a GRR or LRR. General reevaluation studies frequently are similar to feasibility studies in scope and detail. This is reanalysis of a previously completed study using current planning criteria and policies, which is required due to changed conditions and/or assumptions. The results may affirm the previous plan, reformulate and modify the plan, as appropriate, or find that no plan is currently justified. When a project must be reformulated or estimated costs updated, an Engineering Appendix shall be prepared.	Headquarters
Post-Authorization Change Report (PACR)	Section 902 of the WRDA of 1986, as amended, legislates a maximum total project cost. Projects to which this limitation applies, and for which increases in costs exceed the limitations established by Section 902, as amended, will require further authorization by Congress raising the maximum cost established for the project. The maximum project cost allowed by Section 902 includes the authorized cost (adjusted for inflation); the current cost of any studies, modifications, and actions authorized by the WRDA of 1986 or any later law; and 20 percent of the authorized cost (without adjustment for inflation).	Congressional authorization

NOTES: Changes to an authorized project or operations plan may require NEPA documentation. "The scope and nature of the changes in the environmental effects of the project identified as a result of acquisition of new information, of changed conditions, or changes in the project will determine the appropriate type of NEPA documentation. Options include an Environmental Assessment which may result in a Finding of No Significant Impact or a Supplemental Environmental Impact Statement" (USACE, 2000).
[a] Project cost estimates are approved at the district level, unless the cost increase exceeds the project Section 902 limits, which then requires project reauthorization and the preparation of a PACR.
SOURCES: Information from USACE, 1999, 2000.

Project-Level Adaptation During Design and Construction

In this section, the committee examines two projects in which attempts to modify projects post-authorization based on new information during design or construction have been made: Picayune Strand and the Central Everglades Planning Project (CEPP). These case studies are intended to highlight processes that work well and issues that could impact restoration progress.

Picayune Strand

Construction at the Picayune Strand Restoration Project has occurred in phases, moving from east to west. The first construction component (plugging the easternmost canal) was completed in 2007, and as of 2024 construction is focused on the western portion (see Chapter 2: Picayune Strand Restoration Project). Over this time period, modeling tools and scientific understanding have advanced, offering new information about species and system response that could improve the original authorized project plan. Two major post-authorization modifications, based on new information that arose during design and construction, are discussed in this section: (1) mitigation of project impacts to Florida manatees (*Trichechus manatus latirostris*), a subpopulation of the West Indian Manatee, and (2) potential project alterations to reduce harm to endangered red-cockaded woodpeckers (*Dryobates borealis*).

Manatee mitigation. Early assessments of the Picayune Strand Restoration Project during development of the PIR (USACE and SFWMD, 2004) noted uncertainties about downstream flow impacts to cold weather refugium for the Florida manatee, which is protected by the Endangered Species Act and the Marine Mammal Protection Act:

> large numbers of manatees aggregate in the Port of the Islands marina basin in winter . . . in the deepest parts of the (dredged) marina basin. The recommended project will not affect this basin, and is not expected to adversely affect manatee aggregation in winter, but it will affect flows over the weir. . . . It is unknown what role the dynamics of freshwater inflow at the weir play in the balance of marina water temperature or basin stratification.

The PIR stated that the CERP agencies would continue to consult with the U.S. Fish and Wildlife Service (FWS) and that an additional source of groundwater would be added in the vicinity of the Faka Union Canal weir if needed to maintain warm water refugia (USACE and SFWMD, 2004). However, no adaptive management plan was included in the 2004 PIR, because it preceded the USACE requirement for project-level adaptive management plans in the CERP.

A manatee mitigation element was later determined to be needed after a U.S. Geological Survey study concluded that canal plugging would reduce post-project discharge flows compared to the pre-construction condition, impacting existing manatee refugia in the marina basin (Stith et al., 2011). The Manatee Mitigation Feature was designed to create a new area with deep pools that provide a connection to warmer groundwater as refugia for manatees in winter months. The feature, along with other project design modifications, was included in an LRR (USACE, 2015), which required congressional reauthorization (obtained in WRDA 2016) because the total project modification costs exceeded 20 percent of the authorized project costs. The LRR also included an adaptive management plan for the Picayune Strand Restoration Project. The modifications necessitated a revision of the Project Partnership Agreement between the USACE and the SFWMD, which was finalized in 2019 (USACE and SFWMD, 2019).

At its own financial risk, the SFWMD expedited construction prior to congressional approval to prevent further delays in overall project completion; construction was completed in 2016 (Figure 5-5). Manatee and water temperature monitoring has been conducted post-construction to assess the project performance, although the boat basin (the original refugia) is still providing warm water conditions. Once canal filling progresses to the point that the temperature of the original refugia falls below a 68°F threshold, the performance of the manatee mitigation feature can be formally assessed (H. Edwards, Florida Fish and Wildlife Conservation Commission, personal communication, 2024).

The overall adaptive management process from the time that a change in the project design was proposed through construction took approximately 5 years because of the willingness of the SFWMD to expedite the design and construction of the feature at its own financial risk. The potential impact to the manatee created strong regulatory and public interest to address the problem in

FIGURE 5-5 Manatee mitigation feature added to Picayune Strand Restoration Project after new information was discovered about potential impacts.

SOURCE: FWS, https://www.fws.gov/media/pup-manatee-mitigation.

a timely manner, and the SFWMD was able to move quickly to incorporate the new information and minimize delays on the overall project.

Red-cockaded woodpeckers. The LRR that resulted in the addition of the Manatee Mitigation Feature also included a redesign of the Southwest Protection Feature (USACE, 2015). The project footprint was enlarged based on land easements required for this redesign to include a portion of the adjacent South Belle Meade tract to the west containing endangered red-cockaded woodpeckers, which do not occur within the original project area (Figure 2-9). Modeling associated with the redesign of the Southwest Protection Feature indicated that the project would create a flow-way through the area inhabited by red-cockaded woodpeckers, which would likely adversely affect them by increasing hydroperiods enough to convert much of the pine habitat they occupy to unsuitable wet prairie and marsh habitat (USACE, 2020c) (see Chapter 2 for further details).

Adverse anticipated impact on red-cockaded woodpecker habitat is a new issue that represents an unanticipated outcome of the project. It was not among the uncertainties identified in the Picayune Strand adaptive management plan created in the LRR just 4 years prior to its discovery (USACE, 2015). As with the manatee, modifications to prevent harmful effects on this endangered species are being evaluated in consultation with FWS, and RECOVER is not involved. In this case, unlike that of the manatee, the modifications involve no new features and are anticipated to result in very modest additional costs, no cost, or cost savings. Thus, this is an example of an action that can likely be approved administratively at the Jacksonville District level via an EDR (see Table 5-1). However, more than 4 years have already elapsed in analysis of the alternatives and planning of the redesign, and the required NEPA compliance review still lies ahead. How long the process takes, and whether it delays the completion of the Picayune Strand Restoration Project, will be instructive with respect to the speed at which projects can be modified in response to new information post-PIR. This example also illustrates the capacity to modify CERP projects when threatened and endangered species are involved as well as the time required by analysis and redesign.

CEPP

The CEPP is an example where multiple modifications have been proposed, and the challenges and successes provide several lessons for project-level adaptive management. The opportunities for incorporating changes to CEPP design based on new information, however, differ from those for other CERP projects because an extra step—a Validation Report—is required between project authorization and the Project Partnership Agreement for each of the four components of the CEPP. This step was added by the USACE due to concerns that the more rapid USACE planning and approval process first piloted with the CEPP (termed the 3×3×3 process) resulted in greater risk and uncertainty associated with project design for a project as large and complex as the CEPP. The Validation Report creates an added mechanism to obtain approval for some design changes to authorized projects based on new information prior to construction. In this section, the committee examines five modifications being considered or implemented—two for CEPP North, which does not yet have an approved Validation Report, and three for CEPP South, which does.

CEPP North vegetated hammocks. Backfilling the Miami Canal is a central component of CEPP North. The CEPP PIR recommended the addition of "upland landscape (constructed tree islands) approximately every mile along the entire reach of the backfilled Miami canal section (S-8 to I-75) where historic ridges or tree islands once existed" (USACE and SFWMD, 2014). The PIR discussed options such as spoil mounds, vegetated mounds, and tree islands and recommended the involvement of scientific input in the design of these features. To fulfill the latter recommendation, RECOVER convened a workshop in July 2022 to investigate backfilling methods. One outcome was a recommendation to use tear-drop-shaped vegetated hammocks oriented in the direction of flow with peak elevations 2–3 feet above marsh grade, which was higher than the 1.5 feet mentioned in the PIR (RECOVER, 2022o; USACE and SFWMD, 2014). The latest science was used to develop this recommendation, including information from the Loxahatchee Impoundment Landscape Assessment (LILA), a field-based tree-island research project that has been operating for more than 20 years. As of May 2024, the Project Delivery Team was evaluating the RECOVER recommendations in the finalization of the CEPP North design.

Because new information was anticipated in the PIR and significant cost increases are not involved, the proposed modification can be adopted easily and quickly without additional reporting requirements. This case illustrates the value of identifying uncertainties and the need for additional scientific input to inform the project design in a PIR, which undoubtedly expedited the process to convene a RECOVER workshop and enhanced the willingness of the design team to consider its input. By identifying the uncertainties and the plan for additional scientific input, the project team built additional flexibility to incorporate design changes into the CEPP PIR, such that the refinements to the vegetated hammocks did not need additional approval.

CEPP North L-4 Canal. The second CEPP North case—a proposed redesign of features associated with the L-4 Canal in the northwestern corner of Water Conservation Area (WCA)-3A—is more complicated. The original design consisted of degrading the L-4 levee and constructing a culvert (S-8A) within the Miami Canal (Figure 5-6a). Preliminary modeling associated with pre-construction engineer-

Adaptive Management and Use of New Information in Decision Making 191

FIGURE 5-6 Evolution of the CEPP North L-4 Canal features from PIR to more detailed design: top image shows the original design of L-4 Canal features: L-4 levee degrade and S-8A culvert; lower image shows the proposed redesign of L-4 Canal features: L-4 three culverts with spreader canals and S-8A Spillway with connecting canal.

SOURCES: USACE and SFWMD, 2023c; A. Kahn, SFWMD, personal communication, 2024.

ing and design revealed that this design would not produce the desired volume and distribution of flow into northwestern WCA-3A (H. Jarvinen, SFWMD, personal communication, 2023). To meet the project objectives, several modifications of the L-4 conveyance and distribution features have been proposed, including altering the geometry of the levee degrade, constructing three culverts with associated spreader canals in the L-4 levee degrade, and converting the S-8A culvert to a spillway and connecting canal (Figure 5-6b).

Like the Miami Canal backfill, the performance of the L-4 features was identified as an uncertainty and discussed in the PIR (USACE and SFWMD, 2014), but unlike the Miami Canal backfill, the proposed design changes added entirely new structures that were not mentioned in the PIR. The addition of these structures led to staff concerns that the new design might need to be approved by USACE headquarters and potentially Congress. However, it was determined that USACE approval for the proposed changes could be incorporated as a component of the required Validation Report for CEPP North, which can be approved at the (South Atlantic) Division level. Producing the Validation Report and the NEPA analysis necessitated by the changes to the L-4 features is the responsibility of the Project Delivery Team and will not involve RECOVER.

The contrast between this case and the Miami Canal backfill case highlights the challenges associated with adding previously unmentioned features to a project in a timely manner subsequent to its original authorization (i.e., post-PIR), even when the issue involved is identified as an uncertainty in the PIR. Staff noted that amid uncertainty, identifying the potential design alternatives in the PIR can obviate the need for high-level USACE approvals of new structures. Inclusion of Validation Reports in the design phase for the CEPP will facilitate approval of these modifications.

CEPP South. In contrast to CEPP North, CEPP South is further along in its implementation progress (see Chapter 2) and has an approved Validation Report. Changes occurring during the construction would necessitate additional USACE approval (Table 5-1) and may require renegotiation of the Project Partnership Agreement. CEPP South's original design is challenged by new information developed from the Decomp Physical Model (DPM), a pilot project designed to evaluate uncertainties associated with canal backfilling and restoration of sheet flow within the central Everglades. The DPM produced several results relevant to the CEPP that were not anticipated in the initial CEPP adaptive management plan (see Box 5-1).

Based on these findings, the DPM team (Saunders et al., 2021) identified uncertainties meriting further study and recommended four new adaptive management strategies in CEPP South (see Box 5-2). A series of discussions about an earlier draft of the DPM findings led to a memorandum from RECOVER in August 2020 recommending adoption of the first three adaptive management

BOX 5-1
Decomp Physical Model

The DPM is a large-scale pilot project intended to address uncertainties associated with restoring sheet flow to the ridge and slough landscape and to improve understanding of the ecological benefits of different canal backfilling options that could accompany levee removal. The DPM experiment was conducted between L-67A and L-67C, in an area near the border of WCA-3A and WCA-3B known as the "the pocket" (Figure 5-7). The project components include 10 gated culverts on the L-67A Canal (referred to as S-152) and a 3,000-foot gap created in the L-67C levee with three backfill "treatments" in the adjacent canal.

The first 2-month flow experiment was initiated in November 2013 by opening the gated culverts on S-152. Flows through S-152 were initially restricted to November and December, but this period was gradually lengthened over the next 3 years. Beginning in 2017, year-round flows through S-152 were permitted provided that total phosphorus (TP) concentrations in the L-67A inflow waters did not exceed 10 ppb and that other stage constraints for surrounding basins and canals were met.

Analyses of observations from DPM have demonstrated that

- restored surface-water flows are not following historic ridge and slough flow paths;
- sustained flow velocities of 1.5 to 3 cm/s are needed to promote sediment transport and rebuild ridge-and-slough topography;
- the Active Marsh Improvement project demonstrated the ability of vegetation removal in remnant sloughs to redirect surface-water flows and increase spatial extent of elevated flow velocities needed for sediment redistribution from sloughs to ridges;
- extreme high flows (>3 cm/s) can result in localized, damaging phosphorus loading followed by cattail invasion, and excessively high flows (>5 cm/s) may trigger erosion and topographic flattening; and
- canal backfilling improves habitat quality, and canal fill composed of limestone may retain phosphorus and reduce phosphorus transport downstream.

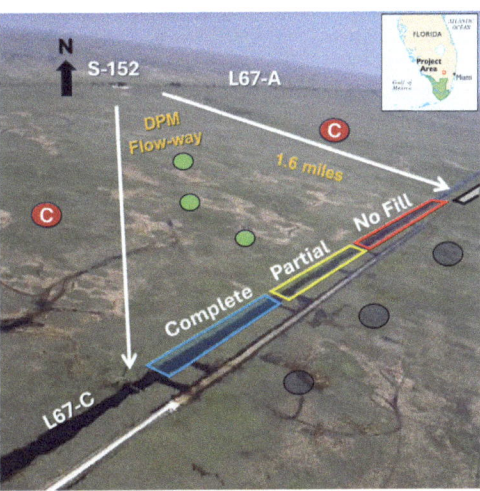

FIGURE 5-7 Map of the DPM located in "the pocket" between L-67A and L-67C.

SOURCES: Modified from Saunders and Newman, 2023. Inset map by International Mapping.

SOURCES: Saunders, 2020; Saunders and Newman, 2023; Sklar et al., 2021.

BOX 5-2
Major Uncertainties and Four Recommended Adaptive Management Strategies for CEPP South

A final draft of the following uncertainties and adaptive management recommendations for CEPP South were provided by the DPM team (Saunders et al., 2021).

Uncertainties

- **Marsh degradation associated with point source discharges.** Even with inflow TP concentrations below 10 µg/L, high TP loading can occur based on high flow velocities, leading to ecosystem disruptions such as cattail growth. "The extent to which increased sediment TP is driven by localized variation in surface water TP, velocity, or P loading or a combination of the two remains uncertain." Uncertainties about the ability of spreader swales to reduce these effects were also identified.
- **Reducing canal-to-marsh flow.** Uncertainties were identified regarding the most effective strategies to reduce the mobilization of nutrients and sediments from the L-67C Canal into the marsh and regarding the effect of different levee removal strategies on water quality and flow patterns.
- **Active Marsh Improvement (AMI) in the Blue Shanty Flow-way.** Several uncertainties were identified regarding the appropriate scale and design of AMI for the CEPP and the effect of AMI on biogeochemical processes and food webs. Additional identified uncertainties related to the design of ditch backfilling efforts and the effects on invasive species and the ecological risks and benefits with increasing flow through the Blue Shanty Flow-way, compared to the S-333 structures.
- **Effects of spoil mound removal on culvert inflows and water quality.** "Although spoil mound removal has the potential to restore some marsh flow in WCA-3A and downstream (the Pocket of WCA-3B), the extent to which removal will reduce inflow water TP or increase discharge performance of the culverts is unknown. Also unknown are the potential benefits of reduced sediment mobilization in and around the L-67A canal."
- **Tree island responses to hydrologic changes associated with the Blue Shanty Flow-way.** Several uncertainties were identified regarding the effect of changes in flow volumes, velocity, water levels, and nutrient loading in CEPP South on tree islands, including nutrient dynamics and uptake, and wildlife activity. Also uncertain is whether management strategies could increase recovery of degraded tree islands.
- **Impacts of CEPP South on invertebrate and fish assemblages.** Uncertainties were identified regarding the optimal flow rates to maximize faunal benefits while minimizing risks as well as effects of strategies such as AMI and levee degradation on fauna, invasive species, and food webs.

Recommended Active Adaptive Management Strategies

1. Use two field tests, including an in situ flume and experimental spreader swale tests, to identify optimal phosphorus loading conditions and strategies to reduce damaging, extreme flow conditions downstream of the inflow culverts (S-631, S-632, S-633).
2. "The testing and application of canal plugs or backfill designs to reduce canal-to-marsh flow and maximize marsh-to-marsh flow across the degraded L-67C levee. In this test, observations will focus on tracking the mobilization, transport, and accumulation of canal-derived, P-enriched sediments into marshes downstream." Testing would be accompanied by large-scale hydrologic modeling to determine cost-effective design alternatives.
3. "Implement active marsh improvement (AMI) to enhance restorative sheet-flow velocities and Ridge-and-Slough (R&S) functions within the BSFW [Big Shanty Flow-way]."
4. "Remove spoil mounds along the western edge of the L-67A canal. The objective of this field test is to evaluate the extent to which ~2000-ft (~600 m) gaps along the canal edge and adjacent to S-63x culverts will increase marsh-to-marsh connectivity between WCA-3A and WCA-3B and improve inflow water quality through the S-63x culverts."

SOURCE: Saunders et al., 2021.

strategies noted in Box 5-2 (RECOVER, 2020b). The examples below represent the progress made in CEPP South in response to the recommendations from the DPM team and RECOVER.

CEPP South active marsh improvement and spoil removal. AMI involves vegetation removal to reconnect historic sloughs and to enhance sheet-flow and ridge-and-slough functions within the Blue Shanty Flow-way, and its effectiveness was documented in the DPM (see Box 5-1). AMI was recommended by both Saunders et al. (2021) and RECOVER (2020b) and addresses an uncertainty identified in the CEPP PIR (#73, USACE and SFWMD, 2014, Annex D). Spoil removal, as recommended by Saunders et al. (2021), involves removal of dredge spoils west of the L-67A Canal to increase marsh inflows and reduce phosphorus concentrations in the canal. Adding spoil removal to CEPP South reduces overall project costs, by providing additional fill material that can be used in other components of CEPP South, while potentially enhancing ecosystem benefits. Both modifications are proposed as active adaptive management projects that could improve the understanding of ecosystem dynamics and potentially inform future management decisions. These modifications were approved at the USACE District level and by SFWMD leadership, and the work is moving forward.

In these two examples, design changes were implemented relatively quickly, with less than 4 years between acquiring new information and implementing the changes. A rapid response was facilitated by the arrival of new information in the form of explicit recommendations for design modifications that were relatively low cost (or cost savings); a fairly high certainty of enhancing restoration outcomes; coverage by existing NEPA; and, in the AMI example, linkage to an uncertainty that had been identified in the PIR.

CEPP South L-67C Canal backfill. Both Saunders et al. (2021) (see Box 5-2) and RECOVER (2020b) recommended an adaptive management approach, based on the DPM findings, to explore applications of plugs or backfilling in the L-67C Canal to reduce the mobilization of nutrient-enriched sediments into WCA-3B in order to prevent adverse ecosystem effects. As of June 2024, the CEPP Project Delivery Team was still assessing the hydrologic and biogeochemical outcomes of various hydrologic modeling scenarios for backfilling the L-67C Canal, and the committee has not reviewed these results. Even though the original purpose of the DPM, which began operations in 2013, was to investigate the ecosystem benefits of complete or partial canal backfilling, this uncertainty was not included in the CEPP adaptive management plan (USACE and SFWMD, 2014, Annex D), nor was canal backfilling included as a management option, in part because its projected costs are substantial. Therefore, any change in design that incorporates canal backfilling, if approved by CERP leadership, will have numerous USACE reporting and approval requirements, perhaps including development of an

LRR with the appropriate level of NEPA and Endangered Species Act analyses; if cost increases exceed 20 percent of original authorized costs across all CEPP components, congressional approval of the modification will be required (see Table 5-1). The ability to consider these changes in 2024 was made easier by contractual delays in other portions of the CEPP project, which delayed the L-67C levee removal contract by several years. If that project component had proceeded, the costs of subsequent canal plugging would increase substantially.

The process to consider this major modification is ongoing, but the time and effort required stand in sharp contrast to the two CEPP South modifications already made. Whereas the new information that drove those two modifications was obtained at the same time as the information driving this one, those two are finished, and this one is just beginning. Several factors likely contribute to the longer time frame of decision making, including that the uncertainty regarding canal backfill and potential management options were not included in the CEPP adaptive management plan, the large potential costs of this change, the potential to delay the implementation of other CEPP components, and the absence of a legal driver such as the Endangered Species Act. These factors highlight the challenges posed by the USACE process in making major changes to an authorized project. Because the change is not driven by a legal requirement, such as the Endangered Species Act, but instead by overall improvement of benefits (and the reduction of adverse impacts), the change may require CERP leadership to prioritize restoration outcomes over the pace of implementation. Executing this modification while continuing the rapid progress of the CEPP will test CERP adaptive management.

Overarching Assessment

At the project level, the objective of adaptation during construction and design generally appears to be to improve restoration outcomes consistent with the original project objectives or to address unforeseen issues in design (Table 5-2). In the two examples from Picayune Strand, expected outcomes were modified to accommodate the needs of endangered species.

CEPP cases illustrate variation in the ease and speed with which projects can be modified based on receipt of new information during post-authorization design and construction. The CERP adaptive management process provides guidance for addressing issues that are included in a project's adaptive management plan. Addressing unanticipated issues or those not included in the initial PIR and adaptive management plan can be more challenging, particularly if the project cost is substantially affected and the primary outcome is an improvement in restoration benefits (rather than a legal constraint such as the Endangered Species Act or the Savings Clause). The administrative level of

TABLE 5-2 CERP Project-Level Adaptation During Design and Construction or Project-Level Adaptation After Operations Begin

Case Study	Timing of Change	Major Drivers	Issue Identified in Project AM Plan?	Approx. Cost of Change	Key Lessons Learned
BBCW: Deering Estate	After operation	Correct project performance issues	No	Low	Even simple changes can take many years if a culture of AM, clear processes, and a sense of urgency are not in place.
Picayune Strand: Manatee Mitigation	During design and construction	Endangered Species Act	No	Moderate	Clear capacity to make major design changes when listed species involved. USACE process for major changes time consuming; SFWMD expedited construction to limit impacts on project schedule.
Picayune Strand: Red-cockaded Woodpeckers	During design and construction	Endangered Species Act	No	Cost savings to low cost	Clear capacity to make design and construction changes when listed species involved.
CEPP North: Vegetated Hammocks	During design and construction	Improve restoration outcomes to better meet project objectives	Yes	Low	Identifying uncertainties and building flexibility into PIR eases process of incorporating new information.
CEPP North: L-4 Canal	During design and construction	Correct project performance issues	No	Moderate	Identifying potential design alternatives in PIR would obviate need for high-level USACE approval of new structures.
CEPP South: Active Marsh Improvement and Spoil Mound Removal	During design and construction	Improve restoration outcomes to better meet project objectives	Yes	Cost savings to low cost	Willingness to make minor change when cost savings involved.
CEPP South: L-67C Canal Backfill	During design and construction	Improve restoration outcomes to better meet project objectives	No	Moderate to high	Lessons limited because early in process of potential change. Lack of inclusion of potential adaptive management strategy in AM plan complicates major change without ESA driver.

NOTES: AM = adaptive management. Costs categorized as low (<$5 million), moderate ($5–20 million), and high (>$20 million) in 2024 dollars.

approval required depends on the magnitude of the proposed modifications and associated costs (Table 5-1). Minor changes can be handled as project redesigns at the district level. More complex and costly decisions require an LRR or GRR, likely triggering additional NEPA and Endangered Species Act analyses, reauthorization by Congress if cost increases exceed 20 percent, and potential renegotiation of the project partnership agreement. As the pace of CERP implementation has quickened under the large budget increases of recent years (see Chapter 2), incorporating important new information into the process can be challenging, because such information requires new analysis and potentially implementation delays while the resulting modification is under consideration. Unless the process can become more nimble, CERP leaders may have to weigh the costs and benefits of modifying a project to improve the restoration outcomes against implementation delays and the increased risk to achieving project objectives.

Difficulties experienced in incorporating new information at the design and construction stages foreshadow difficulties in implementing adaptive management across many operating projects and weaknesses in the implementation of the current adaptive management strategy. Although checks on design changes are appropriate to ensure sound analysis and decision making, the process may include oversight that is excessive for the changes proposed, and there may be ways to alter the process to make it easier to incorporate new information. For example, CERP teams have learned that building flexibility into the PIR (e.g., RECOVER input on Miami Canal plugs) can facilitate incorporation of new information with less burdensome reporting and approval requirements. Not including potentially necessary modifications in an adaptive management plan (e.g., canal plugs in CEPP South) can greatly increase the amount of time and the levels of administrative review needed to effect such a change, delaying restoration benefits (and allowing continued ecosystem decline) or potentially even foreclosing potential actions.

Project-Level Adaptive Management After Operations Begin

To date, there are relatively few examples of changes being made after a CERP project is constructed and fully operational. Broadly, changes can be triggered in two ways under project-level adaptive management. First, system responses to a project's operation may trigger one or more elements of the Management Options Matrix in that project's adaptive management plan. To the committee's knowledge, this scenario has not yet occurred. Second, outside of those anticipated ecosystem response scenarios, other observed project or system responses or research can result in new information that suggests potential changes to a project. An example of the latter is the Deering Estates Flow-way in the Biscayne Bay Coastal Wetlands Project.

Examples of Project Changes After Operations Begin

Deering Estate. Shortly after construction of the Deering Estate Flow-way was completed in 2012, concerns were raised about the effectiveness of short-term, nighttime 100-cfs pumping at the S-700 structure. Staff observed that when the pumps were turned on, the water levels in the freshwater wetland were elevated to the target ranges, but when the pumps were turned off during the day, the stages quickly decreased to pre-pumping levels (Charkhian and Niemeyer, 2023) as shown in Figure 5-8. Therefore, the project did not provide

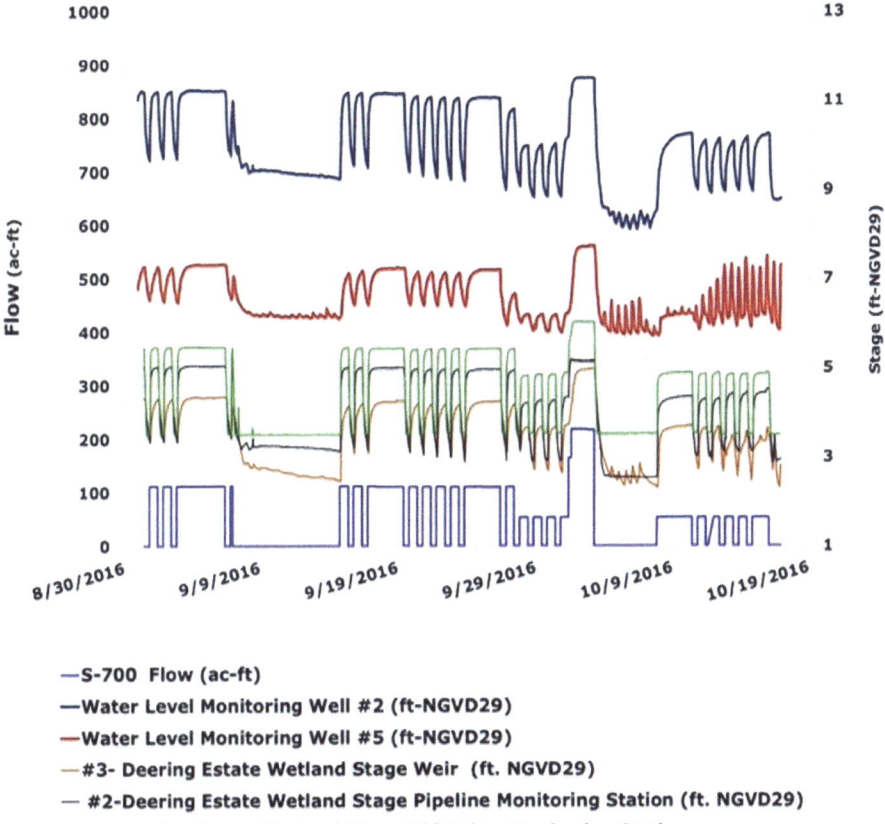

FIGURE 5-8 Deering Estate Flow-way showing flow at S-700 pump station from September and October 2016 and corresponding downstream stage response.

SOURCE: Charkhian and Niemeyer, 2023.

the intended natural wetland hydroperiods. The adaptive management plan did not anticipate this outcome and therefore could not be relied upon for rapid implementation of an alternative management action. In water years (WY) 2015–2017, the SFWMD conducted a series of two wet and two dry season experiments comparing pulse releases to continuous pumping at different rates (Charkhian and Niemeyer, 2023). In June 2017, the SFWMD recommended switching to continuous pumping at 25 cfs, with the lower rate to account for increased operating costs. RECOVER reinforced this recommendation in a July 2017 memo (RECOVER, 2017). After undergoing additional approval processes, the changes were approved by the Florida Department of Environmental Protection in February 2018 and implemented in WY 2019. The time to effect this simple change—from identification of the problem, to analysis of alternatives, to permitting, to final implementation—was 6 years, even with long-term continuity among key staff. At the time, RECOVER had not finalized its SOPs for coordination with projects so whether the process would be completed in less time today is unclear.

Overarching Assessment of Project-Level Adaptive Management

As described earlier in this chapter, project-level adaptive management guidance has been well established, and most projects now include thoughtfully developed adaptive management plans. However, there are limited examples of project-level adaptive management and changes in response to new information after operations have begun (Table 5-2), and the CERP agencies seem challenged to develop timely solutions. This process inherently requires time to analyze the data and potentially design and conduct experiments to tease apart the factors affecting the response (or lack of response). However, because taking actions in the Everglades frequently involves multiple entities and agencies that operate at multiple levels and are governed by multiple laws and regulations, the most pernicious delays appear to be bureaucratic. A culture and clear processes to quickly and effectively transmit, review, and administratively act upon information have not yet been established within the CERP program. Although some changes will involve new engineering design and construction that understandably take time to implement, the Deering Estate example involved a simple operational change and still required 6 years to resolve the issue.

NASEM (2018, 2021) identified several elements of successful application of project-level adaptive management, including tracking of outcomes relative to project objectives and expectations, routine analysis and interpretation of data across long-term trends, and multi-agency data analysis and reporting mechanisms regarding restoration progress for each of the CERP projects. Although several CERP project components are operating, implementation of these routine project-level adaptive management activities has not yet become well established. NASEM (2021) stated that "limitations in monitoring, analysis, and communication of results have impeded quantitative assessment and communication of restoration benefits"—a finding that remains relevant today. Addressing these gaps will enhance the success of adaptive management implementation and inform decisions that can improve restoration outcomes. For example, the C-111 Spreader Canal Western project has been operating since 2012 but has no explicit project-level adaptive management plan and no public reporting mechanism to explain how the project's performance compares to expectations or targets. The Deering Estate component of the Biscayne Bay Coastal Wetlands Project has also been operating since 2012, but analysis of monitoring data is only provided in the SFWMD permit report published in the South Florida Environmental Report, which emphasizes data over the prior water year. Restoration outcomes relative to expectations, long-term trends, and other drivers are extremely difficult to determine from existing reporting, and single-agency reporting limits a full input from CERP partners about the project outcomes. To support adaptive management, the CERP needs to develop a periodic project-level reporting mechanism to communicate multi-agency assessments of restoration progress and implications for decision making.

Significant work has occurred to move the adaptive management program forward, but more work is needed to build a culture and expertise for CERP project-level assessment and reporting to ensure that projects are operating at their full potential and changes are implemented quickly. Such actions will provide confidence to Congress and the taxpayers who support the CERP that their investments are being well spent. Possible strategies for improving adaptive management are discussed later in the chapter.

Adapting Operations at Regional Scales

Adaptation in operations primarily occurs in three forms:

1. **operational flexibility**, which uses the latest scientific information to make near-real time decisions within a pre-determined operations schedule;
2. **operational deviations**, which respond to extreme conditions that necessitate operations that differ from the approved operations schedule; and
3. **periodic revisions to systems operations plans**, to incorporate new information and new projects for improved regional water management.

Water managers have flexibility within system operations plans to make operational adjustments in light of changing conditions, and the leveraging of this flexibility to accommodate new information is a particular strength of

Everglades restoration and ecosystem management in recent years. Beginning in the 2010s, some agency staff began "weekly scientist calls" among various Tribal and governmental agencies to discuss recommended water management actions for the Everglades Restoration Transition Plan (USACE, 2011) relative to hydrologic and ecological conditions in WCA-3 (described in NASEM, 2021). These calls enabled scientists and managers to discuss the latest hydrologic and ecological monitoring data and condition assessments to inform water management decisions (i.e., opening or closing gates). These weekly scientist calls have continued through subsequent operational plans, such as the Combined Operational Plan (COP; see Chapter 2). Appendix C of the COP Biennial Report (USACE et al., 2023a) summarizes issues and recommendations from each call, with input from each participating agency, and each decision. These calls, if appropriately documented, represent opportunities for learning about the system response to improve future management. Recent system operating plans, such as the Lake Okeechobee System Operating Manual, have incorporated more flexibility than prior versions.

Extreme conditions, however, may necessitate deviations to approved operations plans, which must be approved by USACE Jacksonville District leadership, with appropriate levels of NEPA and Endangered Species Act assessments of the impacts of the deviation and coordination with affected parties. Emergency deviations were made in 2016 and 2017 under the incremental testing for the COP in light of high-water conditions. Deviations were also approved in 2023 and 2024 to allow more flow through the S-12s to reduce high-water conditions in WCA-3A that particularly impacted the Miccosukee Tribe (see Box 3-5). Despite the multiple requirements, these deviations can be made relatively quickly to respond to the emergency conditions.

System Operating Manuals require periodic updates to account for new information, new projects, and changing conditions. As noted in Chapter 4, the planned revision cycle for System Operating Manuals represents a useful means to incorporate new knowledge about climate change and lessons learned from recent operations, including extreme events. The committee's last report (NASEM, 2023) and its review of the COP highlight the value of incorporating adaptive management in systems operations to encourage more structured learning. The adaptive management plan included in the COP Environmental Impact Statement (USACE, 2020a) describes a structured approach to learning with well-defined goals and targets, specific triggers, and resulting management actions for each specific uncertainty. Out of 18 total uncertainties, only 2 could be addressed in ways that could change COP operations without major procedural steps, such as NEPA assessment. However, by identifying these uncertainties, project teams could focus monitoring needs and structure data analysis around issues that could inform future System Operating Manual revisions (see Chapter 2: The Combined Operational Plan).

It is important to note the level of commitment in terms of staff time and resources required to incorporate new information into conditions-based decision making and adaptive management of operations. Experts and managers must devote time to regular meetings and associated data analysis and then additional time and resources to address the longer-term uncertainties identified in adaptive management plans. Clear communication, availability of appropriate monitoring data, and timely data analysis, synthesis, and documentation are all essential components of learning and adaptive management. Conditions-based operations may also require upgrades to some older structures to facilitate remote management of gates.

Overarching Assessment of Adapting Operations at Regional Scales in the CERP

Overall, the incorporation of new information in operations is a success story in the CERP. The use of periodic scientist calls has fostered communication between scientists and managers, applied current ecological and hydrological information, and facilitated improved and timely conditions-based operational decision making for the benefit of the ecosystem and the people that depend upon it. Not all regional operations are conditions-based, as described in Box 3-5, but this strategy has been applied successfully in the COP and Lake Okeechobee System Operating Manual. The process of learning from operations at the systems scale has proven valuable and should be continued in the next update to the COP—CEPP 1.0 (see Chapter 2).

Program-Level Adaptive Management

Program-level management involves the integrated coordination of system components to optimize ecosystem restoration benefits while balancing the need for flood control and water supply. Program-level adaptive management represents a structure to learn from new science, monitoring, and modeling to inform system-level decision making. RECOVER's CERP Program-Level Adaptive Management Plan (RECOVER, 2015a) identified 13 "Priority 1 CERP programmatic uncertainties" that were designated "decision critical" and referred to as "showstoppers" (see Box 5-3). The Plan presented an adaptive management strategy to address these uncertainties and management option matrices, which described indicators used to assess performance, thresholds for when corrective action is needed, and actions that could be taken if restoration outcomes are not meeting performance targets. A key element of the Program-Level Adaptive Management Plan (RECOVER, 2015a) is the summary of the project-specific goals, interim goals from RECOVER (2005), "full restoration targets" based on RECOVER's documentation of performance measures (RECOVER, 2007), and triggers for management action. However, for many of the identified uncertainties, the goals and targets have not been defined and are shown in the tables simply as "TBD" (to be determined).

> **BOX 5-3**
> **Priority 1 Mission Critical Uncertainties Identified by the CERP Program-Level Adaptive Management Plan**
>
> Each uncertainty identified in the Program-Level Adaptive Management Plan (RECOVER, 2015a) was evaluated using three criteria:
>
> 1. Knowledge: what is known about the uncertainty (high, medium, low) and how it should be addressed
> 2. Relevance: what is the level of confidence (high, medium, low) that addressing the uncertainty will help improve the design or operation of CERP projects
> 3. Risk: what is the risk (high, medium, low) that CERP goals won't be met if the uncertainty is not addressed
>
> Generally, those uncertainties with high risk, low knowledge, and high relevance were designated as Priority 1 CERP programmatic uncertainties—no further prioritization was conducted within the Priority 1 uncertainties. A total of 13 Priority 1 programmatic uncertainties were identified under topics such as storage capacity, climate change, project targets, and goals. The committee has not examined the rationale behind the identification of these uncertainties or evaluated the priorities established; they are listed here to illustrate a key development in adaptive management.
>
> **Storage**
>
> - Will enough storage be constructed to allow Lake Okeechobee water levels to be kept at ecologically beneficial levels, i.e., reduce extreme high and low periods?
> - Will storage projects (e.g., Aquifer Storage and Recovery) provide enough storage to protect the estuarine resources in both dry and wet seasons?
>
> **Oysters**
>
> - What are the water quality (nutrients and suspended solids) impacts on larval oyster recruitment, given adequate numbers of spawning oysters in the estuaries (i.e., are larvae killed by poor water quality or are they washed downstream)?
>
> **Processes**
>
> - What is the role of flow velocities and flow volumes in maintaining characteristic ridge-and-slough patterns?

NASEM (2016) praised the RECOVER (2015a) Program-Level Adaptive Management Plan for asking highly relevant questions about many important systemwide issues affected by new information since the CERP was launched, such as storage, definition of goals, and climate change. NASEM (2016) also helped to outline some of the steps that should be taken (and when) to answer these questions and inform future decision making. The CERP programmatic adaptive management strategies identified to address the Priority 1 uncertainties included research, modeling, and synthesis, as

Design

- Which areas can be restored quickly (decadal) vs. slowly (century-millennia)? If areas are currently degrading, will waiting to start restoration extend the time of their recovery? Can ridge and slough patterns be reestablished simply by restoring hydrology?
- Is complete backfilling of canals and removal of levees necessary for restoration? Is partial or no backfilling of canals a viable alternative?

Water Quality

- How should restoration projects be designed to deliver increments of clean water to priority restoration areas, i.e., how best can water quantity and water quality goals be balanced to optimized restoration project performance?

Targets

- What water volumes and patterns of flow are required to restore submerged aquatic vegetation, oysters, and fish communities in the coastal Everglades?
- What are the hydrologic needs of the Everglades ecosystem (natural) compared to the human (urban and agricultural) systems? How much of this need is provided by CERP and how much more storage is needed?

Climate Change

- How will sea-level rise affect restoration efforts? How will sea-level rise affect coastal soils and the transition between brackish to oligohaline wetlands?
- What hydrologic/ecological/human changes could be affected by uncertain future demands for water for agriculture and urban users, as well as changes to the system due to climate change (changes in regional water balance) and sea-level rise?
- How will climate change affect the regional water balance? How will this affect the hydrologic assumptions used for CERP projects?

Balance Goals

- If the target lake stage in Lake Okeechobee is achieved, will the discharge problems to the Northern Estuaries be relieved? Which is more damaging to the estuaries, the volume and timing of Lake Okeechobee water releases or the nutrient loads in the water?

well as monitoring. NASEM (2016) noted that although the management options matrix was presented only for assessing monitoring results of "actual performance," those thresholds could also be used to assess the results of forward-looking modeling analyses.

Since its publication, the Program-Level Adaptive Management Plan does not appear to have been used to organize or prioritize data analysis, modeling, or synthesis around the identified uncertainties to support decision making. In fact, the document is rarely mentioned by RECOVER or other CERP agency staff.

RECOVER's attention has been consumed by its basic requirements and available resources (NASEM, 2021, 2023), and an implementation plan was never developed to address these important systemwide questions. The Everglades Science Plan recommended in NASEM (2023) could provide a mechanism to prioritize some of these larger questions, but so far, this recommendation has not been acted upon. Some progress has been made to address a few of the more localized uncertainties (e.g., the role of flow velocities and volumes in maintaining ridge-and-slough patterns), but, largely, the priority uncertainties remain valid and pressing today. The RECOVER 5-year System Status Report (e.g., RECOVER, 2014, 2019) is the primary vehicle to describe systemwide conditions. However, the previous committee's report (NASEM, 2023) stated that the System Status Report "provides only a snapshot of current conditions, failing to provide a synoptic view of how or why the system is changing, why degradation is particularly problematic in specific locations, and what key management questions or knowledge gaps are proving to be obstacles."

The ongoing analysis occurring as part of the Second Periodic CERP Update, which was recommended by NASEM (2016, 2018), may help address some of the pressing program-level questions regarding storage and sea-level rise. The Second Periodic CERP Update is expected to be completed in mid-2025 (Velez, 2024), and details about the ongoing effort are limited. NASEM (2016, 2018) recommended using the required Periodic CERP Update as a mid-course assessment of likely restoration outcomes relative to original expectations based on information gained since the last Periodic CERP Update in 2005. This new information relates to updated modeling tools, the feasibility of CERP storage, and climate change and sea-level rise.

In addition to the Periodic CERP Update, support for program-level adaptive management will necessitate attention toward other system-level issues, such as those identified in RECOVER (2015a) and any new priority uncertainties that may have emerged within the past 9 years. Dedicated agency resources and personnel are needed to address these program-level uncertainties and to build program-level adaptive management capacity. RECOVER funding and staffing have been cut substantially over the past decade, and additional dedicated resources may be necessary to make sufficient progress and ensure systemwide and forward-looking perspectives on these questions.

To support program-level adaptive management, the CERP should develop plans for periodic reporting on the extent of progress made toward restoration goals relative to expectations, whether thresholds for action have been crossed, and the implications of these findings for decision making. The System Status Report (RECOVER, 2014, 2019) provides an existing structure for such reporting; however, it tends to focus on recent trends and current status rather than ecosystem response in terms of expectations for restoration performance and importance to meeting CERP objectives. Timely multi-agency reports or web-based dashboards with clear graphics that convey CERP progress in key metrics, summarize the overall conclusions, and indicate the trends in recovery are needed to communicate progress relative to restoration objectives. A good example of such reporting is the Chesapeake Progress dashboard,[2] which includes an indicator of the trajectory of recent progress in the Chesapeake Bay Program and an assessment of whether the program is on course to meet each specific goal. A geographic information system-based dashboard would facilitate regional and systemwide analyses of project metrics and outcomes.

Overarching Assessment of Program-Level Adaptive Management in the CERP

Program-level adaptive management offers the potential to modify expected CERP outcomes as well as the objectives of individual projects based on new information and maximize the benefits of the restoration as a whole, but program-level adaptive management has yet to be implemented in a meaningful way. RECOVER (2015a) provided an important initial framework for program-level adaptive management, but until recently, little has been done to address system-level uncertainties. More attention is needed to understand evolving system-level risks, such as climate change (see Chapter 4), storage, and changing rates of ecosystem degradation, so that managers and scientists can assess their potential impacts on CERP goals and evaluate potential adaptations that may be needed at a program scale. The ongoing Second Periodic CERP Update is expected to provide useful information about program-level outcomes relative to original expectations and, ideally, will shape future discussions about whether additional program-level changes are needed to reach the desired restoration outcomes. Improved analysis and communication of systemwide restoration progress relative to expectations and targets is also needed to support program-level adaptive management.

REVISITING THE BUILDING BLOCKS OF ADAPTIVE MANAGEMENT

A well-functioning adaptive management program ensures that new information can inform decision making and improve restoration outcomes. However, building and supporting a functional program requires more than sound administrative guidance—it requires developing a culture and sufficient expertise in adaptive management within the CERP, a healthy and integrated science enterprise with adequate resources, effective and timely communication, and a nimble decision process that appropriately balances the level of oversight with the associated risks.

---

[2] See https://www.chesapeakeprogress.com.

A Culture and Expertise in Adaptive Management

Building a culture that embraces adaptive management and the nimble use of new information in decision making is an essential but challenging step for large-scale restoration projects. Evidence-based CERP decisions are built on input from scientific and policy experts from multiple agencies and two Tribes, each having different histories, missions, approaches, protocols, and procedures for decision making that collectively define the culture of each organization. In addition, a myriad of nongovernmental stakeholders is engaged in the restoration effort in some capacity. Developing trust across groups within agencies, as well as across agencies, Tribes, and stakeholders, is a critical component of a culture of adaptive management. Staff have to trust that the data they are collecting and analyzing will be seriously considered, leadership has to trust that the vision they are providing will be implemented, and stakeholders have to trust that their input will be considered fairly and thoroughly. Therefore, the culture for adaptive management depends upon regular communication, transparency (including the sharing of information and data across agencies at multiple levels), and competence to execute. Building trust in the CERP science and decision process will provide the administrative and public support necessary for funding, decision making, and collaboration. See Box 5-4 as an example of the process for building trust in the use of adaptive management in the operations of the Glen Canyon Dam.

Building an effective community for adaptive management takes time and commitment to support the science and operational management collaboration needed by the decision agencies. Even with clear plans and guidance in place for adaptive management, it will not be effectively implemented without leadership commitment to the process across all levels of the CERP agencies and appropriately trained staff responsible for monitoring, analysis, synthesis, and communication. Adaptive management improves as personnel and the public become educated in the process of collaboration and learn to trust that the process will provide the framework for discussion and decision. Within the Sacramento-San Joaquin Delta, a regular multi-agency Adaptive Management Forum has helped to grow awareness and support for adaptive management, with meetings in 2019, 2021, and 2023. The Adaptive Management Forums were recommended by the Delta Science Plan, as well as the Delta Independent Science Board, as a way to engage local staff on current adaptive management issues and challenges. The forum highlights a commitment to using adaptive management throughout the agencies in the Delta and has helped to broaden adoption of adaptive management approaches throughout the region.

Regular adaptive management workshops in the Everglades could bring together thought leaders from other programs and institutions across the country (e.g., Trinity River Restoration Program, Platte River Recovery Implementation

> **BOX 5-4**
> **Building an Adaptive Management Culture in the Glen Canyon**
>
> Developing and maintaining the culture necessary to encourage broad-based participation and data-supported decisions requires an investment in time, resources, governance structure, and leadership. An example of the need to establish a culture of trust early in developing an adaptive management process is provided by the revision of operations of Glen Canyon Dam. Prior to 1982, little to no coordination of the science of the Grand Canyon ecosystem or Glen Canyon Dam operations occurred. The Bureau of Reclamation operations of the Glen Canyon Dam were primarily driven by a defined set of water management rules. Starting in 1982, agencies invested over a decade developing collaborative science, Tribal coordination, and management actions, resulting in subsequent public support.
>
> The Glen Canyon Dam Adaptive Management program was formally instituted in 1996 to improve river management within Glen Canyon through a series of high-flow experiments and broad-scale systemwide monitoring over multiple years to evaluate effects on sand bars, endangered fish, and riparian habitat. The multi-year adaptive management experiments were intended to fine tune high-flow river releases to maximize ecosystem benefits while minimizing costs. As part of the adaptive management program, Congress directed that a structured framework of cooperation be designed, implemented, and supported. Initially, a facilitator was hired to assist the new adaptive management program efforts, including developing meaningful stakeholder communication and multiple levels of outreach and communication tools. The facilitator also worked to establish annual outreach opportunities for decision makers to learn from the science, supported engagement with eight Tribes, and developed and implemented a defined and transparent operational decision-making process. After several years, the Glen Canyon Dam Adaptive Management Work Group developed an adaptive management culture, one with a collective trust of process and information sharing that led to a collaborative decision-making process without yielding any of the congressional-directed control of the dam, natural resources, cultural resources, or recreational resources of the individual federal, state, or Tribal entities.
>
> SOURCE: https://gcdamp.com/index.php/Main_Page.

Program, Chesapeake Bay Restoration) to provide concrete examples of the effective use of adaptive management approaches to addressing challenges/roadblocks (including, but not limited to, programs managed by the USACE). Such efforts would simultaneously grow the adaptive management culture and educate agency staff and increase their capacity to effectively implement adaptive management in the CERP. Given the importance of mutually agreed upon objectives and processes for adaptive management, these issues could be useful initial topics for the proposed adaptive management workshops. Workshops could be used to agree upon the key objectives for both project- and program-level adaptive management; how and when to assess restoration success; how participants including agencies, Tribes, and stakeholders will be engaged in the process; and how learning should be documented and communicated to decision makers.

The process of building a culture and expertise in adaptive management may span many years but can be expedited through targeted hiring of specific scientific, cultural, and physical science expertise to build a critical mass of staff with the necessary skills to support adaptive management. Such skills include statistics; data analysis, visualization, and synthesis; communication; and monitoring. Hiring of practitioners from other large restoration programs with experience in adaptive management can also rapidly build organizational capacity.

An Effective Science Enterprise

Past committee reports have repeatedly emphasized the value of science to support decision making, but the calls for change have become more urgent with the past few reports, as the program transitions from multi-year planning to support of ongoing adaptive management of numerous projects in parallel with ongoing planning efforts. Compounding this issue are the challenges impacting the hydrologic and ecological systems caused by climate change and the increasing variability, extent, and number of extreme weather events (see Chapter 4). NASEM (2021) concluded the following:

> CERP managers face an array of restoration decisions, including adaptive management either at the project or program level based on assessments of restoration performance, near-term operational adjustments, project sequencing, and investments in additional science. The best science should be actively integrated and synthesized to inform these decisions so that restoration benefits are maximized and opportunities for learning across both CERP and non-CERP projects are not lost.

NASEM (2021) called for new and renewed strategies for monitoring, modeling, and synthesis to strengthen the science support for these decisions, thereby strengthening the overall science enterprise. In addition to resources to support these endeavors, a strong science enterprise also requires

- adequate staffing and support of appropriately trained scientists who can respond to management needs by analyzing, synthesizing, and communicating evolving relevant scientific information;
- continuity of expertise to support adaptive management throughout the life cycle of restoration projects, bringing the technical expertise developed during planning to bear on data analysis and assessment of restoration progress toward goals;

- strong science leadership to provide an efficient and direct linkage between scientists who conduct research, modeling, and monitoring and decision makers who need timely summaries of ongoing work and emerging issues; and
- communicators of the science who can interpret the technical information into formats that are understandable and available through multiple media platforms to decision makers, the public, stakeholders, and Congress.

The burden for executing adaptive management has primarily fallen to RECOVER, which is also heavily involved in project planning and bears primary responsibility for systemwide monitoring, assessment, and reporting of actionable issues in CERP performance to decision makers. The need for more resources, particularly staff, for RECOVER has been identified in previous reports (NASEM, 2021, 2023) and remains a problem. As more projects are entering the implementation stage, the shortcoming of staff resources is further exacerbated. A valuable step toward improved coordination between RECOVER and SFWMD decision makers has been the creation of an adaptive management liaison position at the SFWMD. The creation of the SFWMD Environmental Assessment Team, which promises to reduce the demands on RECOVER during project planning, and the recent additions of new positions to RECOVER, are additional notable, initial steps in the right direction. Without sufficient staff commitment and resources, the CERP will not be able to respond in a timely manner to management needs and changing conditions.

As with many other large-scale restoration projects, staff turnover and the loss of long-term knowledge are challenging issues; loss of expertise has been exacerbated in recent years with staff who were active in the development of the CERP reaching retirement age. Gaps in expertise and knowledge of the physical and biological systems can impede the ability of science and adaptive management to keep pace with rapid environmental change. Strategic hiring and training can help ensure sufficient capacity and appropriate expertise and maximize knowledge transfer from long-term staff to new hires.

The committee's previous report (NASEM, 2023) called for the creation of a CERP Science Plan to coordinate science activities and advance essential science activities to support decision making within the Everglades science enterprise. The report highlighted that collaborative identification of science needs for the Everglades would facilitate efficient use of limited science resources and improve science-management linkages. A Science Plan would also improve effective communication of priority science needs to funders and other stakeholders. The report identified the Science Coordination Group as the best-positioned entity to lead the development of a Science Plan. A regularly updated Science Plan to facilitate coordination across agencies would help address mission critical program-level uncertainties, such as those identified in Box 5-3, and would be an integral component of Everglades adaptive management efforts.

Effective Communication

Clear communication is an essential component of an effective and timely adaptive management process. "Closing the loop" in the adaptive management cycle (Figures 5-1 and 5-2) requires effective communication of scientific information to decision makers, who must decide whether a specific change is needed, and stakeholders. Effective communication is needed at a range of levels, from regular meetings among scientists and agency decision makers (e.g., the weekly science calls for the COP), to synthesis and interpretation of project-level or systemwide monitoring data to determine whether a project is meeting its objectives or whether it has crossed any thresholds for corrective action. Improving data synthesis is a common issue for many large-scale restoration and management efforts, because monitoring programs tend to focus on data collection more than data synthesis, interpretation, and communication. Both data collection and synthesis are needed to provide clear communication and direction for adaptive management decisions (Halpern et al., 2020; Ladouceur et al., 2022).

In addition, regular reporting of progress toward quantitative restoration objectives is essential to adaptive management efforts and broader awareness of restoration progress for decision makers, stakeholders, and the public. The Chesapeake Progress website[3] serves as an effective example of timely and transparent communication of large-scale restoration and management efforts and trends in system response, noting whether the restoration is "on course" or "off course" and why. Similarly, the Puget Sound Vital Signs[4] provide regular updates on short- and long-term trends, factors affecting those trends, whether targets are met, and steps needed to achieve the targets, with varying levels of detail depending on the interest of the user. A similar reporting mechanism for the CERP could connect to program-level adaptive management efforts and improve communication about overall restoration, if data were provided to justify the assessments of conditions and trends relative to the objectives. Successful communication and linkages to decision making also require appropriate resources and personnel, which has been a constraint for adaptive management overall as indicated above.

---

[3] See https://www.chesapeakeprogress.com/.
[4] See https://www.psp.wa.gov/evaluating-vital-signs.php.

Timely Decision Processes

Effective adaptive management in the CERP requires a clear framework for timely, structured decision-making processes, if the program is to keep up with new information and lessons learned across scores of projects in design and operation. For the CERP, this involves a process in which CERP leadership with input from RECOVER must approve changes, as well as a USACE-specific framework with multiple paths and requirements for approval depending on the scope of the proposed changes (see Table 5-1). Although various responsibilities for project- and program-level changes have been outlined in guidance documents, no required time frame or process exists for periodic multi-agency assessments of project performance relative to objectives, which limits timely analysis and response. The requirement for a Periodic CERP Update every 5 years most recently took nearly 20 years to implement. In contrast, the Chesapeake Bay Program has established a highly structured biennial Strategy Review System to support analysis of new information and identification of necessary actions to achieve its goals.[5]

To date, the process has mostly not been timely, or nimble, especially for larger-scale changes that require multiple levels of approval. Even lower-level adjustments have taken substantial time to effect, as described in the Deering Estate Flow-way example discussed previously in this chapter. More rapid response has been possible in some project-level adaptation, for example, when effects on a threatened or endangered species prompt an expedited process. Incorporating flexibility in PIRs and well-anticipated adaptive management plans have created pathways for changes that do not require higher-level approvals. However, not all outcomes will be anticipated, and more flexible response mechanisms are needed to handle responses to new information beyond these specific situations. Other initiatives within the USACE have accelerated decision making while managing risk in the planning process (e.g., 3×3×3 process), and such approaches to incorporate new information after project authorization should be considered. The USACE is currently assessing proposed Agency Specific Procedures for the implementation of the Principles, Requirements, and Guidelines for Federal Investments in Water Resources (34 C.F.R. §234 [2024]), which states, "Adaptive management measures shall be clearly identified and evaluated as part of alternatives to the extent that such measures are commensurate with the significance of the proposed activity and available resources." This process could provide a framework for improved incorporation of adaptive management measures into agency activities and more timely implementation.

---

[5] See https://www.chesapeakebay.net/what/what-guides-us/decisions/srs.

CONCLUSIONS AND RECOMMENDATIONS

When the CERP was authorized in 2000, Congress directed the use of adaptive management to ensure continued restoration progress amid uncertainties and to improve restoration outcomes through the incorporation of new information. After 24 years of CERP program development and implementation, the committee evaluates the effectiveness of the incorporation of new information into the CERP at four stages: (1) project-level adaptation during design and construction, (2) project-level adaptive management after operations begin, (3) adaptation of operations at regional scales, and (4) program-level adaptive management. Adaptive management itself is not a stationary process. Increasing input from the Tribes, experience with restoration, and the challenges associated with climate change all demand that the adaptive management program evolve to meet the needs of agency decision makers and stakeholders.

The CERP has developed thorough project-level adaptive management guidance, but the process to incorporate new information is often time-consuming and burdensome, which limits its application and effectiveness at the full scale of the CERP. RECOVER's adaptive management guidance is sound conceptually and sufficiently detailed to be effective and has informed the development of well-crafted project-level adaptive management plans. The guidance promotes a collaborative relationship between RECOVER and Project Delivery Teams that spans design and operation to support adaptive management. However, effective adaptation of authorized designs and operational plans for implemented projects has often been a very lengthy process. The USACE approval processes required to modify a congressionally authorized project contribute to the delays; in some cases, the SFWMD has expedited project components at its own risk to reduce delays. Modifications can be made more quickly when some flexibility is built into the PIRs or when uncertainties are anticipated in project adaptive management plans. As many more projects come online, more timely responses and processes are needed that facilitate adaptation to new information.

Adaptation of operations to new information is a particular strength of water management for restoration, and the USACE and the SFWMD should continue their increased use of conditions-based operations to better adapt to changing conditions. Adaptive approaches have been used effectively to provide and improve operational flexibility, in large part because of regular and transparent communication between scientists and decision makers. Many regional System Operations Manuals, such as the Lake Okeechobee System Operating Manual and the COP, rely on conditions-based operations to provide maximum flexibility and responsiveness to changing conditions, but other parts of the system (e.g., S-12A–D along Tamiami Trail) continue to rely on specific calendar-based operations. Conditions-based operations, where feasible, would enable optimization of water management in light of prior hydrologic and ecological conditions. However, such actions would require modification of existing operational plans and likely require additional resources for staff to review data and make management recommendations or to upgrade control structures to enable remote operations. Structured opportunities for learning from operations, such as the COP Adaptive Management Plan, will further improve future operations.

Program-level adaptive management has not been implemented in detail. The focus of adaptive management to date has instead been at the project level where the objective appears to be to modify projects to achieve the expected outcomes of those projects. At the program level, adaptive management has potential to maximize the benefits of the restoration as a whole. RECOVER identified program-level "mission-critical uncertainties" in 2015, including the implications of climate change and available water storage on restoration outcomes, but no coordinated program-level adaptive management strategy has been implemented. Past System Status Reports have lacked the necessary analysis of how and why the system is changing over time relative to goals and expectations to support program-level decision making. The potential for learning through the long-anticipated Second Periodic CERP Update is substantial and should support renewed attention to program-level adaptive management.

Additional efforts are needed to provide the necessary foundation for successful adaptive management in the CERP. The following three areas are essential to achieve this goal:

1. **Prioritize building expertise and a culture of adaptive management.** Guidance alone is insufficient to support effective implementation of adaptive management. Clear endorsement from leadership of the importance, objectives, and value of adaptive management is essential to promoting a culture of adaptive management. In addition, CERP staff at all levels need additional experience in adaptive management. Building capacity and a culture of adaptive management across agencies and stakeholders can take many years but can be jump-started with strategic hiring of experienced adaptive management practitioners from other restoration programs. Regular workshops have been used in other large-scale restoration efforts across the country to build a culture of adaptive management and should be incorporated into the CERP.
2. **Develop a robust, integrated science enterprise to support adaptive management**, consisting of effective monitoring, modeling, synthesis, and research activities with inclusion of Indigenous Knowledge. This requires adequate staffing of appropriately trained scientists, including a critical mass of expertise in statistics, data analysis, visualization, synthesis, and communication. With the shift from planning to

implementation, responsibilities for adaptive management in both RECOVER and in the agencies have expanded substantially; however, staffing levels have not increased to match these needs. The creation of an Environmental Assessment Team within the SFWMD charged with collecting relevant knowledge and providing it to Project Delivery Teams is a positive step toward supporting adaptive management. A CERP Science Plan would facilitate systemwide coordination of science, which is necessary for effective adaptive management.

3. **Improve communication of restoration performance (relative to expectations and objectives) and implications for decision making.** Effective adaptive management relies on the timely sharing of data and knowledge both within and across projects and the integration of this information into management decision making at project and program scales. Multi-agency reporting on project outcomes, such as Project Delivery Team biennial reports and/or a CERP dashboard built on key metrics that are collectively developed, would improve tracking of project outcomes and communication to management and the public.

All of these efforts will require strong direction from USACE and SFWMD leadership that adaptive management is a CERP priority that adds value and improves efficiency toward restoration success. Such communication should include not only messaging but also provision of the necessary resources for adaptive management implementation, including staffing to support the rapid increase in the number of CERP projects.

USACE headquarters should review required approval processes associated with incorporating new information into design, construction, and project-level adaptive management processes to ensure timely use of new information. Such an effort would benefit not only the CERP but all USACE restoration projects. Under the current framework many decisions related to effective implementation of adaptive management are cumbersome, slow, and often not timely enough to address current or future challenges. Implementation of existing processes has resulted in delays in restoration improvements and could deter staff from making valuable changes because of the administrative burden necessary to affect that change or the implication of delays on other projects. As a result, new information that could substantially improve restoration outcomes may not be used if it is unrelated to legal drivers such as the Endangered Species Act. As more CERP projects come online, each will have its own adaptive management plan. Therefore, a more timely and effective administrative process for incorporating new information into decision making is needed. The USACE has used the 3×3×3 approach in the project planning process to reduce the time associated with review and decision making. The committee encourages USACE headquarters to identify lessons learned from this approach that could be applied to project-level adaptive management and the incorporation of new information. By reviewing when decisions must be elevated to the district, division, or headquarters for action, the USACE can determine whether the effort for these approvals is appropriate to the risks posed, including impacts to ecosystem integrity and Tribal resources associated with lengthy delays. The USACE headquarters review should consider opportunities to incorporate flexibility into the decision-making process and empower decision making at the lowest reasonable level. A more efficient process for incorporating new information while managing operational and ecosystem risk would benefit USACE restoration projects in the CERP and nationally. USACE headquarters should also explore mechanisms to increase post-authorization flexibility within the constraints of existing processes, such as incorporating more flexibility into PIRs and operation manuals.

References

Adams, M. M. 2016. *Who Belongs?: Race, Resources, and Tribal Citizenship in the Native South*. Oxford, UK: Oxford University Press.

AIHC (American Indian Health Commission). 2023. Model Tribal Data Sharing Agreement. American Indian Health Commission. https://aihc-wa.com/tribal-public-health-emergency-resources (accessed May 29, 2024).

Alfonso, S., M. Gesto, and B. Sadoul. 2021. Temperature increase and its effects on fish stress physiology in the context of global warming. *Journal of Fish Biology* 98(6):1496–1508, https://doi.org/10.1111/jfb.14599 (accessed June 7, 2024).

Armstrong, C., T. Piccone, and J. Dombrowski. 2023. The largest constructed treatment wetland project in the world: the story of the Everglades stormwater treatment areas. *Ecological Engineering*. 193:107005, https://doi.org/10.1016/j.ecoleng.2023.107005 (accessed November 24, 2024).

Atkinson, D., and R. M. Sibly. 1997. Why are organisms usually bigger in colder environments? Making sense of a life history puzzle. *Trends in Ecology & Evolution* 12(6):235–239, https://doi.org/10.1016/S0169-5347(97)01058-6 (accessed June 7, 2024).

Ator, S. W., J. D. Blomquist, J. S. Webber, and J. G. Chanat. 2020. Factors driving nutrient trends in streams of the Chesapeake Bay watershed. *Journal of Environmental Quality* 49(4):812–834, https://doi.org/10.1002/jeq2.20101.

Austin, M. 2007. Species distribution models and ecological theory: A critical assessment and some possible new approaches. *Ecological Modelling* 200(1-2):1–19, https://doi.org/10.1016/j.ecolmodel.2006.07.005 (accessed June 7, 2024).

Ban, N. C., A. Frid, M. Reid, B. Edgar, D. Shaw, and P. Siwallace. 2018. Incorporate Indigenous perspectives for impactful research and effective management. *Nature Ecology & Evolution* 2(11):1680–1683.

Barney, J. N., G. N. Ripa, J. Drake, D. Franusich, M. C. Mims, and G. O'Malley. 2024. A silent spring, or a new cacophony? Invasive plants as maestros of modern soundscapes. *Frontiers in Ecology and the Environment* 22(3):e2729, https://doi.org/10.1002/fee.2729 (accessed May 28, 2024).

Bartlett, D., P. Ramos, K. Burchett, Col. J. Booth, and A. Blalock. 2023. *Principals' Concurrence with S-333 Working Group "Phase 1" Recommendations*. Memorandum dated November 2023. https://www.sfwmd.gov/sites/default/files/documents/Memo_to_TOC_from_Principals_S333_WG_Nov_2023.pdf (accessed June 21, 2024).

Bartoszek, I. A., M. J. Schuman, D. S. Addison, K. B. Worley, and J. R. Schmid. 2007. *Biological Monitoring of Aquatic and Terrestrial Fauna for the Picayune Strand Restoration Project (2005-2007)*. Final Report for South Florida Water Management District. Naples, FL: Conservancy of Southwest Florida. https://conservancy.org/wp-content/uploads/2021/01/PSRP-Biological-Monitoring-Final-Report-for-web.pdf (accessed June 21, 2024).

Beerens, J. M., E. G. Noonburg, and D. E. Gawlik. 2015. Linking dynamic habitat selection with wading bird foraging distributions across resource gradients. *PLoS ONE* 10(6):e0128182, http://dx.plos.org/10.1371/journal.pone.0128182 (accessed June 13, 2024).

Bellard, C., C. Bertelsmeier, P. Leadley, W. Thuiller, and F. Courchamp. 2012. Impacts of climate change on the future of biodiversity. *Ecology Letters* 15(4):365–377, https://doi.org/10.1111/j.1461-0248.2011.01736.x (accessed June 7, 2024).

Bennett, J. M., P. Calosi, S. Clusella-Trullas, B. Martínez, J. Sunday, A. C. Algar, M. B. Araújo, B. A. Hawkins, S. Keith, I. Kühn, C. Rahbek, L. Rodríguez, A. Singer, F. Villalobos, M. Ángel Olalla-Tárraga, and I. Morales-Castilla. 2018. GlobTherm, a global database on thermal tolerances for aquatic and terrestrial organisms. *Scientific Data* 5(180022), https://doi.org/10.1038/sdata.2018.22 (accessed June 7, 2024).

Berkes, F. 1998. Indigenous knowledge and resource management systems in the Canadian subarctic. In *Linking Social and Ecological Systems: Management Practices and Social Mechanisms for Building Resilience,* edited by F. Berkes and C. Folke. Cambridge, UK: Cambridge University Press. Pp. 98–128.

Berkes, F. 2017. Environmental governance for the Anthropocene? Social-ecological systems, resilience, and collaborative learning. *Sustainability* 9(7):1232, https://doi.org/10.3390/su9071232 (accessed May 29, 2024).

Berkes, F., D. Feeny, B. J. McCay, and J. M. Acheson. 1989. The benefits of the commons. *Nature* 340:91–93, https://doi.org/10.1038/340091a0 (accessed May 29, 2024).

Berkes, F., J. Colding, and C. Folke. 2000. Rediscovery of traditional ecological knowledge as adaptive management. *Ecological Applications* 10(5):1251–1262, https://doi.org/10.1890/1051-0761(2000)010[1251:ROTEKA]2.0.CO;2 (accessed May 29, 2024).

Betts, A., Z. Welch, and P. Jones. 2024. Chapter 8B: Lake Okeechobee Watershed Protection Plan annual progress report. In *2024 South Florda Environmental Report*, Vol. 1. West Palm Beach, FL: South Florida Water Management District.

Billie, J. 2016. Response from Hon. James Billie, Chairman Seminole Tribe of Florida, dated September 13, 2016, to letter from Colonel Jason A. Kirk, District Commander, dated September 2, 2016. Appendix C.3: Pertinent Correspondence Information. WERP Draft PIR/EIS.

Blake, N. 1980. *Land into Water—Water into Land: A History of Water Management in Florida*. Tallahassee, FL: University Press of Florida.

Blaustein, A. R., L. K. Belden, D. H. Olson, D. M. Green, T. L. Root, and J. M. Kiesecker. 2001. Amphibian breeding and climate change. *Conservation Biology* 15(6):1804–1809, https://doi.org/10.1046/j.1523-1739.2001.00307.x (accessed June 7, 2024).

Blaustein, A. R., S. C. Walls, B. A. Bancroft, J. J. Lawler, C. L. Searle, and S. S. Gervasi. 2010. Direct and indirect effects of climate change on amphibian populations. *Diversity* 2(2):281–313, https://doi.org/10.3390/d2020281 (accessed June 7, 2024).

Bled, F., M. J. Cherry, E. P. Garrison, K. V. Miller, L. M. Conner, H. N. Abernathy, W. H. Ellsworth, L. L. S. Margenau, D. A. Crawford, K. N. Engebretsen, B. D. Kelly, D. B. Shindle, and R. B. Chandler. 2022. Balancing carnivore conservation and sustainable hunting of a key prey species: A case study on the Florida panther and white-tailed deer. *Journal of Applied Ecology* 59(8):2010-2022, https://doi.org/10.1111/1365-2664.14201 (accessed May 24, 2024).

Bock, S. L., R. H. Lowers, T. R. Rainwater, E. Stolen, J. M. Drake, P. M. Wilkinson, S. Weiss, B. Back, L. Guillette, and B. B. Parrott. 2020. Spatial and temporal variation in nest temperatures forecasts sex ratio skews in a crocodilian with environmental sex determination. *Proceedings of the Royal Society B* 287(1926), https://doi.org/10.1098/rspb.2020.0210 (accessed June 7, 2024).

Boltaña, S., N. Sanhueza, A. Aguilar, C. Gallardo-Escarate, G. Arriagada, J. A. Valdes, D. Soto, and R. A. Quiñones. 2017. Influences of thermal environment on fish growth. *Ecology and Evolution* 7(17):6814–6825, https://doi.org/10.1002%2Fece3.3239 (accessed June 7, 2024).

Booth, J. 2023. Commentary on the C-44 Reservoir. *Lake Okeechobee News, November 30*. https://www.lakeonews.com/indiantown/stories/commentary-on-the-c-44-reservoir,61738? (accessed June 17, 2024).

Brandt, L. A., V. Briggs-Gonzalez, J. A. Browder, M. Cherkiss, S. Farris, P. Frederick, E. Gaiser, D. Gawlik, C. Hackett, A. Huebner, C. Kelble, J. Kline, K. Kotun, J. Lorenz, C. Madden, F. J. Mazzotti, M. Parker, L. Rodgers, A. Rodusky, D. Rudnick, R. Sobszak, J. Spencer, J. Trexler, and I. Zink. 2018. *System-wide Indicators for Everglades Restoration: 2018 Report*. Science Coordination Group, South Florida Ecosystem Restoration Task Force. https://static1.squarespace.com/static/5d5179e7e42ca1000117872f/t/604f51fab2eba6767a7f6861/161 5811069215/2018_system_wide_ecological_indicators.pdf (accessed September 2, 2024).

Breithaupt, J. L., J. M. Smoak, V. H. Rivera-Monroy, E. Casteñeda-Moya, R. P. Moyer, M. Simard, and C. J. Sanders. 2017. Partitioning the relative contributions of organic matter and mineral sediment to accretion rates in carbonate platform mangrove soils. *Marine Geology* 390:170–180, https://doi.org/10.1016/j.margeo.2017.07.002 (accessed August 26, 2024).

Brias, A., and S. B. Munch. 2021. Ecosystem based multi-species management using Empirical Dynamic Programming. *Ecological Modelling* 441:109423, https://doi.org/10.1016/j.ecolmodel.2020.109423 (accessed May 7, 2024).

Brook, B. W., M. A. Burgman, and R. Frank. 2000. Differences and congruencies between PVA packages: The importance of sex ratio for predictions of extinction risk. *Conservation Ecology* 4(1):6, https://www.doi.org/10.5751/es-00179-040106 (accessed June 7, 2024).

Brooke, M. 2004. *Albatrosses and Petrels Across the World*. Oxford, UK: Oxford University Press.

Bucklin, D. N., M. Basille, A. M. Benscoter, L. A. Brandt, F. J. Mazzotti, S. S. Romañach, C. Speroterra, and J. I. Watling. 2015. Comparing species distribution models constructed with different subsets of environmental predictors. *Diversity and Distributions*, 21(1):23–35, https://doi.org/10.1111/ddi.12247 (accessed August 13, 2024).

Bulbul Ali, A., and A. Mishra. 2022. Effects of dissolved oxygen concentration on freshwater fish: A review. *International Journal of Fisheries and Aquatic Studies* 10(4):113–127, https://doi.org/10.22271/fish.2022.v10.i4b.2693 (accessed June 7, 2024).

Caneja, E. 2024. Presentation to CISRERP: Lake Okeechobee Watershed Restoration Project and Aquifer Storage and Recovery Science Plan Update. January 10, 2024. Miami, FL.

Carboneras, C., F. Jutglar, and G. M. Kirwan. 2020. Sooty Shearwater (Ardenna grisea), version 1.0. In *Birds of the World,* edited by J. del Hoyo, A. Elliott, J. Sargatal, D. A. Christie, and E. de Juana. Ithaca, NY: Cornell Lab of Ornithology. https://doi.org/10.2173/bow.sooshe.01 (accessed May 29, 2024).

Carr, R. S. 2002. Chapter 6: The archaeology of Everglades tree islands. In *Tree Islands of the Everglades,* edited by F. H. Sklar and A. Van Der Valk. Dordrecht, Netherlands: Springer. https://doi.org/10.1007/978-9009-0001-1_6 (accessed May 29, 2024).

Carrick, H. J., and A. D. Steinman. 2001. Variation in periphyton biomass and species composition in Lake Okeechobee, Florida (USA): Distribution of algal guilds along environmental gradients. *Archiv für Hydrobiologie* 152(3):411–438, https://doi.org/10.1127/archiv-hydrobiol/152/2001/411 (accessed May 24, 2023).

Carroll, S. R., I. Garba, O. L. Figueroa-Rodriguez, J. Holbrook, R. Lovett, S. Materechera, M. Parsons, K. Raseroka, D. Rodriguez-Lonebar, R. Rowe, R. Sara, J. D. Walker, J. Anderson, and M. Hudson. 2020. The CARE principles for Indigenous data governance. *Data Science Journal* 19(43), https://doi.org/10.5334/dsj-2020-043 (accessed May 29, 2024).

Catano, C. P., S. S. Romañach, J. M. BEreens, L. G. Pearlstine, L. A. Brandt, K. M. Hart, F. J. Mazzotti, and J. C. Trexler. 2015. Using scenario planning to evaluate the impacts of climate change on wildlife populations and communities in the Florida Everglades. *Environmental Management* 55:807–823, https://doi.org/10.1007/s00267-014-0397-5 (accessed June 7, 2024).

Ceilley, D. W. 2008. *Picayune Strand Restoration Project Baseline Assessment of Inland Aquatic Fauna*. Final Report for South Florida Water Management District, West Palm Beach, FL.

Ceilley, D. W. 2022. *Third Post-Restoration Picayune Strand Restoration Project Area Aquatic Faunal Monitoring*. Final Report for South Florida Water Management District, West Palm Beach, FL.

Ceilley, D. W., S. E. Clem, L. Martin, E. M. Everham, G. Diaz, and P. E. Clark. 2020. *Third Year Post-Construction Aquatic Fauna Monitoring in the Picayune Strand Restoration Project Area*. Final Report for South Florida Water Management District, West Palm Beach, FL. https://apps.sfwmd.gov/sfwmd/SFER/2023_sfer_final/v3/appendices/v3_app2-1.pdf (accessed May 24, 2024).

Chadès, I., S. Nicol, T. M. Rout, M. Péron, Y. Dujardin, J-B. Pichancourt, A. Hastings, and C. E. Hauser. 2017. Optimization methods to solve adaptive management problems. *Theoretical Ecology* 10:1–20, https://doi.org/10.1007/s12080-016-0313-0 (accessed June 7, 2024).

Chambers, R. M., A. L. Gorsky, E. Castañeda-Moya, and V. H. Rivera-Monroy. 2021. Evaluating a steady-state model of soil accretion in Everglades mangroves (Florida, USA). *Estuaries and Coasts* 44(5):1469–1476, https://doi.org/10.1007/s12237-020-00883-1 (accessed June 7, 2024).

Charkhian, B. 2023. Appendix 2-3: Annual Permit Report for the Biscayne Bay Coastal Wetlands Project. Permit Report (May 1, 2021-April 30, 2022). Permit Number: 0271729. *2023 South Florida Environmental Report* Vol. 3.

Charkhian, B. 2024. Appendix 2-2: Annual Permit Report for the Biscayne Bay Coastal Wetlands Project. Permit Report (May 1, 2022-April 30, 2023). Permit Number: 0271729. *2024 South Florida Environmental Report* Vol. 3.

Charkhian, B., and N. Niemeyer. 2023. Presentation to CISRERP: Biscayne Bay Coastal Wetlands Phase 1 Project – Adapting CERP Projects in Light of New Information. November 6, 2023.

Cherkiss, M. S., J. I. Watling, L. A. Brandt, F. J. Mazzotti, J. Lindsay, J. S. Beauchamp, J. Lorenz, J. A. Wasilewski, I. Fujisaki, and K. M. Hart. 2020. Shifts in hatching date of American crocodile (*Crocodylus acutus*) in southern Florida. *Journal of Thermal Biology* 88:102521, https://doi.org/10.1016/j.jtherbio.2020.102521 (accessed June 7, 2024).

Chimney, M. J. 2020. Chapter 5b: Performance and operation of the Everglades stormwater treatment areas. In *South Florida Environmental Report*. West Palm Beach, FL: South Florida Water Management District.

Chimney, M. J. 2021. Chapter 5b: Performance and operation of the Everglades stormwater treatment areas. In *South Florida Environmental Report*. West Palm Beach, FL: South Florida Water Management District.

Chimney, M. J. 2022. Chapter 5b: Performance and operation of the Everglades stormwater treatment areas. In *South Florida Environmental Report*. West Palm Beach, FL: South Florida Water Management District.

Chimney, M. J. 2023. Chapter 5b: Performance and operation of the Everglades stormwater treatment areas. In *South Florida Environmental Report*. West Palm Beach, FL: South Florida Water Management District.

Chimney, M. J. 2024. Chapter 5b: Performance and operation of the Everglades stormwater treatment areas. In *South Florida Environmental Report*. West Palm Beach, FL: South Florida Water Management District.

Chuirazzi, K., W. Abtew, V. Ciuca, B. Gu, N. Iricanin, C. Mo, N. Niemeyer, and J. Starnes. 2018. Appendix 2-1: Annual Permit Report for the Picayune Strand Restoration Project. In *2018 South Florida Environmental Report* – Volume III. SFWMD. https://apps.sfwmd.gov/sfwmd/SFER/2018_sfer_final/v3/appendices/v3_app2-1.pdf (accessed August 30, 2024).

Clark, P. E. 2020. Monitoring the success of the Picayune Strand Restoration Project in Collier County, FL. Master of Science requirement, Florida Gulf Coast University.

Clutton-Brock, T., and B. C. Sheldon. 2010, The seven ages of Pan. *Science* 327(5970):1207–1208, https://doi.org/10.1126/science.1187796 (accessed May 29, 2024).

Cohen, J. M., M. J. Lajeunesse, and J. R. Rohr. 2018. A global synthesis of animal phenological responses to climate change. *Nature Climate Change* 8(3):224–228, https://doi.org/10.1038/s41558-018-0067-3 (accessed June 7, 2024).

Cotton, P. A. 2003. Avian migration phenology and global climate change. *Proceedings of the National Academy of Sciences* 100(21):12219–12222, https://doi.org/10.1073/pnas.1930548100 (accessed June 7, 2024).

Cottrell, C. 2022. Avoiding a new era in biopiracy: Including Indigenous and local knowledge in nature-based solutions to climate change. *Environmental Science & Policy* 135:162–168, https://doi.org/10.1016/j.envsci.2022.05.003 (accessed May 29, 2024).

Covington, J. W. 1993. *The Seminoles of Florida*. Gainesville, FL: University Press of Florida.

Craft, C. B., and C. J. Richardson. 1998. Recent and long-term organic soil accretion and nutrient accumulation in the Everglades. *Soil Science Society of America Journal* 62(3):834–843, https://doi.org/10.2136/sssaj1998.03615995006200030042x (accessed June 7, 2024).

CTKW (Climate and Traditional Knowledges Workgroup). 2014. *Guidelines for Considering Traditional Knowledges in Climate Change Initiatives.* http://climatetkw.wordpress.com (accessed May 29, 2024).

Cuda, J. P., D. H. Habeck, W. A. Overholt, J. C. Medal, J. H. Pedrosa-Macedo, M. D. Vitorino, and C. R. Minteer. 2023. Classical biological control of Brazilian peppertree (Schinus terebinthifolia) in Florida. University of Florida, IFAS Extension ENY-820. https://edis.ifas.ufl.edu/publication/IN114 (accessed May 24, 2024).

Cypress, B. 2020. Letter from Hon Billy Cypress, Chairman Miccosukee Tribe of Indians of Florida, to Col. Andrew Kelly, District Commander, dated January 21, 2020. Appendix C.3: Pertinent Correspondence Information. WERP Draft PIR/EIS.

Cypress, T. 2022. Letter from Hon. Talbert Cypress, Chairman Miccosukee Tribe of Indians of Florida, to Hon. Michael L. Connor, Assistant Secretary of the Army for Civil Works, dated December 15, 2022. Appendix C.3: Pertinent Correspondence Information. WERP Draft PIR/EIS.

Cypress, T. 2023. Letter from Hon. Talbert Cypress, Chairman Miccosukee Tribe of Indians of Florida, to Martha Williams, Director, U.S. Fish and Wildlife Service, dated March 28, 2023.

D'Acunto, L. E., L. Pearlstine, and S. S. Romañach. 2021. Joint species distribution models of Everglades wading birds to inform restoration planning. *PLoS ONE* 16(1):e0245973, https://doi.org/10.1371/journal.pone.0245973 (accessed June 11, 2024).

D'Acunto, L. E., L. Pearlstine, S. M. Haider, C. E. Hackett, D. Shinde, and S. S. Romañach. 2023. The Everglades vulnerability analysis: Linking ecological models to support ecosystem restoration. *Frontiers in Ecology and Evolution* 11, https://doi.org/10.3389/fevo.2023.1111551 (accessed June 11, 2024).

Darby, P. C., D. L. DeAngelis, S. S. Romañach, K. Suir, and J. Bridevaux. 2015. Modeling apple snail population dynamics on the Everglades landscape. *Landscape Ecology* 30:1497–1510, https://doi.org/10.1007/s10980-015-0205-5 (accessed June 11, 2024).

Daufresne, M., K. Lengfellner, and U. Sommer. 2009. Global warming benefits the small in aquatic ecosystems. *Proceedings of the National Academy of Sciences* 106(31):12788–12793.

David-Chavez, D. M., and M. C. Gavin. 2018. A global assessment of Indigenous community engagement in climate research. *Environmental Research Letters* 13(12):123005.

Davis, S. M., and J. C. Ogden, eds. 1994. *Everglades: The Ecosystem and Its Restoration.* Delray Beach: St. Lucie Press.

De Solla, S. R., K. J. Fernie, G. C. Barrett, and C. A. Bishop. 2006 Population trends and calling phenology of anuran populations surveyed in Ontario estimated using acoustic surveys. *Topics in Biodiversity & Conservation* 4:113–129, https://doi.org/10.1007/978-1-4020-5734-2_9 (accessed May 24, 2024).

Department of Climate Change, Energy, the Environment and Water. 2023. *Conservation Advice for Ardenna grisea (sooty shearwater).* Canberra, Australia.

Desjonquères, C., T. Gifford, and S. Linke. 2020. Passive acoustic monitoring as a potential tool to survey animal and ecosystem processes in freshwater environments. *Freshwater Biology* 65(1):7–19, https://doi.org/10.1111/fwb.13356 (accessed May 24, 2024).

Doering, P. H. 1996. Temporal variability of water quality in the St. Lucie Estuary, South Florida. *Journal of the American Water Resources Association* 32(6):1293–1306.

Doering, P. H., and R. H. Chamberlain. 1999. Water quality and source of freshwater discharge to the Caloosahatchee Estuary, Florida. *Journal of the American Water Resources Association* 35(4):793–806.

DOI (U.S. Department of the Interior). 2014. Order No. 3335: *Reaffirmation of the Federal Trust Responsibility to Federally Recognized Indian Tribes and Individual Indian Beneficiaries.* Washington, DC: Secretary of the Interior. https://www.doi.gov/sites/doi.gov/files/migrated/news/pressreleases/upload/Signed-SO-3335.pdf (accessed May 24, 2024).

DOI. 2024. Interior Department Reaches Landmark Agreement with Klamath Basin Tribes, Project Irrigators to Collaborate on Ecosystem Restoration and Water Reliability. February 14, 2024. https://www.doi.gov/pressreleases/interior-department-reaches-landmark-agreement-klamath-basin-tribes-project (accessed May 24, 2024).

DOI and DOC (U.S. Department of Commerce). 2009. Responding to Climate Change in the Chesapeake Bay Watershed. A Draft Report Fulfilling Section 202d of Executive Order 13508. Washington, DC: DOI and DOC. https://federalleadership.chesapeakebay.net/files/2009/11/202f%20Scientific%20Support%20Report. pdf (accessed August 29, 2024).

DOI and USACE (U.S. Army Corps of Engineers). 2005. *Central and Southern Florida Project Comprehensive Everglades Restoration Plan: 2005 Report to Congress*. https://www.nrc.gov/docs/ML1219/ML12193A328.pdf (accessed September 2, 2024).

Dorcas, M. E., S. J. Price, S. C. Walls, and W. J. Barichivich. 2009. Chapter 16: Auditory monitoring of anuran populations. In *Amphibian Ecology and Conservation: A Handbook of Techniques*, edited by C. K. Dodd. New York: Oxford University Press. Pp. 281–298.

Doren, R. F., J. C. Trexler, A. D. Gottlieb, and M. C. Harwell. 2009. Ecological indicators for system-wide assessment of the Greater Everglades Ecosystem Restoration program. *Ecological Indicators* 9(6):S2–S16, https://doi.org/10.1016/j.ecolind.2008.08.009 (accessed May 29, 2024).

Dray, F. A., M. Rayamajhi, M. Smith, D. Halbritter, and J. Aquino-Thomas. 2021. *Biological Control Program Comprehensive Everglades Restoration Program, FY'22 Annual Plan*. Fort Lauderdale, FL: USDA/ARS Invasive Plant Research Laboratory.

Dray, F. A., M. Rayamajhi, M. Smith, D. Halbritter, and J. Aquino-Thomas. 2022. *Biological Control Program Comprehensive Everglades Restoration Program, FY'23 Annual Plan*. Fort Lauderdale, FL: USDA/ARS Invasive Plant Research Laboratory.

Dray, F. A., M. Rayamajhi, M. Smith, D. Halbritter, and J. Aquino-Thomas. 2023. *Biological Control Program Comprehensive Everglades Restoration Program, FY'23 Combined 4th Quarter (1 Jul 2023 – 30 Sep 2023)/Annual Report*. Fort Lauderdale, FL: USDA/ARS Invasive Plant Research Laboratory.

DuBose, T. P., V. Catalan, C. E. Moore, V. R. Farallo, A. L. Benson, J. L. Dade, W. A. Hopkins, and M. C. Mims. 2024. Thermal traits of Anurans database for the Southeastern United States (TRAD): A database of thermal trait values for 40 anuran species. *Ichthyology & Herpetology* 112(1):21–30, https://doi.org/10.1643/h2022102 (accessed May 11, 2024).

Duever, M. 2023. Hydrologic Restoration in the Picayune Strand Restoration Project and Adjacent Fakahatchee Strand Preserve State Park. Presentation at the Greater Everglades Ecosystem Restoration Conference, April 18, 2023. https://conference.ifas.ufl.edu/geer/past/geer2023 (accessed August 26, 2024).

Dunn, P. O., and D. W Winkler. 2010. Chapter 10: Effects of climate change on timing of breeding and reproductive success in birds. In *Effects of Climate Change on Birds*, edited by A. P. Moller, W. Fielder, and P. Berthold. New York: Oxford University Press. Pp. 113–128.

Ehlinger, G. S. 2023. Letter from Gretchen Ehlinger, Chief, Environmental Branch, U. S. Army Corps of Engineers, Jacksonville District, to Larry Williams, State Supervisor of Ecological Services, Fish and Wildlife Service, dated October 20, 2023.

Eitzel, M. V., D. Sarna-Wojcicki, S. Hogan, J. Sowerwine, M. Mucioki, K. McCovey, S. Bourque, L. Hillman, L. Morehead-Hillman, F. Lake, and V. Preston. 2024. Using mixed-method analytical historical ecology to map land use and land cover change for ecocultural restoration in the Klamath River Basin (Northern California). *Ecological Informatics* 81:102552.

Elith, J., M. A. Burgman, and H. M. Regan. 2002. Mapping epistemic uncertainties and vague concepts in predictions of species distribution. *Ecological Modelling* 157(2-3):313–329, https://doi.org/10.1016/S0304-3800(02)00202-8 (accessed June 11, 2024).

Elmore, R. D., J. M. Carroll, E. P. Tanner, T. J. Hovick, B. A. Grisham, S. D. Fuhlendorf, and S. K. Windels. 2017. Implications of the thermal environment for terrestrial wildlife management. *Wildlife Society Bulletin* 41(2):183–193, https://doi.org/10.1002/wsb.772 (accessed June 10, 2024).

EOP (Executive Office of the President). 2000. Executive Order 13175 of November 6, 2000: Consultation and Coordination with Indian Tribal Governments. *U.S. Federal Register* 65(218):67249–67252, https://www.federalregister.gov/documents/2000/11/09/00-29003/consultation-and-coordination-with-indian-tribal-governments (accessed May 29, 2024).

EOP. 2021a. Executive Order 13990 of January 20, 2021: Protecting Public Health and the Environment and Restoring Science to Tackle the Climate Crisis. *U.S. Federal Register* 86(14):7037–7043, https://www.federalregister.gov/documents/2021/01/25/2021-01765/protecting-public-health-and-the-environment-and-restoring-science-to-tackle-the-climate-crisis (accessed August 26, 2024).

EOP. 2021b. Executive Order 14049 of October 11, 2021: White House Initiative on Advancing Educational Equity, Excellence, and Economic Opportunity for Native Americans and Strengthening Tribal Colleges and Universities. *U.S. Federal Register* 86(196):57313–57318, https://www.federalregister.gov/documents/2021/10/14/2021-22588/white-house-initiative-on-advancing-educational-equity-excellence-and-economic-opportunity-for (accessed August 26, 2024).

EOP. 2021c. Memorandum for the Heads of Departments and Agencies of November 15, 2021: Indigenous Traditional Ecological Knowledge and Federal Decision Making. Office of Science and Technology Policy, Council on Environmental Quality. https://www.whitehouse.gov/wp-content/uploads/2021/11/111521-OSTP-CEQ-ITEK-Memo.pdf (accessed August 26, 2024).

EOP. 2022a. Memorandum for the Heads of Departments and Agencies of December 1, 2022: Guidance for Federal Departments and Agencies on Indigenous Knowledge. Office of Science Technology Policy and White House Council on Environmental Quality. https://www.whitehouse.gov/wp-content/uploads/2022/12/OSTP-CEQ-IK-Guidance.pdf (accessed August 26, 2024).

EOP. 2022b. Memorandum for the Heads of Departments and Agencies of December 1, 2022: Implementation of Guidance for Federal Departments and Agencies on Indigenous Knowledge. Office of Science Technology Policy and White House Council on Environmental Quality. https://www.whitehouse.gov/wp-content/uploads/2022/12/IK-Guidance-Implementation-Memo.pdf (accessed August 26, 2024).

EOP. 2022c. Executive Order 14072 of April 22, 2022: Strengthening the Nation's Forests, Communities and Local Economies. *U.S. Federal Register* 87(81):24851–24855, https://www.federalregister.gov/documents/2022/04/27/2022-09138/strengthening-the-nations-forests-communities-and-local-economies (accessed August 26, 2024).

EOP. 2022d. Memorandum for the Heads of Departments and Agencies of November 30, 2022: Uniform Standards for Tribal Consultation. Office of Science and Technology Policy and White House Council on Environmental Quality. https://www.whitehouse.gov/briefing-room/presidential-actions/2022/11/30/memorandum-on-uniform-standards-for-tribal-consultation (accessed August 26, 2024).

EOP. 2023a. Executive Order 14096 of April 21, 2023: Revitalizing our Nation's Commitment to Environmental Justice for All. *U.S. Federal Register* 88(80):25251–25261, https://www.federalregister.gov/documents/2023/04/26/2023-08955/revitalizing-our-nations-commitment-to-environmental-justice-for-all (accessed August 26, 2024).

EOP. 2023b. Executive Order of December 6, 2023: Reforming Federal Funding and Support for Tribal Nations to Better Embrace Our Trust Responsibilities and Promote the Next Era of Tribal Self-Determination. *U.S. Federal Register* 88(236):86021–86025, https://www.federalregister.gov/documents/2023/12/11/2023-27318/reforming-federal-funding-and-support-for-tribal-nations-to-better-embrace-our-trust (accessed August 26, 2024).

EPA (U.S. Environmental Protection Agency). 2010. United States Environmental Protection Agency Amended Determination. https://www.epa.gov/sites/default/files/2014-01/documents/1-ad-final-version-09-03-10.pdf (accessed September 2, 2024).

EPA. 2024. *Water Quality Standards: Regulations and Resources: EPA Actions on Tribal Water Quality Standards and Contacts*. https://www.epa.gov/wqs-tech/epa-actions-tribal-water-quality-standards-and-contacts (accessed January 19, 2024).

Evans, T. G., S. E. Diamond, and M. W. Kelly. 2015. Mechanistic species distribution modelling as a link between physiology and conservation. *Conservation Physiology* 3(1):cov056, https://doi.org/10.1093/conphys/cov056 (accessed June 11, 2024).

Fagherazzi, S., G. Mariotti, N. Leonardi, A. Canestrelli, W. Nardin, and W. S. Kearney. 2020. Salt marsh dynamics in a period of accelerated sea level rise. *Journal of Geophysical Research: Earth Surface* 125(8):e2019JF005200, https://doi.org/10.1029/2019JF005200 (accessed June 11, 2024).

FDEP (Florida Department of Environmental Protection). 2010. *Loxahatchee River National Wild and Scenic River Management Plan: Update 2010*. Florida Department of Environmental Protection and South Florida Water Management District.

FDEP. 2012a. *STA Everglades Forever Act (EFA) Watershed Permit (no. 0311207) and Associated Consent Order OGC 12-1149.* Tallahassee, FL: Florida Department of Environmental Protection.

FDEP. 2012b. *STA National Pollutant Discharge Elimination System Watershed Permit Number FL0778451 and Associated Consent Order OGC 12-1148.* Tallahassee, FL: Florida Department of Environmental Protection.

FDEP. 2015. *Report on the Beneficial Use of Reclaimed Water, Stormwater, and Excess Surface Water (Senate Bill 536).* Office of Water Policy. December 1, 2015. https://floridadep.gov/sites/default/files/SB536%20Final%20Report.pdf (accessed September 2, 2024).

FDEP. 2017. *Everglades Forever Act Permit Number 031-102006.7* September 10, 2017. https://floridadep.gov/sites/default/files/2018-01-31_Watershed%20EFA_Consolidated%20Permit_0.pdf (accessed September 1, 2024).

FDEP. 2020. *Implementation of Chapter 2008-232, Laws of Florida Domestic Wastewater Ocean Outfalls 2020 Progress Report.* July 2020. https://floridadep.gov/sites/default/files/July%202020%20Ocean%20Outfall%20Report.pdf (accessed September 1, 2024).

Feher, L. C., M. J. Osland, G. H. Anderson, W. C. Vervaeke, K. W. Krauss, K. R. T. Whelan, K. M. Balentine, G. Tiling-Range, T. J. Smith III, and D. R. Cahoon. 2020. The long-term effects of hurricanes Wilma and Irma on soil elevation change in Everglades mangrove forests. *Ecosystems* 23(5):917–931, https://doi.org/10.1007/s10021-019-00446-x (accessed June 21, 2024).

Feit, H. A. 1986. James Bay Cree Indian management and moral considerations of fur bearers. In *Native People and Renewable Resource Management.* 1986 Symposium of the Alberta Society of Professional Biologists. Edmonton, Alberta: Alberta Society of Professional Biologists. Pp. 49–65.

Ficke, A. D., C. A. Myrick, and L. J. Hanson. 2007. Potential impacts of global climate change on freshwater fisheries. *Reviews in Fish Biology and Fisheries* 17:581–613, https://doi.org/10.1007/s11160-007-9059-5 (accessed June 11, 2024).

Fixico, D. L. 2017. *"That's What They Used to Say": Reflections on American Indian Oral Traditions.* Norman, OK: University of Oklahoma Press.

Flaig, E. G., and K. R. Reddy. 1995. Fate of phosphorus in the Lake Okeechobee watershed, Florida, USA: Overview and recommendations. *Ecological Engineering* 5(2-3):127–142.

Folk, M. 2018. *Summary Report of 2018 Activities for Red-Cockaded Woodpecker Management on Picayune Strand State Forest.* Kissimmee, FL: Wild Folk, LLC.

Fordham, D. A., H. Resit Akçakaya, M. B. Araújo, J. Elith, D. A. Keith, R. Pearson, T. D. Auld, C. Mellin, J. W. Morgan, T. J. Regan, and M. Tozer. 2012. Plant extinction risk under climate change: Are forecast range shifts alone a good indicator of species vulnerability to global warming? *Global Change Biology* 18(4):1357–1371, https://doi.org/10.1111/j.1365-2486.2011.02614.x (accessed August 14, 2024).

François, B., A. Dufour, T. N. K. Nguyen, A. Bruce, D. K. Park, and C. Brown. 2024. From many futures to one: Climate-informed planning scenario analysis for resource-efficient deep climate uncertainty analysis. *Climatic Change* 177(7):1–23.

Franklin, J., H. M. Regan, and A. D. Syphard. 2014. Linking spatially explicit species distribution and population models to plan for the persistence of plant species under global change. *Environmental Conservation* 41(2):97–109, https://doi.org/10.21425/F5FBG51254 (accessed August 14, 2024).

Fujisaki, I., K. M. Hart, F. J. Mazzotti, M. S. Cherkiss, A. R. Sartain, B. M. Jeffery, J. S. Beauchamp, and M. Denton. 2014. Home range and movements of American alligators (*Alligator mississippiensis*) in an estuary habitat. *Animal Biotelemetry* 2:8, https://doi.org/10.1186/2050-3385-2-8 (accessed September 2, 2024).

FWS (U.S. Fish and Wildlife Service). 2016. *Biological Opinion for the Everglades Restoration Transition Plan—2016.* Vero Beach, FL: U.S. Fish and Wildlife Service. https://www.eenews.net/assets/2016/07/22/document_pm_02.pdf (accessed May 29, 2024).

FWS. 2020. *Species Status Assessment for the Florida Panther.* Vero Beach, FL: U.S. Fish and Wildlife Service. https://www.researchgate.net/publication/355080952_Species_Status_Assessment_for_the_Florida_P anther (accessed March 14, 2024).

FWS. 2024. Biological Opinion and Conference Opinion for the Western Everglades Restoration Project. May 30, 2024.

Gadgil, M., F. Berkes, and C. Folke. 1993. Indigenous knowledge for biodiversity conservation. *Ambio* 22(2/3):151–156, https://www.jstor.org/stable/4314060 (accessed June 21, 2024).

Gaglia, T. K. 2022. Monitoring the Picayune Strand restoration project using aquatic invertebrates as bioindicators. M.S. thesis. Florida Gulf Coast University.

Gaiser, E. E. 2023. FCE LTER- Water Total Phosphorus Changes, Update 11-30-23. PowerPoint.

Gan, H., J. Zhang, M. Towsey, A. Truskinger, D. Stark, B. J. van Rensberg, Y. Li, and P. Roe. 2021. A novel frog chorusing recognition method with acoustic indices and machine learning. *Future Generation Computer Systems* 125:485–495, https://doi.org/10.1016/j.future.2021.06.019 (accessed May 24, 2024).

Gerber, L. R., and E. R. White. 2014. Two-sex matrix models in assessing population viability: When do male dynamics matter? *Journal of Applied Ecology* 51(1):270–278, https://doi.org/10.1111/1365-2664.12177 (accessed June 11, 2024).

Gitau, M. W., W. J. Gburek, and A. R. Jarrett. 2005. A tool for estimating best management practice effectiveness for phosphorus pollution control. *Journal of Soil and Water Conservation* 60(1):1–10, https://www.jswconline.org/content/60/1/1 (accessed May 24, 2024).

Godfrey, M. C., and T. Catton. 2011. *River of Interests: Water Management in South Florida and the Everglades, 1948-2010*. Jacksonville, FL: U.S. Army Corps of Engineers.

Goode, A. B. C., P. W. Tipping, C. R. Minteer, E. N. Pokorny, B. K. Knowles, J. R. Foley, and R. J. Valmonte. 2021. *Megamelus scutellaris* (Berg) (Hemiptera: Delphacidae) biology and population dynamics in the highly variable landscape of southern Florida. *Biological Control* 160:104679, https://doi.org/10.1016/j.biocontrol.2021.104679 (accessed May 24, 2024).

Gorelick, D. E., L. Lin, H. B. Zeff, Y. Kim, J. M. Vose, J. W. Coulston, D. N. Wear, L. E. Band, P. M. Reed, and G. W. Characklis. 2020. Accounting for adaptive water supply management when quantifying climate and land cover change vulnerability. *Water Resources Research* 56:e2019WR025614.

Goss, J. A. 1995. *Usual and Customary Use and Occupancy by the Miccosukee and Seminole Indians in Big Cypress National Preserve, Florida*. Report to the Southeast Region National Park Service Atlanta, GA. http://npshistory.com/publications/bicy/goss-1995.pdf (accessed May 30, 2024).

Gottlieb, A., A. Melligon, and C. Qiu. 2024. Appendix 2-3: Annual Permit Report for the C-111 Spreader Canal. *2024 South Florida Environmental Report*, Vol. III, https://apps.sfwmd.gov/sfwmd/SFER/2024_sfer_final/v3/appendices/v3_app2-3.pdf (accessed June 13, 2024).

Grieger, R., S. J. Capon, W. L. Hadwen, and B. Mackey. 2020. Between a bog and a hard place: A global review of climate change effects on coastal freshwater wetlands. *Climatic Change* 163(1):161–179, https://doi.org/10.1007/s10584-020-02815-1 (accessed June 11, 2024).

Grunwald, M. 2006. *The Swamp: The Everglades, Florida, and the Politics of Paradise*. New York: Simon and Schuster.

Halpern, B. S., E. Berlow, R. Williams, E. T. Borer, F. W. Davis, A. Dobson, B. J. Enquist, H. E. Froehlich, L. R. Gerber, C. J. Lortie, M. I. O'Connor, H. Regan, D. P. Vázquez, and G. Willard. 2020. Ecological synthesis and its role in advancing knowledge. *BioScience* 70(11):1005–1014, https://doi.org/10.1093/biosci/biaa105 (accessed June 17, 2024).

Harding, A., B. Harper, D. Stone, C. O'Neill, P. Berger, S. Harris, and J. Donatuto. 2012. Conducting research with tribal communities: Sovereignty, ethics, and data-sharing issues. *Environmental Health Perspectives* 120(1):6–10, https://doi.org/10.1289/ehp.1103904 (accessed May 30, 2024).

Harvell, C. D., C. E. Mitchell, J. R. Ward, S. Altizer, A. P. Dobson, R. S. Ostfeld, and M. D. Samuel. 2002. Climate warming and disease risks for terrestrial and marine biota. *Science* 296(5576):2158–2162, https://doi.org/10.1126/science.1063699 (accessed June 10, 2024).

Havens, K. E., and W. W. Walker Jr. 2002. Development of a total phosphorus concentration goal in the TMDL process for Lake Okeechobee, Florida (USA). *Lake and Reservoir Management* 18(3):227–238, https://doi.org/10.1080/07438140209354151 (accessed May 24, 2024).

Havens, K. E., C. Hanlon, and R. T. James. 1994. Seasonal and spatial variation in algal bloom frequencies in Lake Okeechobee, Florida, USA. *Lake and Reservoir Management* 10(2):139–148, https://doi.org/10.1080/07438149409354185 (accessed June 11, 2024).

Heisler, L., D. T. Towles, L. A. Brandt, and R. T. Pace. 2002. Tree island vegetation and water management in the central Everglades. In *Tree Islands of the Everglades*. Dordrecht, Netherlands: Springer. Pp. 283–309. https://doi.org/10.1007/978-94-009-0001-1_9 (accessed May 30, 2024).

Herman, J. D., J. D. Quinn, S. Steinschneider, M. Giuliani, and S. Fletcher. 2020. Climate adaptation as a control problem: Review and perspectives on dynamic water resources planning under uncertainty. *Water Resources Research* 56(2):e24389, https://doi.org/10.1029/2019WR025502 (accessed June 11, 2024).

Hernandez, J., and K. A. Vogt. 2020. Indigenizing restoration: Indigenous lands before urban parks. *Human Biology* 92(1):37–44, https://doi.org/10.13110/humanbiology.92.1.02 (accessed May 30, 2024).

Hiers, J. K., R. J. Mitchell, A. Barnett, J. R. Walters, M. Mack, B. Williams, and R. Sutter. 2012. The dynamic reference concept: Measuring restoration success in a rapidly changing no-analogue future. *Ecological Restoration* 30(1):27–36.

Hopkins, J. M., W. Edwards, and L. Schwarzkopf. 2022. Invading the soundscape: Exploring the effects of invasive species' calls on acoustic signals of native wildlife. *Biological Invasions* 24(11):3381–3393, https://doi.org/10.1007/s10530-022-02856-w (accessed May 24, 2024).

Howes, A. L., M. Martine, and C. A. McAlpine. 2010. Bayesian networks and adaptive management of wildlife habitat. *Conservation Biology* 24(4):974–983, https://www.jstor.org/stable/40864197 (accessed June 11, 2024).

Hughes, B. B., R. Beas-Luna, A. K. Barner, K. Brewitt, D. R. Brumbaugh, E. B. Cerny-Chipman, S. L. Close, K. E. Coblentz, K. L. de Nesnera, S. T. Drobnitch, J. D. Figurski, B. Focht, M. Friedman, J. Freiwald, K. K. Heady, W. N. Heady, A. Hettinger, A. Johnson, K. A. Karr, B. Mahoney, M. M. Moritsch, A.-M. K. Osterback, J. Reimer, J. Robinson, T. Rohrer, J. M. Rose, M. Sabal, L. M. Segui, C. Shen, J. Sullivan, R. Zuercher, P. T. Raimondi, B. A. Menge, K. Grorud-Colvert, M. Novak, and M. H. Carr. 2017. Long-term studies contribute disproportionately to ecology and policy. *BioScience* 67(3):271–281, https://doi.org/10.1093/biosci/biw185 (accessed May 30, 2024).

Humphries, G. R. W., and H. Moller. 2017. Fortune telling seabirds: Sooty shearwaters (Puffinus griseus) predict shifts in Pacific climate. *Marine Biology* 164(6):150, https://doi.org/10.1007/s00227-017-3182-1 (accessed May 30, 2024).

Hutchins, J. 2024. S333 Working Group Update. Presentation to the Technical Oversight Committee Meeting, June 24, 2024. https://www.sfwmd.gov/sites/default/files/documents/S333_WG_Update_TOC_Meeting_June2024.pdf (accessed August 30, 2024).

IPBES (Intergovernmental Science-Policy Platform on Biodiversity and Ecosystem Services). 2022. Summary for policymakers of the Thematic Assessment Report on the Sustainable Use of Wild Species of the Intergovernmental Science-Policy Platform on Biodiversity and Ecosystem Services. *Zenodo* Version 1, https://doi.org/10.5281/zenodo.6425599 (accessed May 30, 2024).

Irizarry-Ortiz, M. M., and J. F. Stamm. 2022. Change Factors to Derive Future Precipitation Depth-Duration-Frequency (DDF) Curves at 174 National Oceanic and Atmospheric Administration (NOAA) Atlas 14 Stations in Central and South Florida. United States Geological Survey Data Release. https://doi.org/10.5066/P935WRTG (accessed August 15, 2024).

Irizarry-Ortiz, M. M., J. F. Stamm, C. Maran, and J. Obeysekera. 2022. *Development of Projected Depth-Duration-Frequency Curves (2050–89) for South Florida*. Scientific Investigations Report. https://doi.org/10.3133/sir20225093 (accessed August 26, 2024).

Jackson, J. B. 2014. Seminole histories of the Calusa: Dance, narrative, and historical consciousness. *Native South* 7(1):122–142, https://doi.org/10.1353/nso.2014.0000 (accessed May 30, 2024).

James, R., R. Tsosie, P. Sahota, M. Parker, D. Dillard, I. Sylvester, J. Lewis, J. Klejka, L. Muzquiz, P. Olsen, R. Whitener, and W. Burke. 2014. Exploring pathways to trust: A tribal perspective on data sharing. *Genetics in Medicine* 16(11):820–826, https://doi.org/10.1038/gim.2014.47 (accessed May 30, 2024).

James, R. J., T. Piccone, J. King, S. Bornhoeft, S. Mason, and O. Villapando. 2024a. Chapter 5C: Restoration Strategies Science Plan. In *South Florida Environmental Report*, Vol. 1. West Palm Beach, FL: South Florida Water Management District. Pp. 5C1–5C69.

James, R. T., M. J. Chimney, C. Armstrong, T. Piccone, J. King, J. R. White, and K. R. Reddy. 2024b. Everglades Stormwater Treatment Area research: Synthesis, conclusions, and potential management options. *Ecological Engineering* 204:107256, https://doi.org/10.1016/j.ecoleng.2024.107256 (accessed May 28, 2024).

Jennings, L., T. Anderson, A. Martinez, R. Sterling, D. D. Chavez, I. Garba, M. Hudson, N. A. Garrison, and S. R. Carroll. 2023. Applying the "CARE Principles for Indigenous Data Governance" to ecology and biodiversity research. *Nature Ecology & Evolution* 7(10):1547–1551, https://doi.org/10.1038/s41559-023-02161-2 (accessed May 30, 2024).

Jessen, T. D., N. C. Ban, N. X. Claxton, and C. T. Darimont. 2021. Contributions of Indigenous Knowledge to ecological and evolutionary understanding. *Frontiers in Ecology and the Environment* 20(2):93–101, https://doi.org/10.1002/fee.2435 (accessed May 30, 2024).

Joanen, T. E. D., and L. L. McNease. 1989. Ecology and physiology of nesting and early development of the American alligator. *American Zoologist* 29(3):987–998, https://doi.org/10.1093/icb/29.3.987 (accessed June 11, 2024).

Johnson, R. 2020. Presentation on Tamiami Trail Next Steps Phase II Roadway improvements. February 25, 2020. https://static1.squarespace.com/static/5d5179e7e42ca1000117872f/t/5f2a0da6a21b6a77b2e5be33/159 6591527976/3_Tamiami_Trail_Next_Steps_Phase_2_Update.pdf.

Jones, G. M., A. J. Shirk, Z. Yang, R. J. Davis, J. L. Ganey, R. J. Gutiérrez, S. P. Healey, S. J. Hedwall, S. J. Hoagland, R. Maes, K. Malcolm, K. S. McKelvey, J. S. Sanderlin, M. K. Schwartz, M. E. Seamans, H. Yi Wan, and S. A. Cushman. 2023. Spatial and temporal dynamics of Mexican spotted owl habitat in the southwestern US. *Landscape Ecology* 38(1):23–37, https://doi.org/10.1007/s10980-022-01418-8 (accessed June 11, 2024).

J-Tech. 2020. *C-43 West Basin Storage Reservoir Water Quality Feasibility Study*. Deliverable 4.3.1: Final Feasibility Study Update. https://www.sfwmd.gov/sites/default/files/DEL%204.3.1_C-43%20Final%20Feasibility%20Study%20Update_11202020.pdf (accessed May 28, 2024).

Julian, P. and Davis, S. E. 2024. Evaluating water quality change with hydrologic restoration in the Western Everglades (Florida, USA), an application of WAM. *Watershed Ecology and the Environment* 6:70-83, https://doi.org/10.1016/j.wsee.2024.05.001 (accessed November 22, 2024).

Julian II, P., and L. Reidenbach. 2024. Upstream water management and its role in estuary health, evaluation of freshwater management and subtropical estuary function. *Watershed Ecology and the Environment* 6:84–94.

Kearney, M., and W. Porter. 2009. Mechanistic niche modelling: Combining physiological and spatial data to predict species' ranges. *Ecology Letters* 12(4):334–350, https://doi.org/10.1111/j.1461-0248.2008.01277.x (accessed June 11, 2024).

Kiker, C. F., J. W. Milon, and A. W. Hodges. 2001. South Florida: The reality of change and the prospects for sustainability: Adaptive learning for science-based policy: The Everglades restoration. *Ecological Economics* 37(3):403–416.

Kimmerer, R. 2013. *Braiding Sweetgrass: Indigenous Wisdom, Scientific Knowledge and the Teachings of Plants*. Minneapolis, MN: Milkweed editions. https://sps.berkeley.edu/static/documents/EnI/Week_3_2.pdf (accessed August 30, 2024).

Kimmerer, R. W. 2017. Chapter 26: The covenant of reciprocity. In *The Wiley Blackwell Companion to Religion and Ecology*, edited by J. Hart. Hoboken, NJ: John Wiley & Sons Ltd. Pp. 368–381. https://doi.org/10.1002/9781118465523.ch26 (accessed May 30, 2024).

Kirchmeier-Young, M. C., and X. Zhang. 2020. Human influence has intensified extreme precipitation in North America. *Proceedings of the National Academy of Sciences* 117:13308–13313.

Koch, M. S., S. A. Schopmeyer, O. I. Nielsen, C. Kyhn-Hansen, and C.J. Madden. 2007. Conceptual model of seagrass die-off in Florida Bay: Links to biogeochemical processes. *Journal of Experimental Marine Biology and Ecology* 350(1-2):73–88.

Koebel, J. W., S. G. Bousquin, D. H. Anderson, C. Carroll, D. Marois, R. Botta, C. Hanlon, L. Spencer, T. Beck, E. Tate-Boldt, B. Anderson, D. Nelson, and R. Ford. 2024. Chapter 9: Kissimmee River Restoration and other Kissimmee Basin initiatives. In *South Florida Environmental Report 2024*, Vol. 1. West Palm Beach, FL: South Florida Water Management District. Pp. 1–99. https://apps.sfwmd.gov/sfwmd/SFER/2024_sfer_final/v1/chapters/v1_ch9.pdf (accessed August 30, 2024).

Koons, D. N., D. T. Iles, M. Schaub, and H. Caswell. 2016. A life-history perspective on the demographic drivers of structured population dynamics in changing environments. *Ecology Letters* 19(9):1023–1031, https://doi.org/10.1111/ele.12628 (accessed June 11, 2024).

Kuebbing, S. E., A. P. Reimer, S. A. Rosenthal, G. Feinberg, A. Leiserowitz, J. A. Lau, and M. A. Bradford. 2018. Long-term research in ecology and evolution: A survey of challenges and opportunities. *Ecological Monographs* 88(2):245–258, https://doi.org/10.1002/ecm.1289 (accessed May 30, 2024).

Kukutai, T., and J. Taylor. 2016. *Indigenous Data Sovereignty: Toward an Agenda*. Canberra, Australia: The Australian National University Press.

Kunkel, K. E., T. R. Karl, M. F. Squires, X. Yin, S. T. Stegall, and D. R. Easterling. 2020. Precipitation extremes: Trends and relationships with average precipitation and precipitable water in the contiguous United States. *Journal of Applied Meteorology and Climatology* 59(1):125–142, https://doi.org/10.1175/JAMC-D-19-0185.1 (accessed June 11, 2024).

Kushlan, J. A., and T. Jacobsen. 1990. Environmental variability and the reproductive success of Everglades alligators. *Journal of Herpetology* 24(2):176–184, https://www.jstor.org/stable/1564225 (accessed June 11, 2024).

Ladouceur, E., N. Shackelford, K. Bouazza, L. Brudvig, A. Bucharova, T. Conradi, T. E. Erickson, M. Garbowski, K. Garvy, W. S. Harpole, H. P. Jones, T. Knight, M. M. Nsikani, G. Paterno, K. Suding, V. M. Temperton, P. Török, D. E. Winkler, and J. M. Chase. 2022. Knowledge sharing for shared success in the decade on ecosystem restoration. *Ecological Solutions and Evidence* 3(1):e12117, https://doi.org/10.1002/2688-8319.12117 (accessed June 17, 2024).

Lancaster, L. T., and A. M. Humphreys. 2020. Global variation in the thermal tolerances of plants. *Proceedings of the National Academy of Sciences* 117(24):13580–13587, https://doi.org/10.1073/pnas.1918162117 (accessed June 11, 2024).

Landis, W. G., A. J. Markiewicz, K. K. Ayre, A. F. Johns, M. J. Harris, J. M. Stinson, and H. M. Summers. 2017. A general risk-based adaptive management scheme incorporating the Bayesian Network Relative Risk Model with the South River, Virginia, as case study. *Integrated Environmental Assessment and Management* 13(1):115–126, https://doi.org/10.1002/ieam.1800 (accessed June 11, 2024).

Lapp, S., T. Wu, C. Richards-Zawacki, J. Voyles, K. M. Rodriguez, H. Shamon, and J. Kitzes. 2021. Automated detection of frog calls and choruses by pulse repetition rate. *Conservation Biology* 35(5):1659–1668, https://doi.org/10.1111/cobi.13718 (accessed May 28, 2024).

Lawson, A. J., B. A. Strickland, and A. E. Rosenblatt. 2018. Patterns, drivers and effects of alligator movement behavior and habitat use. In *American Alligators: Habitats, Behaviors, and Threats*, edited by C. B. Eversole and S. E. Henke. Hauppauge, NY: Nova Science Publishers. Pp. 47–77.

LeBrasseur, M. M., and E. S. Freark. 1982. Touch a child—they are my people: Ways to teach American Indian children. *Journal of American Indian Education* 21(3):6–12, https://www.jstor.org/stable/24397306 (accessed May 30, 2024).

Lee, K. N. 1999. Appraising adaptive management. *Conservation Ecology* 3(2):2, http://www.consecol.org/vol3/iss2/art2 (accessed June 17, 2024).

Light, S., and J. Dineen. 1994. Water control in the Everglades: A historical perspective. In *Everglades: The Ecosystem and Its Restoration*, edited by S. Davis and J. Ogden. Delray Beach, FL: St. Lucie Press. Pp. 47–84.

Lopez, M. S., L. Hudson, G. G. Payne, and S. K. Xue. 2024. Chapter 3: Water quality in the Everglades Protection Area. In *South Florida Environmental Report 2024*, Vol. 1. West Palm Beach, FL: South Florida Water Management District. Pp. 1–64. https://apps.sfwmd.gov/sfwmd/SFER/2024_sfer_final/v1/chapters/v1_ch3.pdf (accessed August 26, 2024).

Lord, L. A. 1993. *Guide to Florida Environmental Issues and Information*. Winter Park, FL: Florida Conservation Foundation.

Lorenz, J. J. 2014. A review of the effects of altered hydrology and salinity on vertebrate fauna and their habitats in Northeastern Florida Bay. *Wetlands* 34(1):189–200, https://doi.org/10.1007/s13157-013-0377-1 (accessed June 12, 2024).

Love, R. P., B. J. Hardy, C. Hefferman, A. Heyd, M. Cardinal-Grant, L. Sparling, B. Healy, J. Smylie, and R. Long. 2022. Developing data governance agreements with Indigenous communities in Canada: Toward equitable tuberculosis programming, research, and reconciliation. *Health and Human Rights* 24(1):21–33.

Lukawiecki, J., R. Gagnon, C. Dokis, D. Walters, and L. Molot. 2021. Meaningful engagement with Indigenous peoples: A case study of Ontario's Great Lakes Protection Act. *International Journal of Water Resources Development* 37(4):603–618, https://doi.org/10.1080/07900627.2019.1681261 (accessed May 30, 2024).

Lurgi, M., B. C. López, and J. M. Montoya. 2012. Novel communities from climate change. *Philosophical Transactions of the Royal Society B: Biological Sciences* 367(1605):2913–2922, https://doi.org/10.1098/rstb.2012.0238 (accessed June 12, 2024).

Lynch, J. C., J. R. Meriwether, B. A. McKee, F. Vera-Herrera, and R. R. Twilley. 1989. Recent accretion in mangrove ecosystems based on ^{137}Cs and ^{210}Pb. *Estuaries* 12:284–299, https://doi.org/10.2307/1351907 (accessed June 12, 2024).

Lyver, P. 2002. Use of traditional knowledge by Rakiura Maori to guide sooty shearwater harvests. *Wildlife Society Bulletin* 30(1):29–40, http://dx.doi.org/10.2307/3784632 (accessed May 30, 2024).

Lyver, P., H. Moller, and C. Thomas. 1999. Changes in sooty shearwater Puffinus griseus chick production and harvest precede ENSO events. *Marine Ecology Progress Series* 188:237–248, https://doi.org/10.3354/meps188237 (accessed May 30, 2024).

Lyver, P. O'B., A. Akins, H. Phipps, V. Kahui, D. R. Towns, and H. Moller. 2016. Key biocultural values to guide restoration action and planning in New Zealand. *Research Ecology* 24(3):314–323, https://doi.org.ezproxy.lib.vt.edu/10.1111/rec.12318 (accessed May 30, 2024).

MacDonald-Beyers, K., and R. F. Labisky. 2005. Influence of flood waters on survival, reproduction, and habitat use of white-tailed deer in the Florida Everglades. *Wetlands* 25(3):659–666, https://doi.org/10.1672/0277-5212(2005)025[0659:IOFWOS]2.0.CO;2 (accessed May 28, 2024).

Mangione, T., and J. Spickler. 2022. *Red-cockaded Woodpecker Monitoring Project on Picayune Strand State Forest & Wildlife Management Area*. 2022 Summary Report. Tallahassee, FL: Florida Fish & Wildlife Conservation Commission.

Mantyka-Pringle, C. S., T. G. Martin, and J. R. Rhodes. 2012. Interactions between climate and habitat loss effects on biodiversity: A systematic review and meta-analysis. *Global Change Biology* 18(4):1239–1252, https://doi.org/10.1111/j.1365-2486.2011.02593.x (accessed June 12, 2024).

Margalida, A. 2017. Importance of long-term studies to conservation practice: The case of the bearded vulture in the Pyrenees. In *High Mountain Conservation in a Changing World. Advances in Global Change Research*, Vol 62, edited by J. Catalan, J. Ninot, and M. Aniz. Pp. 343–383. https://doi.org/10.1007/978-3-319-55982-7_15 (accessed May 30, 2024).

Marshall, C., Jr., R. Pielke, Sr., L. Steyaert, and D. Willard. 2004. The impact of anthropogenic land cover change on the Florida peninsula sea breezes and warm season sensible weather. *Monthly Weather Review* 132(1):28–52.

Maul, G. A., and D. M. Martin. 1993. Sea level rise at Key West, Florida, 1846–1992: America's longest instrument record? *Geophysical Research Letters* 20:1955–1958.

Mazzotti, F. J. 1989. Factors affecting the nesting success of the American crocodile, Crocodylus acutus, in Florida Bay. *Bulletin of Marine Science* 44(1):220–228.

McCartney, A. M., J. Anderson, L. Liggins, M. L. Hudson, M. Z. Anderson, B. TeAika, J. Geary, R. Cook-Deegan, H. R. Patel, and A. M. Phillipy. 2022. Balancing openness with Indigenous data sovereignty: An opportunity to leave no one behind in the journey to sequence all of life. *Proceedings of the National Academy of Sciences* 119(4):e2115860119, https://doi.org/10.1073/pnas.2115860119 (accessed May 31, 2024).

McLaughlin, J. F., J. J. Hellmann, C. L. Boggs, and P. R. Ehrlich. 2002. Climate change hastens population extinctions. *Proceedings of the National Academy of Sciences* 99(9):6070–6074, https://doi.org/10.1073/pnas.052131199 (accessed June 12, 2024).

McLean, A. R. 2010. The Conceptual Ecological Model for Everglades Tree Islands. Presentation at the Greater Everglades Restoration Conference 2010, July 2010.

McLean, A. R., K. Jacobs, and J. C. Ogden. 2004. Chapter 7: RECOVER Activities. In *South Florida Water Management District, 2004 Everglades Consolidated Report*. https://apps.sfwmd.gov/sfwmd/SFER/2004_ECR/final/index.html (accessed May 31, 2024).

McPherson, B. F., and R. Halley. 1996. *The South Florida Environment: A Region Under Stress*. USGS Circular 1134. Washington, DC: U.S. Government Printing Office.

Measey, G. J., B. C. Stevenson, T. Scott, R. Altwegg, and D. L. Borchers. 2017. Counting chirps: Acoustic monitoring of cryptic frogs. *Journal of Applied Ecology* 54(3):894–902, https://doi.org/10.1111/1365-2664.12810 (accessed May 28, 2024).

Meeder, J. F., R. W. Parkinson, P. L. Ruiz, and M. S. Ross. 2017. Saltwater encroachment and prediction of future ecosystem response to the Anthropocene Marine Transgression, Southeast Saline Everglades, Florida. *Hydrobiologia* 803:29–48, https://doi.org/10.1007/s10750-017-3359-0 (accessed June 12, 2024).

Meshaka, Jr., W. E., R. Snow, O. L. Bass, and W. B. Robertson. 2002. Occurrence of wildlife on tree islands in the southern Everglades. In *Tree Islands of the Everglades*. Dordrecht, Netherlands: Springer. Pp. 391–427.

Miccosukee Tribe of Indians of Florida. 2021. *Miccosukee Environmental Protection Code Subtitle B: Water Quality Standards for Surface Waters of the Miccosukee Tribe of Indians of Florida*. https://www.epa.gov/sites/default/files/2014-12/documents/miccosukee.pdf (accessed May 31, 2024).

Mo, C., V. Ciuca, and C. Qui. 2022. *Settlement Agreement Report: Fourth Quarter 2021, October-December*. May 2, 2022. https://www.sfwmd.gov/sites/default/files/documents/SA_4th_qtr_2021_report_final.pdf (accessed August 29, 2024).

Mo, C., V. Ciuca, and C. Qui. 2024. *Settlement Agreement Report: Third Quarter 2023, July-September*. February 15, 2024. https://www.sfwmd.gov/sites/default/files/documents/SA_3rd_qtr_2023_report_final.pdf (accessed August 26, 2024).

Moller, H., F. Berkes, P. O. Lyver, and M. Kislalioglu. 2004. Combining science and traditional ecological knowledge: Monitoring populations for co-management. *Ecology and Society* 9(3):2, http://www.ecologyandsociety.org/vol9/iss3/art2 (accessed May 31, 2024).

Moody, P., and C. Brown. 2013. Robustness indicators for evaluation under climate change: Application to the upper Great Lakes. *Water Resources Research* 49(6):3576–3588, https://doi.org/10.1002/wrcr.20228 (accessed June 12, 2024).

Morris, J. T., P. V. Sundareshwar, C. T. Nietch, B. Kjerfve, and D. R. Cahoon. 2002. Responses of coastal wetlands to rising sea level. *Ecology* 83(10):2869–2877.

NASEM (National Academies of Sciences, Engineering, and Medicine). 2016. *Progress Toward Restoring the Everglades: The Sixth Biennial Review—2016*. Washington, DC: The National Academies Press. https://doi.org/10.17226/23672 (accessed May 31, 2024).

NASEM. 2018. *Progress Toward Restoring the Everglades: The Seventh Biennial Review—2018*. Washington, DC: The National Academies Press. https://doi.org/17226/25198 (accessed August 29, 2024).

NASEM. 2021. *Progress Toward Restoring the Everglades: The Eighth Biennial Review—2020*. Washington, DC: The National Academies Press. https://doi.org/17226/25853 (accessed August 29, 2024).

NASEM. 2023. *Progress Toward Restoring the Everglades: The Ninth Biennial Review—2022*. Washington, DC: The National Academies Press. https://doi.org/17226/26706 (accessed August 29, 2024).

NMFS (National Marine Fisheries Service). 2023. *Endangered Species Act – Section 7 Consultation Biological Opinion Implementation of the Lake Okeechobee System Operation Manual (LOSOM)*. Miami, FL: National Oceanic and Atmospheric Administration. https://repository.library.noaa.gov/view/noaa/56367 (accessed June 24, 2024).

NOAA (National Oceanic and Atmospheric Administration). 2017. *Global and Regional Sea Level Rise Scenarios for the United States*. Silver Spring, MD: Center for Operational Oceanographic Products and Services. https://tidesandcurrents.noaa.gov/publications/techrpt83_Global_and_Regional_SLR_Scenarios_for_t he_US_final.pdf (accessed September 2, 2024).

Nocentini, A., J. Redwine, E. Gaiser, T. Hill, S. Hoffman, J. S. Kominoski, J. Sah, D. Shinde, and D. Surratt. 2024. Rehydration of degraded wetlands: Understanding drivers of vegetation community trajectories. *Ecosphere* 15:e4813, https://doi.org/10.1002/ecs2.4813 (accessed June 21, 2024).

NRC (National Research Council). 1996. *Upstream*. Washington, DC: National Academy Press. https://doi.org/10.17226/4976 (accessed September 16, 2024).

NRC. 1999. *New Directions for Water Resources Planning for the U.S. Army Corps of Engineers*. Washington, DC: National Academy Press. https://doi.org/10.17226/6128 (accessed September 16, 2024).

NRC. 2001. *Aquifer Storage and Recovery in the Comprehensive Everglades Restoration Plan: A Critique of the Pilot Projects and Related Plans for ASR in the Lake Okeechobee and Western Hillsboro Areas*. Washington, DC: National Academy Press. https://doi.org/10.17226/10061 (accessed September 16, 2024).

NRC. 2002a. *Florida Bay Research Programs and Their Relation to the Comprehensive Everglades Restoration Plan*. Washington, DC: The National Academies Press. https://doi.org/10.17226/10479 (accessed September 16, 2024).

NRC. 2002b. *Regional Issues in Aquifer Storage and Recovery for Everglades Restoration*. Washington, DC: The National Academies Press. https://doi.org/10.17226/10521 (accessed September 16, 2024).

NRC. 2003a. *Adaptive Monitoring and Assessment for the Comprehensive Everglades Restoration Plan*. Washington, DC: The National Academies Press. https://doi.org/10.17226/10663 (accessed September 16, 2024).

NRC. 2003b. *Does Water Flow Influence Everglades Landscape Patterns?* Washington, DC: The National Academies Press. https://doi.org/10.17226/10758 (accessed September 16, 2024).

NRC. 2003c. *Science and the Greater Everglades Ecosystem Restoration: An Assessment of the Critical Ecosystem Studies Initiative*. Washington, DC: The National Academies Press. https://doi.org/10.17226/10589 (accessed September 16, 2024).

NRC. 2004a. *River Basins and Coastal Systems Planning Within the U.S. Army Corps of Engineers*. Washington, DC: The National Academies Press. https://doi.org/10.17226/10970 (accessed September 16, 2024).

NRC. 2004b. *Adaptive Management for Water Resources Project Planning*. Washington, DC: The National Academies Press. https://doi.org/10.17226/10972 (accessed September 16, 2024).

NRC. 2005. *Re-Engineering Water Storage in the Everglades: Risks and Opportunities*. Washington, DC: The National Academies Press. https://doi.org/10.17226/11215 (accessed September 16, 2024).

NRC. 2007. *Progress Toward Restoring the Everglades: The First Biennial Review—2006*. Washington, DC: The National Academies Press. https://doi.org/10.17226/11754 (accessed September 16, 2024).

NRC. 2008. *Progress Toward Restoring the Everglades: The Second Biennial Review—2008*. Washington, DC: The National Academies Press. https://doi.org/10.17226/12469 (accessed September 16, 2024).

NRC. 2010. *Progress Toward Restoring the Everglades: The Third Biennial Review—2010*. Washington, DC: The National Academies Press. https://doi.org/10.17226/12988 (accessed September 16, 2024).

NRC. 2012. *Progress Toward Restoring the Everglades: The Fourth Biennial Review—2012*. Washington, DC: The National Academies Press. https://doi.org/10.17226/13422 (accessed September 16, 2024).

NRC. 2014. *Progress Toward Restoring the Everglades: The Fifth Biennial Review—2014*. Washington, DC: The National Academies Press. https://doi.org/10.17226/18809 (accessed September 16, 2024).

NRC. 2015. *Review of the Everglades Aquifer Storage and Recovery Regional Study*. Washington, DC: The National Academies Press. https://doi.org/10.17226/21724 (accessed September 16, 2024).

Nungesser, M. K., C. McVoy, Y. Wu, and N. Wang. 2003. Quantifying the current landscape patterns of the Everglades ridge and slough. GEER conference, Palm Harbor, FL, April 12-18. Abstract.

Obeysekera, J., J. Barnes, and M. Nungesser. 2015. Climate sensitivity runs and regional hydrologic modeling for predicting the response of the greater Florida Everglades ecosystem to climate change. *Environmental Management* 55(4):749–762, https://doi.org/10.1007/s00267-014-0315-x (accessed June 21, 2024).

OMB (Office of Management and Budget). 2024. *Budget of the U.S. Government Fiscal Year 2025*. Washington, DC: Office of Management and Budget.

Orem, W. H., C. Gilmour, D. Axelrad, D. P. Krabbenhoft, D. Scheidt, P. I. Kalla, P. McCormick, M. Gabriel, and G. Aiken. 2011. Sulfur in the South Florida ecosystem: Distribution, sources, biogeochemistry, impacts, and management for restoration. *Critical Reviews in Environmental Science and Technology* 41(S1):249–288.

Ornstein, E. R. 2024. Indigenous Knowledge as evidence in federal rule-making. *University of Miami Law Review* 78(2):409, https://repository.law.miami.edu/umlr/vol78/iss2/5 (accessed May 31, 2024).

Osceola, M., Jr. 2019. Letter to Dr. Gretchen Ehlinger. Seminole Tribe Comments Regarding the Lake Okeechobee Watershed Restoration Project's Revised Draft Integrated Project Implementation Report and Environmental Impact Statement.

Osceola, M., Jr. 2020. Letter from Hon. Marcellus Osceola Jr., Chairman Seminole Tribe of Florida, to Col. Andrew Kelly, District Commander. Appendix C.3: Pertinent Correspondence Information. WERP Draft PIR/EIS.

Osceola, M., Jr. 2022. Letter to Col. James L. Booth, District Commander, Jacksonville District: Project Benefits, the Incorporation of ITEK into WERP Planning, Operations, Adaptive Management, and Water Quality Monitoring. Appendix C.3: Draft PIR/EIS.

Osceola, M., Jr. 2023. Letter from Hon. Marcellus Osceola Jr., Chairman Seminole Tribe of Florida, to Hon. Michael L. Connor, Assistant Secretary of the Army for Civil Works. Appendix C.3: Pertinent Correspondence Information. WERP Draft PIR/EIS.

Osland, M. J., B. Chivoiu, N. M. Enwright, K. M. Thorne, G. R. Guntenspergen, J. B. Grace, L. L. Dale, W. Brooks, N. Herold, J. W. Day, F. H. Sklar, and C. M. Swarzenzki. 2022. Migration and transformation of coastal wetlands in response to rising seas. *Science Advances* 8(26), https://doi.org/10.1126/sciadv.abo5174 (accessed June 12, 2024).

Overholt, W. A., M. Rayamajhi, E. Rohrig, S. Hight, F. A. Dray, E. Lake, M. Smith, K. Hibbard, G. P. Bhattarai, K. Bowers, and R. Poffenberger. 2016. Release and distribution of *Lilioceris cheni* (Coleoptera: Chrysomelidae), a biological control agent of air potato (*Dioscorea bulbifera*: Dioscoreaceae), in Florida. *Biocontrol Science and Technology* 26(8):1087–1099, https://doi.org/10.1080/09583157.2016.1185090 (accessed June 21, 2024).

Parrott, M. 2024. Program and Project Update: South Florida Ecosystem Restoration Task Force Working Group/Science Coordination Group. Presentation on March 20, 2024. https://www.evergladesrestoration.gov/s/9_SFWMD-Prog-and-Project-Update_-WGSCG_20March_2024.pdf (accessed September 2, 2024).

Patterson, K., and R. Finck. 1999. *Tree Islands of the WCA3 Aerial Photointerpretation and Trend Analysis Project Summary Report*. Report to the South Florida Water Management District.

Perez, L., J. I. Watling, D. Bucklin, M. Basille, F. J. Mazzotti, S. Romañach, and L. Brandt. 2017. *Climate Changes, Shifting Ranges: Climate Change Effects on Wildlife in the Florida Everglades and Keys*. WEC383, Wildlife Ecology and Conservation Department: University of Florida/IFAS Extension. https://edis.ifas.ufl.edu/publication/UW428 (accessed June 12, 2024).

Perry, W. 2004. Elements of south Florida's comprehensive Everglades restoration plan. *Ecotoxicology* 13:185–193.

Peters, D. P., K. M. Havstad, J. Cushing, C. Tweedie, O. Fuentes, and N. Villanueva-Rosales. 2014. Harnessing the power of big data: Infusing the scientific method with machine learning to transform ecology. *Ecosphere* 5(6):1–15, https://doi.org/10.1890/ES13-00359.1 (accessed June 12, 2024).

Peterson, G., G. A. De Leo, J. J. Hellman, M. A. Janssen, A. Kinzig, J. R. Malcolm, K. L. O'Brien, S. E. Pope, D. S. Rothman, E. Shevliakova, and R. R. T. Tinch. 1997. Uncertainty, climate change, and adaptive management. *Conservation Ecology* 1(2):4, http://www.consecol.org/vol1/iss2/art4 (accessed June 12, 2024).

Phlips, E. J., S. Badylak, N. G. Nelson, and K. E. Havens. 2020. Hurricanes, El Niño and harmful algal blooms in two sub-tropical Florida estuaries: Direct and indirect impacts. *Scientific Reports* 10(1):1910, https://doi.org/10.1038/s41598-020-58771-4 (accessed May 28, 2024).

Poret, N., R. R. Twilley, V. H. Rivera-Monroy, and C. Coronado-Molina. 2007. Below ground decomposition of mangrove roots in Florida coastal everglades. *Estuaries and Coasts* 30:491–496, https://doi.org/10.1007/BF02819395 (accessed August 30, 2024).

Pottier, P., H-Y. Lin, R. R. Y. Oh, P. Pollo, A. N. Rivera-Villanueva, J. O. Valdenito, Y. Yang, T. Amano, S. Burke, S. M. Drobniak, and S. Nakagawa. 2022. A comprehensive database of amphibian heat tolerance. *Scientific Data* 9(1):600, https://doi.org/10.1038/s41597-022-01704-9 (accessed June 12, 2024).

Prior, C. 2013. Permitting problems: Environmental justice and the Miccosukee Indian Tribe. *Environmental and Earth Law Journal* 3(1).

Qiu, C. 2024a. *Settlement Agreement Quarterly Report July-September 2023*. South Florida Water Management District, Technical Oversight Committee. https://www.sfwmd.gov/sites/default/files/documents/SA_TOC_Q3_2023_Presentation_final_0.pdf (accessed May 28, 2024).

Qiu, C. 2024b. *Shark River Slough Final WY202 Annual Compliance Results*. South Florida Water Management District, Technical Oversight Committee. https://www.sfwmd.gov/sites/default/files/documents/SA_TOC_SRS_WY2023_final_Presentation.pdf (accessed August 29, 2024).

Qiu, C., J. Godin, B. Gu, and J. Shaffer. 2018. Appendix 2-4: Annual Permit Report for the C-111 Spreader Canal Phase 1 (Western) Project. *2016 South Florida Environmental Report—Volume III*. http://apps.sfwmd.gov/sfwmd/SFER/2018_sfer_final/v3/appendices/v3_app2-4.pdf (accessed August 29, 2024).

Rainie, S. C., T. Kukutai, M. Walter, O. L. Figueroa-Rodriguez, J. Walker, and P. Alexsson. 2019. Indigenous data sovereignty. In *The State of Open Data: Histories and Horizons*, edited by T. Davies, S. B. Walker, M. Rubinstein, and F. Perini. Cape Town and Ottawa: African Minds and International Development Research Centre. Pp. 300–319.

Ramesh, V., P. Hariharan, V. A. Akshay, P. Choksi, S. Khanwilkar, R. Defries, and V. V. Robin. 2023. Using passive acoustic monitoring to examine the impacts of ecological restoration on faunal biodiversity in the Western Ghats. *Biological Conservation* 282:110071, https://doi.org/10.1016/j.biocon.2023.110071 (accessed May 28, 2024).

Ray, P., S. Wi, A. Schwarz, M. Correa, M. He, and C. Brown. 2020. Vulnerability and risk: Climate change and water supply from California's Central Valley water system. *Climatic Change* 161:177–199, https://doi.org/10.1007/s10584-020-02655-z (accessed June 21, 2024).

Rayamajhi, M. B., E. Rohrig, J. Leidi, C. Kerr, E. Salcedo, R. Poffenberger, M. Smith, E. Lake, F. A. Dray Jr., P. Pratt, P. Tipping, and T. Center. 2019. Herbivory by the biocontrol agent *Lilioceris cheni* suppresses propagule production and smothering ability of the invasive vine *Dioscorea bulbifera*. *Biological Control* 130:1–8, https://doi.org/10.1016/j.biocontrol.2018.12.001 (accessed June 21, 2024).

RECOVER (REstoration COordination and VERification). 2004. *2004 RECOVER Monitoring and Assessment Plan*. Jacksonville, FL: U.S. Army Corps of Engineers.

RECOVER. 2005. *The RECOVER Team's Recommendations for Interim Goals and Interim Targets for the Comprehensive Everglades Restoration Plan*. Jacksonville, FL: U.S. Army Corps of Engineers and West Palm Beach, FL: South Florida Water Management District.

RECOVER. 2006a. *Comprehensive Everglades Restoration Plan Adaptive Management Strategy. Restoration Coordination and Verification*. Jacksonville, FL: U.S. Army Corps of Engineers and West Palm Beach, FL: South Florida Water Management District.

RECOVER. 2006b. *Monitoring and Assessment Plan (MAP) Part 2: 2006 Assessment Strategy for the MAP*. Jacksonville, FL: U.S. Army Corps of Engineers and West Palm Beach, FL: South Florida Water Management District.

RECOVER. 2007. *Development and Application of Comprehensive Everglades Restoration Plan System-wide Performance Measures*. Jacksonville, FL: U.S. Army Corps of Engineers and West Palm Beach, FL: South Florida Water Management District.

RECOVER. 2009. *2009 RECOVER Monitoring and Assessment Plan*. Jacksonville, FL: U.S. Army Corps of Engineers and West Palm Beach, FL: South Florida Water Management District.

RECOVER. 2011. *Adaptive Management Integration Guide: The Comprehensive Everglades Restoration Plan March 2011*. Jacksonville, FL: U.S. Army Corps of Engineers and West Palm Beach, FL: South Florida Water Management District.

RECOVER. 2014. *2014 System Status Report*. Jacksonville, FL: U.S. Army Corps of Engineers and West Palm Beach: South Florida Water Management District. https://usace.contentdm.oclc.org/utils/getfile/collection/p16021coll7/id/8694 (accessed June 17, 2024).

RECOVER. 2015a. *Program-Level Adaptive Management Plan: Comprehensive Everglades Restoration Plan*. Jacksonville, FL: U.S. Army Corps of Engineers and West Palm Beach, FL: South Florida Water Management District. https://usace.contentdm.oclc.org/digital/collection/p16021coll7/id/8184 (accessed June 17, 2024).

RECOVER. 2015b. *RECOVER Southern Coastal Systems Performance Measure: American Crocodile Growth and Survival*. Jacksonville, FL: REstoration COordination and VERification. https://www.saj.usace.army.mil/Portals/44/docs/Environmental/RECOVER/Southern_Coastal_Syste ms_Crocodile_Performance_Measure_Final_102015.pdf?ver=2016-10-26-112403-007 (accessed May 31, 2024).

RECOVER. 2017. *Adaptive Management Recommendations for the BBCW Project During Project Implementation*. RECOVER Memorandum. Jacksonville, FL: U.S. Army Corps of Engineers and West Palm Beach, FL: South Florida Water Management District.

RECOVER. 2019. *2019 Everglades System Status Report: Assessment Period of 2012-2017*. Jacksonville, FL: REstoration COordination and VERification. https://usace.contentdm.oclc.org/utils/getfile/collection/p16021coll7/id/11519 (accessed May 31, 2024).

RECOVER. 2020a. *The RECOVER Team's Recommendations for Revisions to the Interim Goals and Interim Targets for the Comprehensive Everglades Restoration Plan: 2020*. REstoration COordination and VERification. Jacksonville, FL: U.S. Army Corps of Engineers and West Palm Beach, FL: South Florida Water Management District. https://usace.contentdm.oclc.org/utils/getfile/collection/p16021coll7/id/14710.

RECOVER. 2020b. *RECOVER's Consideration of Findings from the Decompartmentalization and Sheetflow Enhancement Physical Model (DPM) Findings for CEPP Presented in Project Memorandum Dated April 3, 2020*. August 24, 2020, Memorandum from RECOVER Executive Committee to CEPP Project Managers. Jacksonville, FL: U.S. Army Corps of Engineers and West Palm Beach, FL: South Florida Water Management District.

RECOVER. 2022a. *Project Delivery Team-RECOVER Project Planning Interaction Point #1: RECOVER Project Management Plan Input*. Jacksonville, FL: U.S. Army Corps of Engineers and West Palm Beach, FL: South Florida Water Management District.

RECOVER. 2022b. *Project Delivery Team-RECOVER Project Planning Interaction Point #2: RECOVER Overview Presentation*. Jacksonville, FL: U.S. Army Corps of Engineers and West Palm Beach, FL: South Florida Water Management District.

RECOVER. 2022c. *Project Delivery Team-RECOVER Project Planning Interaction Point #3: System-wide Modeling Assumptions*. Jacksonville, FL: U.S. Army Corps of Engineers and West Palm Beach, FL: South Florida Water Management District.

RECOVER. 2022d. *Project Delivery Team-RECOVER Project Planning Interaction Point #4: RECOVER Consistency Review of Project-Level Performance Measures*. Jacksonville, FL: U.S. Army Corps of Engineers and West Palm Beach, FL: South Florida Water Management District.

RECOVER. 2022e. *Project Delivery Team-RECOVER Project Planning Interaction Point #5: RECOVER Evaluation of Project Alternative Plans*. Jacksonville, FL: U.S. Army Corps of Engineers and West Palm Beach, FL: South Florida Water Management District.

RECOVER. 2022f. *Project Delivery Team-RECOVER Project Planning Interaction Point #6: RECOVER Review of Monitoring Plan*. Jacksonville, FL: U.S. Army Corps of Engineers and West Palm Beach, FL: South Florida Water Management District.

RECOVER. 2022g. *Project Delivery Team-RECOVER Project Planning Interaction Point #7: RECOVER Input to Draft Operating Manual*. Jacksonville, FL: U.S. Army Corps of Engineers and West Palm Beach, FL: South Florida Water Management District.

RECOVER. 2022h. *Project Delivery Team-RECOVER Project Implementation Interaction Point #1: Scoping Project/RECOVER Communication Begins*. Jacksonville, FL: U.S. Army Corps of Engineers and West Palm Beach, FL: South Florida Water Management District.

RECOVER. 2022i. *Project Delivery Team-RECOVER Project Implementation Interaction Point #2: Design RECOVER Involvement in Preconstruction, Engineering and Design (PED) and/or the Value Engineering (VE) Study*. Jacksonville, FL: U.S. Army Corps of Engineers and West Palm Beach, FL: South Florida Water Management District.

RECOVER. 2022j. *Project Delivery Team-RECOVER Project Implementation Interaction Point #3: Design-Crosswalk with MAP/Review Monitoring Plan*. Jacksonville, FL: U.S. Army Corps of Engineers and West Palm Beach, FL: South Florida Water Management District.

RECOVER. 2022k. *Project Delivery Team-RECOVER Project Implementation Interaction Point #4: Design Review of Analysis for Ecological Monitoring of Operational Testing Phase*. Jacksonville, FL: U.S. Army Corps of Engineers and West Palm Beach, FL: South Florida Water Management District.

RECOVER. 2022l. *Project Delivery Team-RECOVER Project Implementation Interaction Point #5: Construction-Incorporation of Project Monitoring into RECOVER Assessments and Reports*. Jacksonville, FL: U.S. Army Corps of Engineers and West Palm Beach, FL: South Florida Water Management District.

RECOVER. 2022m. *Project Delivery Team-RECOVER Project Implementation Interaction Point #6: Construction-RECOVER and Project-Level Monitoring Data Feeds into Adaptive Management*. Jacksonville, FL: U.S. Army Corps of Engineers and West Palm Beach, FL: South Florida Water Management District.

RECOVER. 2022n. *Project Delivery Team-RECOVER Project Implementation Interaction Point #7: Operations, Maintenance, Repair, Rehabilitation & Replacement (OMRR&R) RECOVER and Project-Level Monitoring Data and Evaluation of Project Benefits Feed into Adaptive Management*. Jacksonville, FL: U.S. Army Corps of Engineers and West Palm Beach, FL: South Florida Water Management District.

RECOVER. 2022o. Miami Canal Engineered Vegetated Hammock Design: Individual Hammocks (S-8 to S-339). Product from the July 2022 RECOVER Topic Workshop. Jacksonville, FL: U.S. Army Corps of Engineers and West Palm Beach, FL: South Florida Water Management District.

RECOVER. 2022p. Project Performance Measure Documentation Sheet: Draft Adaptive Foundational Resilience (AFR). June 7, 2022.

RECOVER. 2023. Performance Measure Documentation: Adaptive Foundational Resilience (AFR). June 5, 2023.

Reddy, K. R., W. F. DeBusk, R. D. DeLaune, and M. S. Koch. 1993. Long-term nutrient accumulation rates in the Everglades. *Soil Science Society of America Journal* 57(4):1147–1155, https://doi.org/10.2136/sssaj1993.03615995005700040044x (accessed June 12, 2024).

Reddy, K. R., C. Armstrong, M. J. Chimney, R. T. James, and J. R. White. 2024. Stormwater Treatment Areas of the Everglades ecosystem: Science and applications-Everglades STAs special issue. *Ecological Engineering* 203:107253, https://doi.org/10.1016/j.ecoleng.2024.107253 (accessed May 28, 2024).

Regan, H. M., M. Colyvan, and M. A. Burgman. 2002. A taxonomy and treatment of uncertainty for ecology and conservation biology. *Ecological Applications* 12(2):618–628, https://doi.org/10.1890/1051-0761(2002)012[0618:ATATOU]2.0.CO;2 (accessed June 12, 2024).

Reo, N. J., K. P. Whyte, D. McGregor, M. A. Smith, and J. F. Jenkins. 2017. Factors that support Indigenous involvement in multi-actor environmental stewardship. *AlterNative: An International Journal of Indigenous Peoples* 13(2):58–68.

Research Data Alliance International Indigenous Data Sovereignty Interest Group. 2019. CARE Principles for Indigenous Data Governance. The Global Indigenous Data Alliance. https://www.gida-global.org/care.

Reséndez, A. 2016. *The Other Slavery: The Uncovered Story of Indian Enslavement in America*. Boston, MA: Mariner Books, Houghton Mifflin Harcourt.

Reynolds, J. 2022. *CEPP New Water Seepage Barrier Wall Project*. https://d3n9y02raazwpg.cloudfront.net/sfwmd/9d1ee7f6-069c-11ed-baa3-0050569183fa-72fe29af-2810-417d-9194-0eba8f-28d3a3-1660319670.pdf (accessed August 29, 2024).

Ribeiro, Jr, J. W., K. Harmon, G. A. Leite, T. N. de Melo, J. LeBien, and M. Campos-Cerqueria. 2022. Passive acoustic monitoring as a tool to investigate the spatial distribution of invasive alien species. *Remote Sensing* 14(18):4565, https://doi.org/10.3390/rs14184565 (accessed May 28, 2024).

Riedlinger, D., and F. Berkes. 2001. Contributions of traditional knowledge to understanding climate change in the Canadian Arctic. *Polar Record* 37(203):315–328.

Rizzardi, K. W. 2001. Translating science into law: Phosphorus standards in the Everglades. *Journal of Land Use and Environmental Law* 17:149–168.

Robinson, J. M., N. Gellie, D. MacCarthy, J. G. Mills, K. O'Donnell, and N. Redvers. 2021. Traditional ecological knowledge in restoration ecology: A call to listen deeply, to engage with, and respect Indigenous voices. *Restoration Ecology* 29(4):e13381, https://doi.org/10.1111/rec.13381 (accessed May 31, 2024).

Romañach, S. S., S. M. Haider, C. Hackett, M. KcKelvy, and L. G. Pearlstine. 2022. Managing multiple species with conflicting needs in the Greater Everglades. *Ecological Indicators* 136:108669, https://doi.org/10.1016/j.ecolind.2022.108669 (accessed June 12, 2024).

Romañach, S. S., S. M. Haider, and A. M. Benscoter. 2023. Sea level rise may pose conservation challenges for the endangered Cape Sable seaside sparrow. *Frontiers in Ecology and Evolution* 10, https://doi.org/10.3389/fevo.2022.1085970 (accessed June 12, 2024).

Rosner, A., R. M. Vogel, and P. H. Kirshen. 2014. A risk-based approach to flood management decisions in a nonstationary world. *Water Resources Research* 50(3):1928–1942, https://doi.org/10.1002/2013WR014561 (accessed June 12, 2024).

Ross, M. S., J. F. Meeder, J. P. Sah, P. L. Ruiz, and G. J. Telesnicki. 2000. The southeast saline Everglades revisited: 50 years of coastal vegetation change. *Journal of Vegetation Science* 11:101–112.

Ruiz, P. L., T. N. Schall, R. B. Shamblin, and K. R. T. Whelan. 2021. *The Vegetation of Everglades National Park*. National Park Service Natural Resource Report NPS/SFCN/NRR—2021/2256. Washington, DC: National Park Service.

Rumpff, L., D. H. Duncan, P. A. Vesk, D. A. Keith, and B. A. Wintle. 2011. State-and-transition modelling for adaptive management of native woodlands. *Biological Conservation* 144(4):1224–1236, https://doi.org/10.1016/j.biocon.2010.10.026 (accessed June 12, 2024).

Ruuskanen, S., B-Y. Hsu, and A. Nord. 2021. Endocrinology of thermoregulation in birds in a changing climate. *Molecular and Cellular Endocrinology* 519:111088, https://doi.org/10.1016/j.mce.2020.111088 (accessed June 12, 2024).

S-333 Working Group. 2023. *Multiagency Recommendations-Phase 1*. Technical Oversight Committee Meeting. https://www.sfwmd.gov/sites/default/files/documents/S333_WG_Recommendations_TOC_Meeting_Dec2023.pdf (accessed September 2, 2024).

Saintilan, N., N. S. Khan, E. Ashe, L. L. Kelleway, K. Rodgers, C. D. Woodroffe, and B. P. Horton. 2020. Thresholds of mangrove survival under rapid sea level rise. *Science* 368(6495):1118–1121, https://doi.org/10.1126/science.aba2656 (accessed June 12, 2024).

Salmón, E. 2000. Kincentric ecology: Indigenous perceptions of the human-nature relationship. *Ecological Applications* 10(5):1327–1332, https://doi.org/10.1890/1051-0761(2000)010[1327:KEIPOT]2.0.CO;2 (accessed May 31, 2024).

Samuel, G. 2020. *Independent Review of the EPBC Act—Interim Report.* Department of Agriculture, Water and the Environment, Canberra. https://epbcactreview.environment.gov.au/resources/interim-report (accessed June 4, 2024).

Saunders, C. (ed.). 2020. Appendix 6-1: Decomp Physical Model research. In *2016 South Florida Environmental Report*, Vol 1. West Palm Beach, FL: South Florida Water Management District.

Saunders, C., and S. Newman. 2023. Presentation to CISRERP on the Decomp Physical Model and Its Potential to Inform CEPP Projects. August 2, 2023.

Saunders, C., S. Newman, E. Tate-Boldt, C. Hansen, J. Harvey, L. Larsen, B. Buskirk, J. Lewis, J. Gomez-Velez, A. Swartz, R. Jaffé, B. Rosen, J. Choi, E. Cline, C. Coronado-Molina, C. Zweig, and F. Sklar. 2016. Appendix 6-1: Interim findings of the Decomp Physical Model. In *2016 South Florida Environmental Report*, Vol. 1. West Palm Beach, FL: South Florida Water Management District.

Saunders, C., S. Newman, C. Zweig, E. Tate-Boldt, D. Y. Lee, L. Jackson, C. Coronado-Molina, J. Redwine, N. Dorn, J. Harvey, M. Manna, and J. Choi. 2021. *Central Everglades Planning Project-South, Adaptive Management Recommendations.*

Schile, L. M., J. C. Callaway, J. T. Morris, D. Stralberg, V. T. Parker, and M. Kelly. 2014. Modeling tidal marsh distribution with sea-level rise: Evaluating the role of vegetation, sediment, and upland habitat in marsh resiliency. *PLoS ONE* 9(2):e88760, https://doi.org/10.1371/journal.pone.0088760 (accessed June 12, 2024).

Scofield, P. 2000. Declines in sooty shearwaters *Puffinus griseus* on their breeding grounds: The effects of bycatch or global warming? *Marine Ornithology* 146.

Scofield, R. P., and D. Christie. 2002. Beach patrol records indicate a substantial decline in sooty shearwater (*Puffinus griseus*) numbers. *Notornis* 49(3):158–165.

SCT (Science Coordination Team). 2003. *The Role of Flow in the Everglades Ridge and Slough Landscape.*

Sena, P. H., T. Gonçalves-Souza, P. H. Gonçalves, P. S. Ferreira, R. A. Gusmao, and F. P. Melo. 2022. Biocultural restoration improves delivery of ecosystem services in social-ecological landscapes. *Restoration Ecology* 30(5):e13599.

SFERTF (South Florida Ecosystem Restoration Task Force). 2000. *Coordinating Success: Strategy for Restoration of the South Florida Ecosystem.*

SFERTF. 2012. *July 2010-June 2012 Strategy and Biennial Report.* https://static1.squarespace.com/static/5d5179e7e42ca1000117872f/t/5ebd8cc6f1bf6b0b891adf0e/1589480660514/2012+Biennial+Report.pdf (accessed August 28, 2024).

SFERTF. 2014. South Florida Ecosystems Restoration Task Force Meeting May 7: Draft Meeting Minutes. Royal Palm Beach, FL: Royal Palm Beach Cultural Center. https://www.evergladesrestoration.gov/task-force-1/may-6-7-2014-task-force-meeting (accessed November 25, 2024).

SFERTF. 2016. South Florida Ecosystems Restoration Task Force Meeting June 29: Draft Meeting Minutes. West Palm Beach, FL. https://static1.squarespace.com/static/5d5179e7e42ca1000117872f/t/5f21c56b509f742614e2da66/159 6048748110/June+29+2016+TF+Meeting+Summary.pdf.

SFERTF. 2023. *2023 Integrated Financial Plan.* Davie, FL: South Florida Ecosystem Restoration Task Force. https://static1.squarespace.com/static/5d5179e7e42ca1000117872f/t/65272df19b18242a1956a892/16 97066492596/2023_IFP.pdf (accessed May 28, 2024).

SFERTF. 2024. *Fiscal Year 2024 Cross Cut Budget.* Davie, FL: South Florida Ecosystem Restoration Task Force. https://www.evergladesrestoration.gov/s/2024-Cross-Cut-Budget.pdf (accessed August 30, 2024).

SFWMD. 2008. *Greater Everglades Performance Measure - Sheet flow in the Everglades Ridge and Slough Landscape.* CERP System-wide Performance Measure Greater Everglades Sheet Flow Documentation Sheet. West Palm Beach, FL: South Florida Water Management District.

SFWMD. 2009. *Settlement Agreement Report Third Quarter July–September 2009.* Prepared for the Technical Oversight Committee. https://www.sfwmd.gov/sites/default/files/documents/sa_rpt_12_08_2009_revised.pdf (accessed August 28, 2024).

SFWMD. 2012. Regional Strategies Regional Water Quality Plan. April 27, 2012. https://www.sfwmd.gov/sites/default/files/documents/rs_waterquality_plan_042712_final.pdf.

SFWMD. 2022. *Technical Memorandum: Adoption of Future Extreme Rainfall Change Factors for Flood Resiliency Planning in South Florida*. https://apps.sfwmd.gov/sfwmd/gsdocs/TPubs/2022_SFWMD_TM_Adoption_of_Future_Extreme_Ra infall_Change_Facotrs_for_Resiliency_Planning_in_South_Florida_rev2.0.pdf (accessed August 28, 2024).

SFWMD. 2024a. *South Florida Environmental Report*. West Palm Beach, FL: South Florida Water Management District. https://www.sfwmd.gov/science-data/scientific-publications-sfer (accessed August 29, 2024).

SFWMD. 2024b. *Lake Okeechobee Storage Reservoir Section 203 Study*. West Palm Beach, FL: South Florida Water Management District. https://www.sfwmd.gov/sites/default/files/documents/2024%20LOCAR%20Section%20203%20Feas ibility%20Study.pdf (accessed May 31, 2024).

Shaffer, S. A., Y. Tremblay, H. Weimerskirch, D. Scott, D. R. Thompson, P. M. Sagar, H. Moller, G. A. Taylor, D. G. Foley, B. A. Block, and D. P. Costa. 2006. Migratory shearwaters integrate oceanic resources across the Pacific Ocean in an endless summer. *Proceedings of the National Academy of Sciences* 103(34):12799–12802.

Sharpe, L. M., M. C. HArwell, and C. A. Jackson. 2021. Integrated stakeholder prioritization criteria for environmental management. *Journal of Environmental Management* 282:111719.

Shelter, Support and Housing Administration. 2019. *Meeting in the Middle: Protocols and Practices for Meaningful Engagement with Indigenous Partners and Communities*. Toronto, Canada: Shelter, Support and Housing Administration.

Shinde, D., L. Pearlstine, L.A. Brandt, F. J. Mazzotti, M. W. Parry, B. Jeffery, and A. LoGalbo. 2014. *Alligator Production Suitability Index Model (GATOR–PSIM v. 2.0): Ecological and Design Documentation*. South Florida Natural Resources Center, Homestead, FL: Everglades National Park. http://npshistory.com/publications/ever/sfnrc/2014-1.pdf (accessed June 12, 2024).

Shiva, V. 2016. *Biopiracy: The Plunder of Nature and Knowledge*. Berkeley, CA: North Atlantic Books.

Shore, J., and J. C. Straus. 1990. The Seminole Water Rights Compact and the Seminole Indian Land Claims Settlement Act of 1987. *Journal of Land Use and Environmental Law* 6:1.

Shuford, R., J. Otero, T. Solaiman, and J. Smith. 2024. Chapter 5A: Restoration Strategies – Design and construction status of water quality improvement projects. In *South Florida Environmental Report*. West Palm Beach, FL: South Florida Water Management District.

Sklar, F. H. (ed.). 2020. Chapter 6: Everglades research and evaluation. In *2020 South Florida Environmental Report*, Vol. 1. West Palm Beach, FL: South Florida Water Management District.

Sklar, F. H., and A. van der Valk. 2002. Tree islands of the Everglades: An overview. In *Tree Islands of the Everglades*. Dordrecht, Netherlands: Springer. Pp. 1–18. https://doi.org/10.1007/978-94-009-0001-1_1 (accessed May 31, 2024).

Sklar, F. H., and A. van der Valk. 2012. *Tree Islands of the Everglades*. Springer Science & Business Media.

Sklar, F. H., M. J. Chimney, S. Newman, P. McCormick, D. Gawlik, S. Miao, C. McVoy, W. Said, J. Newman, C. Coronado, and G. Crozier. 2005. The ecological–societal underpinnings of Everglades restoration. *Frontiers in Ecology and the Environment* 3(3):161–169.

Sklar, F. H., L. Heisler, A. McLean, C. Coronado-Molina, C. Saunders, G. Kiker, and P. Fletcher. 2011. "The" Conceptual Ecological Model for Everglades Tree Islands. Presentation at the 4th National Conference on Ecosystem Restoration, August 1-5, 2011.

Sklar, F. H., C. Carlson, C. Coronado-Molina, and A. C. Moran. 2021. Coastal ecosystem vulnerability and sea level rise (SLR) in south Florida: A mangrove transition projection. *Frontiers in Ecology and Evolution* 9:646083, https://doi.org/10.3389/fevo.2021.646083 (accessed June 12, 2024).

Smith, E. N. 1979. Behavioral and physiological thermoregulation of crocodilians. *American Zoologist* 19(1):239–247, https://doi.org/10.1093/icb/19.1.239 (accessed August 14, 2024).

Smith, M. C. 2022. Chapter 29: Successful Melaleuca biological control in the Florida Everglades. In *Contributions of Classical Biological Control to the U.S. Food Security, Forestry, and Biodiversity*, edited by R. G. Van Driesche, R. L. Winston, T. M. Perring, and V. M. Lopez. FHAAST-2019-05. Morgantown, WV: USDA Forest Service. Pp. 356–366. https://bugwood-cloud.org/resource/files/23194.pdf (accessed May 28, 2024).

Smith, T. J., G. H. Anderson, K. Balentine, G. Tiling, G. A. Ward, and K. R. T. Whelan. 2009. Cumulative impacts of hurricanes on Florida mangrove ecosystems: Sediment deposition, storm surges and vegetation. *Wetlands* 29:24–34, https://doi.org/10.1672/08-40.1 (accessed June 12, 2024).

Smoak, J. M., J. L. Breithaupt, T. J. Smith, and C. J. Sanders. 2013. Sediment accretion and organic carbon burial relative to sea-level rise and storm events in two mangrove forests in Everglades National Park. *CATENA* 104:58–66, https://doi.org/10.1016/j.catena.2012.10.009 (accessed June 12, 2024).

Society for Ecological Restoration International Science & Policy Working Group. 2004. *The SER International Primer on Ecological Restoration*. Tucson, AZ: Society for Ecological Restoration International.

SSG (Science Sub-Group). 1993. *Federal Objectives for the South Florida Restoration by the Science Sub-Group of the South Florida Management and Coordination Working Group*. Miami, FL.

State of Florida. 2023. Fiscal Year 2023-24 Framework for Freedom Budget: Statewide Overview and Taxes https://www.flgov.com/wp-content/uploads/2023/06/FY-23-24-Budget-Highlights-Draft-1.pdf (accessed August 30, 2024).

State of Florida. 2024. Focus on Florida's Future Budget: Overview, Debt Reduction, and Reserves. https://www.flgov.com/wp-content/uploads/2024/06/2024-25-GAA-Highlights.pdf (accessed August 30, 2024).

Statham, P. J. 2012. Nutrients in estuaries—An overview and the potential impacts of climate change. *Science of the Total Environment* 434:213–227, https://doi.org/10.1016/j.scitotenv.2011.09.088 (accessed June 12, 2024).

Steinschneider, S., and C. Brown. 2013. A semiparametric multivariate, multisite weather generator with low-frequency variability for use in climate risk assessments. *Water Resources Research* 49(11):7205–7220.

Steinschneider, S., P. Ray, S. H. Rahat, and J. Kucharski. 2019. A weather-regime-based stochastic weather generator for climate vulnerability assessments of water systems in the western United States. *Water Resources Research* 55(8):6923–6945, https://doi.org/10.1029/2018WR024446 (accessed June 12, 2024).

Stevens, C. E., A. W. Diamond, and T. S. Gabor. 2002. Anuran call surveys on small wetlands in Prince Edward Island, Canada restored by dredging of sediments. *Wetlands* 22:90–99, https://doi.org/10.1672/02775212(2002)022[0090:ACSOSW]2.0.CO;2 (accessed May 28, 2024).

Stith, B. M., J. P. Reid, C. A. Langtimm, E. D. Swain, T. J. Doyle, D. H. Slone, J. D. Decker, and L. E. Soderqvist. 2011. Temperature inverted haloclines provide winter warm-water refugia for manatees in southwest Florida. *Estuaries and Coasts* 34:106–119, https://doi.org/10.1007/s12237-010-9286-1 (accessed June 17, 2024).

Stupariu, M. S., S. A. Cushman, A-I. Pleşoianu, I. Pătru-Stupariu, and C. Fürst. 2022. Machine learning in landscape ecological analysis: A review of recent approaches. *Landscape Ecology* 37(5):1227–1250, https://doi.org/10.1007/s10980-021-01366-9 (accessed June 12, 2024).

Sturtevant, W. C., and J. R. Cattelino 2004. Florida Seminole and Miccosukee. *Handbook of North American Indians (Southeast)* 14:429–449.

Stys, B., T. Foster, M. M. P. B. Fuentes, B. Glazer, K. Karish, N. Montero, and J. S. Reece. 2017. Climate change impacts on Florida's biodiversity and ecology. In *Florida's Climate: Changes, Variations & Impacts,* edited by E. P. Chassignet, J. W. Jones, V. Misra, and J. Obesekera. Florida Climate Institute. Pp. 339–389. https://doi.org/10.17125/fci2017.ch12 (accessed June 12, 2024).

The White House, Office of the Press Secretary. 2022. *White House Releases First-of-a-Kind Indigenous Knowledge Guidance for Federal Agencies* [Press Release]. https://www.whitehouse.gov/ceq/news-updates/2022/12/01/white-house-releases-first-of-a-kind-indigenous-knowledge-guidance-for-federal-agencies (accessed August 28, 2024).

Thompson, K. L., T. C. Lantz, and N. C. Ban. 2020. A review of Indigenous knowledge and participation in environmental monitoring. *Ecology & Society* 25(2).

Thompson, K. L., N. Reece, N. Robinson, H. J. Fisher, N. C. Ban, and C. R. Picard. 2019. "We monitor by living here": Community-driven actualization of a social-ecological monitoring program based in the knowledge of Indigenous harvesters. *Facets* 4(1):293–314.

Thompson, S., M. Vehkaoja, J. Pellikka, and P. Nummi. 2021. Ecosystem services provided by beavers *Castor* spp. *Mammal Review* 51(1):25–39, https://doi.org/10.1111/mam.12220 (accessed May 31, 2024).

Tipping, P. W., M. R. Martin, E. N. Pokorny, K. R. Nimmo, D. L. Fitzgerald, F. A. Dray, Jr., and T. D. Center. 2014. Current levels of suppression of waterhyacinth in Florida USA by classic biological control agents. *Biological Control* 71:65–69, https://doi.org/10.1016/j.biocontrol.2014.01.008 (accessed May 28, 2024).

Tipping, P. W., L. A. Gettys, C. R. Minteer, J. R. Foley, and S. N. Sardes. 2017. Herbivory by biological control agents improves herbicidal control of Waterhyacinth (*Eichhornia crassipes*). *Invasive Plant Science and Management* 10:271–276, https://doi.org/10.1017/inp.2017.30 (accessed May 28, 2024).

Todd, B. D., D. E. Scott, J. H. K. Pechmann, and J. W. Gibbons. 2011. Climate change correlates with rapid delays and advancements in reproductive timing in an amphibian community. *Proceedings of the Royal Society B: Biological Sciences* 278(1715):2191–2197, https://doi.org/10.1098/rspb.2010.1768 (accessed June 12, 2024).

Uhlmann, S. S., and J. M. Jeschke. 2011. Comparing factors associated with total and dead sooty shearwater bycatch in New Zealand trawl fisheries. *Biological Conservation* 144:1859–1865, https://doi.org/10.1016/j.biocon.2011.02.025 (accessed May 31, 2024).

United Nations. 2007. *United Nations Declaration on the Rights of Indigenous Peoples*. New York: United Nations. https://www.un.org/development/desa/indigenouspeoples/wp-content/uploads/sites/19/2018/11/UNDRIP_E_web.pdf (accessed May 31, 2024).

University of Florida Water Institute. 2015. *Options to Reduce High Volume Freshwater Flows to the St. Lucie and Caloosahatchee Estuaries and Move More Water South from Lake Okeechobee to the Southern Everglades*. Gainesville, FL: University of Florida. https://waterinstitute.ufl.edu/faculty/graham/wp-content/uploads/UF-Water-Institute-Final-Report-March-2015.pdf (accessed October 20, 2016).

USACE (U.S. Army Corps of Engineers). 1999. *Engineering and Design for Civil Works Projects. Engineering Regulation 1100-2-1150*. Jacksonville, FL: U.S. Army Corps of Engineers.

USACE. 2000. *Planning Guidance Notebook. Engineering Regulation 1105-2-100*. Jacksonville, FL: U.S. Army Corps of Engineers.

USACE. 2007. *Memorandum for Director of Civil Works on Comprehensive Everglades Restoration Plan, Water Quality Improvements*. Washington, DC: USACE.

USACE. 2009. *CECW-PB. Implementation Guidance for Section 2039 of the Water Resources Development Act of 2007 (WRDA 2007)—Monitoring Ecosystem Restoration*. Memorandum for Commanders. Jacksonville, FL: U.S. Army Corps of Engineers. https://usace.contentdm.oclc.org/digital/api/collection/p16021coll11/id/2925/download (accessed June 17, 2024).

USACE. 2011. *Everglades Restoration Transition Plan Final Environmental Impact Statement*. Jacksonville, FL: U.S. Army Corps of Engineers. https://news.caloosahatchee.org/docs/Ertp_Final_Eisreport_130201.pdf (accessed June 21, 2024).

USACE. 2015. *Picayune Strand Restoration Project: Limited Reevaluation Report and Environmental Assessment*. Jacksonville, FL: U.S. Army Corps of Engineers. https://usace.contentdm.oclc.org/utils/getfile/collection/p16021coll7/id/14224 (accessed June 17, 2024).

USACE. 2019. *Procedures to Evaluate Sea Level Change: Impacts, Responses, and Adaptation. Engineer Pamphlet 1100-2-1*. Jacksonville, FL: U.S. Army Corps of Engineers. https://www.publications.usace.army.mil/Portals/76/Users/182/86/2486/EP-1100-2-1.pdf?ver=2019-09-12- (accessed June 21, 2024).

USACE. 2020a. *Combined Operational Plan Final Environmental Impact Statement*. Jacksonville, FL: U.S. Army Corps of Engineers Jacksonville District. https://usace.contentdm.oclc.org/utils/getfile/collection/p16021coll7/id/15765 (accessed May 28, 2024).

USACE. 2020b. *Central and Southern Florida, Everglades Agricultural Area (EAA), Florida, Everglades Agricultural Area Southern Reservoir and Stormwater Treatment Area, Final Environmental Impact Statement*. Jacksonville, FL: U.S. Army Corps of Engineers. https://usace.contentdm.oclc.org/digital/collection/p16021coll7/id/13245 (accessed June 21, 2024).

USACE. 2020c. *Supplemental Environmental Assessment and Proposed Finding of No Significant Impact. Picayune Strand Restoration Project Southwest Protection Feature, Additional Conveyance Features, and Partial Plugging of the Faka Union Canal.* Jacksonville, FL: U.S. Army Corps of Engineers. https://erdc-library.erdc.dren.mil/jspui/bitstream/11681/36955/1/SEA%20 Picayune%20Strand%20Restoration_20 20.pdf (accessed June 17, 2024).

USACE. 2022a. *Integrated Delivery Schedule (IDS) 2022 Update.* Jacksonville, FL: U.S. Army Corps of Engineers. https://usace.contentdm.oclc.org/utils/getfile/collection/p16021coll11/id/5989 (accessed May 29, 2024).

USACE. 2022b. Draft Environmental Impact Statement, Lake Okeechobee System Operating Manual. https://usace.contentdm.oclc.org/utils/getfile/collection/p16021coll7/id/21271.

USACE. 2022c. *C-111 South Dade Project: Facts and Information.* September 2022. Jacksonville, FL: U.S. Army Corps of Engineers. https://usace.contentdm.oclc.org/utils/getfile/collection/p16021coll11/id/5908 (accessed June 24, 2024).

USACE. 2022d. *Facts and Information: Indian River Lagoon – South.* Jacksonville, FL: U.S. Army Corps of Engineers. https://usace.contentdm.oclc.org/utils/getfile/collection/p16021coll11/id/5921 (accessed June 21, 2024).

USACE. 2023a. *U.S. Army Corps of Engineers-Civil Works Tribal Consultation Policy.* https://api.army.mil/e2/c/downloads/2023/12/06/f10ab368/dec2023-usace-tribal-consultation-policy.pdf (accessed August 31, 2024).

USACE. 2023b. *Facts and Information: Lake Okeechobee Watershed Restoration Project (LOWRP).* Jacksonville, FL: U.S. Army Corps of Engineers. https://usace.contentdm.oclc.org/utils/getfile/collection/p16021coll11/id/6030 (accessed September 1, 2024).

USACE. 2023d. *South Florida Ecosystem Restoration Comprehensive Federal/State of Florida Cost-Share Transparency.* Jacksonville, FL: U.S. Army Corps of Engineers. https://www.evergladesrestoration.gov/s/Cost-Share-Transparency-Report-FY2023.pdf (accessed June 24, 2024).

USACE. 2023e. *Integrated Delivery Schedule (IDS) 2023 Update.* Jacksonville, FL: U.S. Army Corps of Engineers. https://usace.contentdm.oclc.org/utils/getfile/collection/p16021coll11/id/6589 (accessed May 29, 2024).

USACE. 2023f. *Biannual Assessment Report: Report 1, March 2023, Combined Operational Plan (COP).* Jacksonville, FL: U.S. Army Corps of Engineers.

USACE. 2023g. *Biannual Assessment Report: Report 2, October 2023, Combined Operational Plan (COP).* Jacksonville, FL: U.S. Army Corps of Engineers.

USACE. 2024a. *Facts and Information: Central Everglades Planning Project.* Jacksonville, FL: U.S. Army Corps of Engineers. https://usace.contentdm.oclc.org/utils/getfile/collection/p16021coll11/id/6733 (accessed June 21, 2024).

USACE. 2024b. *Final Environmental Impact Statement: Lake Okeechobee System Operating Manual.* Jacksonville, FL: U.S. Army Corps of Engineers. https://usace.contentdm.oclc.org/utils/getfile/collection/p16021coll7/id/25886 (accessed August 29, 2024).

USACE. 2024c. *Final Environmental Impact Statement: North of Lake Okeechobee Storage Reservoir Section 203 Study.* Jacksonville, FL: U.S. Army Corps of Engineers. https://usace.contentdm.oclc.org/utils/getfile/collection/p16021coll7/id/25174 (accessed May 29, 2024).

USACE. 2024d. *Congressional Fact Sheet: CERP C-43 Caloosahatchee West Basin Storage Reservoir.* Jacksonville, FL: U.S. Army Corps of Engineers. https://www.saj.usace.army.mil/About/Congressional-Fact-Sheets-2024/CERP-C-43-Caloosahatchee-West-Basin-Storage-ReservoirC/3- (accessed June 17, 2024).

USACE. 2024e. *Congressional Fact Sheet: CERP Indian River Lagoon-South.* Jacksonville: FL: U.S. Army Corps of Engineers. https://www.saj.usace.army.mil/About/Congressional-Fact-Sheets-2024/CERP-Indian-River-Lagoon-South-C (accessed May 29, 2024).

USACE. 2024f. *CERP Lake Okeechobee Watershed Restoration Project Congressional Fact Sheet.* Jacksonville, FL: U.S. Army Corps of Engineers. https://www.saj.usace.army.mil/About/Congressional-Fact-Sheets-2024/CERP-Lake-Okeechobee-Watershed-C (accessed May 29, 2024).

USACE and DOI. 2020. *2015-2020 MOMENTUM: Report to Congress – Comprehensive Everglades Restoration Plan, Central & Southern Florida Project.* https://issuu.com/usace_saj/docs/final_2020_report_to_congress_on_cerp_progress_hig (accessed August 28, 2024).

USACE, DOI, and the State of Florida. 2007. *Intergovernmental Agreement Among the United States Department of the Army, the United States Department of the Interior, and the State of Florida Establishing Interim Restoration Goals for the Comprehensive Everglades Restoration Plan.*

USACE, ENP (Everglades National Park), and SFWMD. 2023a. *Combined Operational Plan (COP) Biennial Report.* Jacksonville, FL: U.S. Army Corps of Engineers. https://usace.contentdm.oclc.org/utils/getfile/collection/p16021coll7/id/23364 (accessed May 28, 2024).

USACE and SFWMD. 1999. *Central and Southern Florida Project Comprehensive Review Study, Final Integrated Feasibility Report and Programmatic Environmental Impact Statement.* Jacksonville, FL: U.S. Army Corps of Engineers and West Palm Beach, FL: South Florida Water Management District. https://www.sfwmd.gov/sites/default/files/documents/CENTRAL_AND_SOUTHERN_FLORIDA_P ROJECT_COMPREHENSIVE_REVIEW_STUDY.pdf (accessed June 17, 2024).

USACE and SFWMD. 2004. *Comprehensive Everglades Restoration Plan Picayune Strand Restoration (Formerly Southern Golden Gate Estates Ecosystem Restoration) Final Integrated Project Implementation Report and Environmental Impact Statement.* Jacksonville, FL: U.S. Army Corps of Engineers and West Palm Beach, FL: South Florida Water Management District. https://usace.contentdm.oclc.org/utils/getfile/collection/p16021coll7/id/23368 (accessed May 24, 2024).

USACE and SFWMD. 2010. *Caloosahatchee River (C-43) West Basin Storage Reservoir Final Integrated Project Implementation Report and Environmental Impact Statement.* Jacksonville, FL: U.S. Army Corps of Engineers and West Palm Beach, FL: South Florida Water Management District. http://141.232.10.32/pm/projects/docs_04_c43_pir_final.aspx (accessed June 21, 2024).

USACE and SFWMD. 2012. *Broward County Water Preserve Areas: Revised Final Integrated Project Implementation Report and Environmental Impact Statement.* Jacksonville, FL: U.S. Army Corps of Engineers and West Palm Beach, FL: South Florida Water Management District. https://usace.contentdm.oclc.org/utils/getfile/collection/p16021coll7/id/12462 (accessed August 29, 2024).

USACE and SFWMD. 2014. *CERP – Central Everglades Planning Project: Final Integrated Project Implementation Report and Environmental Impact Statement.* Jacksonville, FL: U.S. Army Corps of Engineers and West Palm Beach, FL: South Florida Water Management District. https://usace.contentdm.oclc.org/digital/collection/p16021coll7/id/16157 (accessed June 17, 2024).

USACE and SFWMD. 2018. *CERP Guidance Memorandum 66: RECOVER Assistance to Projects During Implementation.* Jacksonville, FL: U.S. Army Corps of Engineers and West Palm Beach, FL: South Florida Water Management District. https://usace.contentdm.oclc.org/utils/getfile/collection/p16021coll11/id/2245 (accessed June 17, 2024).

USACE and SFWMD. 2019. *Amendment No. 1 to Project Partnership Agreement Between the Department of the Army and South Florida Water Management District for Constructing and Operating, Maintaining, Repairing, Replacing and Rehabilitating the Picayune Strand Restoration Project.* Jacksonville, FL: U.S. Army Corps of Engineers and West Palm Beach, FL: South Florida Water Management District. https://usace.contentdm.oclc.org/utils/getfile/collection/p16021coll7/id/11944 (accessed June 17, 2024).

USACE and SFWMD. 2020a. *Loxahatchee River Watershed Restoration Project: Final Integrated Project Implementation Report and Environmental Impact Statement.* Jacksonville, FL: U.S. Army Corps of Engineers and West Palm Beach, FL: South Florida Water Management District. https://usace.contentdm.oclc.org/utils/getfile/collection/p16021coll7/id/13409 (accessed May 28, 2024).

USACE and SFWMD. 2020b. *Lake Okeechobee Watershed Restoration Project Final Integrated Project Implementation Report and Environmental Impact Statement.* Jacksonville, FL: U.S. Army Corps of Engineers. https://usace.contentdm.oclc.org/utils/getfile/collection/p16021coll7/id/15175 (accessed May 29, 2024).

USACE and SFWMD. 2020c. *Central and Southern Florida Project Comprehensive Everglades Restoration Plan Biscayne Bay Southeastern Everglades Ecosystem Restoration Project Management Plan.* Jacksonville, FL: U.S. Army Corps of Engineers and West Palm Beach, FL: South Florida Water Management District. https://usace.contentdm.oclc.org/utils/getfile/collection/p16021coll7/id/15573 (accessed June 21, 2024).

USACE and SFWMD. 2021. *2021 Aquifer Storage and Recovery Science Plan*. Jacksonville, FL: U.S. Army Corps of Engineers. https://www.sfwmd.gov/sites/default/files/documents/2021_ASR_Science_Plan_Final_062121.pdf (accessed May 28, 2024).

USACE and SFWMD. 2022a. *Lake Okeechobee Watershed Restoration Project Third Revised Draft Integrated Project Implementation Report and Environmental Impact Statement*. Jacksonville, FL: U.S. Army Corps of Engineers. https://usace.contentdm.oclc.org/utils/getfile/collection/p16021coll7/id/20830 (accessed May 29, 2024).

USACE and SFWMD. 2022b. *2022 Aquifer Storage and Recovery Science Plan – Draft October 2022*. Jacksonville, FL: U.S. Army Corps of Engineers. https://www.sfwmd.gov/sites/default/files/2022_ASR_science_plan_draft.pdf (accessed May 29, 2024).

USACE and SFWMD. 2023a. *Western Everglades Restoration Project: Project Implementation Report and Environmental Impact Statement*. Jacksonville, FL: U.S. Army Corps of Engineers.

USACE and SFWMD. 2023b. *Biscayne Bay and Southeastern Everglades Ecosystem Restoration (BB-SEER) – Project Delivery Team Meeting, August 14, 2023*. Jacksonville, FL: U.S. Army Corps of Engineers and West Palm Beach, FL: South Florida Water Management District. https://usace.contentdm.oclc.org/utils/getfile/collection/p16021coll11/id/6498 (accessed June 21, 2024).

USACE and SFWMD. 2023c. Untitled presentation on CEPP North to the CISRERP. December 4, 2023.

USACE and SFWMD. 2024. Central Everglades Planning Project Public Information Meeting. Jacksonville, FL: U.S. Army Corps of Engineers and West Palm Beach, FL: South Florida Water Management District. https://usace.contentdm.oclc.org/utils/getfile/collection/p16021coll11/id/6906 (accessed June 21, 2024).

USACE, SFWMD, and FWS. 2023b. Presentation to CISRERP: Adapting CERP Projects in Light of New Information: Picayune Strand Restoration Project. November 11, 2023.

USGCRP (U.S. Global Change Research Program). 2023. *Fifth National Climate Assessment*, edited by A. R. Crimmins, C. W. Avery, D. R. Easterling, K. E. Kunkel, B. C. Stewart, and T. K. Maycock. Washington, DC: U.S. Global Change Research Program. https://doi.org/10.7930/NCA5.2023 (accessed August 28, 2024).

Vega-Liriano, Z. 2024. Presentation to CISRERP: Lake Okeechobee Watershed Restoration Project. January 10, 2024.

Veit, R., J. A. McGowan, D. G. Ainley, T. R. Wahl, and P. J. Pyle. 1997. Apex marine predator declines ninety percent in association with changing oceanic climate. *Global Change Biology* 3(1):23–28.

Velez, E. 2024. *South Florida Ecosystem Restoration Program: USACE Program Update*, Supplemental Slides. Presentation to the South Florida Ecosystem Restoration Task Force on April 25, 2024. https://static1.squarespace.com/static/5d5179e7e42ca1000117872f/t/66211aeb2e60507cf47817bc/1713445614912/7b_USACE+SFER+Program_Supplemental+Slides.pdf (accessed June 21, 2024).

Visintin, C., N. J. Briscoe, S. N. C. Woolley, P. E. Lentini, R. Tingley, B. A. Wintle, and N. Golding. 2020. STEPS: Software for spatially and temporally explicit population simulations. *Methods in Ecology and Evolution* 11(4):596–603, https://doi.org/10.1111/2041-210X.13354 (accessed June 12, 2024),

Walker, G. S., E. C. Lake, E. Mattison, and G. F. Sutton. 2024. Host range, biology, and climate suitability of *Callopistria exotica*, a potential biological control agent of Old World climbing fern (*Lygodium microphyllum*) in the USA. *Biological Control* 188:105410, https://doi.org/10.1016/j.biocontrol.2023.105410 (accessed May 29, 2024).

Walters, C. J., and C. S. Holling. 1990. Large-scale management experiments and learning by doing. *Ecology* 71(6):2060–2068, https://doi.org/10.2307/1938620 (accessed June 17, 2024).

Walther, G.R., E. Post, P. Convey, A. Menzel, C. Parmesan, T. J. C. Beebee, J-M. Fromentin, O. Hoegh-Guldberg, and F. Bairlein. 2002. Ecological responses to recent climate change. *Nature* 416(6879):389–395, https://doi.org/10.1038/416389a (accessed June 12, 2024).

Wang, Y., and M. Mahmoudi. 2024. Appendix 4-3: Supplementation Information for Other Tributary Basins. *2024 South Florida Environmental Report*, Vol. I. https://apps.sfwmd.gov/sfwmd/SFER/2024_sfer_final/v1/appendices/v1_app4-3.pdf (accessed August 28, 2024).

Wdowinski, S., R. Bray, B. P. Kirtman, and Z. Wu. 2016. Increasing flooding hazard in coastal communities due to rising sea level: Case study of Miami Beach, Florida. *Ocean & Coastal Management* 126:1–8, https://doi.org/10.1016/j.ocecoaman.2016.03.002 (accessed June 12, 2024).

Wetzel, P. R., A. G. van der Valk, S. Newman, D. E. Gawlik, T. T. Gann, C. A. Coronado-Molina, D. L. Childers, and F. H. Sklar. 2005. Maintaining tree islands in the Florida Everglades: Nutrient redistribution is the key. *Frontiers in Ecology and the Environment* 3(7):370–376.

Whateley, S., S. Steinschneider, and C. Brown. 2016. Selecting stochastic climate realizations to efficiently explore a wide range of climate risk to water resource systems. *Journal of Water Resources Planning and Management* 142(6):06016002.

Williams, A. B. 2023. Fired: Caloosahatchee reservoir contractor out; district evaluates next steps. *News-Press*, May 2. https://www.news-press.com/story/news/environment/2023/05/02/fired-caloosahatchee-reservoir-contractor-out-what-to-know-as-project-continues/70173832007 (accessed May 23, 2024).

Williams, B. K., and E. D. Brown. 2014. Adaptive management: From more talk to real action. *Environmental Management* 53:465–479, https://doi.org/10.1007/s00267-013-0205-7 (accessed August 29, 2024).

Williams, L. 2023. Letter from Larry Williams, State Supervisor of Ecological Services, Fish and Wildlife Service, to Hon. Talbert Cypress, Chairman Miccosukee Tribe of Indians of Florida, dated October 20, 2023.

Williams, M. 2024. C-44 reservoir hampered by seepage after pumps turned on in 2021. *WPTV*, April 1. https://www.wptv.com/news/protecting-paradise/c-44-reservoir-hampered-by-seepage-3-years-after-pumps-turned-on (accessed June 17, 2024).

Williams, P. J., and W. L. Kendall. 2017. A guide to multi-objective optimization for ecological problems with an application to cackling goose management. *Ecological Modelling* 343:54–67, https://doi.org/10.1016/j.ecolmodel.2016.10.010 (accessed June 12, 2024).

Williamson, B., S. Provost, and C. Price. 2023. Operationalising Indigenous data sovereignty in environmental research and governance. *Environment and Planning F* 2(1-2):281–304.

Wilsey, C. B., J. J. Lawler, E. P. Maurer, D. McKenzie, P. A. Townsend, R. Gwozdz, J. A. Freund, K. Hagmann, and K. M. Hutten. 2013. Tools for assessing climate impacts on fish and wildlife. *Journal of Fish and Wildlife Management* 4(1):220–241.

Winter, K. B., T. Ticktin, and S. A. Quazi. 2020. Biocultural restoration in Hawai'i also achieves core conservation goals. *Ecology and Society* 25(1):26, https://doi.org/10.5751/ES-11388-250126 (accessed May 31, 2024).

Wolkovich, E. M., B. I. Cook, K. K. McLauchlan, and T. J. Davies. 2014. Temporal ecology in the Anthropocene. *Ecology Letters* 17(11):1365–1379, https://doi.org/10.1111/ele.12353 (accessed June 12, 2024).

Wright, J. P., C. G. Jones, and A. S. Flecker. 2002. An ecosystem engineer, the beaver, increases species richness at the landscape scale. *Oecologia* 132:96–101, https://doi.org/10.1007/s00442-002-0929-1 (accessed May 31, 2024).

Wu, Y., M. K. Nungesser, N. Wang, and C. McVoy. 2003. A tool for measuring landscape changes (ridge and slough) in the Everglades. GEER conference, Palm Harbor, FL, April 12-18. Abstract.

Zedler, J. B., and M. L. Stevens. 2018. Western and traditional ecological knowledge in ecocultural restoration. *San Francisco Estuary and Watershed Science* 16:1–18.

Appendix A

The Restoration Plan in Context

This chapter sets the stage for the tenth of this committee's biennial assessments of restoration progress in the South Florida ecosystem. Background for understanding the project is provided through descriptions of the ecosystem decline, restoration goals, the needs of a restored ecosystem, and the specific activities of the restoration project.

BACKGROUND

The Everglades once encompassed about 3 million acres of slow-moving water and associated biota that stretched from Lake Okeechobee in the north to the Florida Keys in the south (Figures 1-1a and A-1a). The conversion of the Everglades wilderness into an area of high agricultural productivity and cities was a dream of 19th-century investors, and projects begun between 1881 and 1894 affected the flow of water in the watershed north and west of Lake Okeechobee. These early projects included dredging canals in the Kissimmee River Basin and constructing a channel connecting Lake Okeechobee to the Caloosahatchee River and, ultimately, the Gulf of Mexico. By the late 1800s, more than 50,000 acres north and west of the lake had been drained and cleared for agriculture (Grunwald, 2006). In 1907, Governor Napoleon Bonaparte Broward created the Everglades Drainage District to construct a vast array of ditches, canals, dikes, and "improved" channels. By the 1930s, Lake Okeechobee had a second outlet, through the St. Lucie Canal, leading to the Atlantic Ocean, and 440 miles of other canals altered the hydrology of the Everglades (Blake, 1980). After hurricanes in 1926 and 1928 resulted in disastrous flooding from Lake Okeechobee due to failures of the earthen dike that bordered the southern edge of the lake, the U.S. Army Corps of Engineers (USACE) replaced the small berm with the massive Herbert Hoover Dike, which was eventually expanded in the 1960s to encircle the lake. The hydrologic end product of these drainage

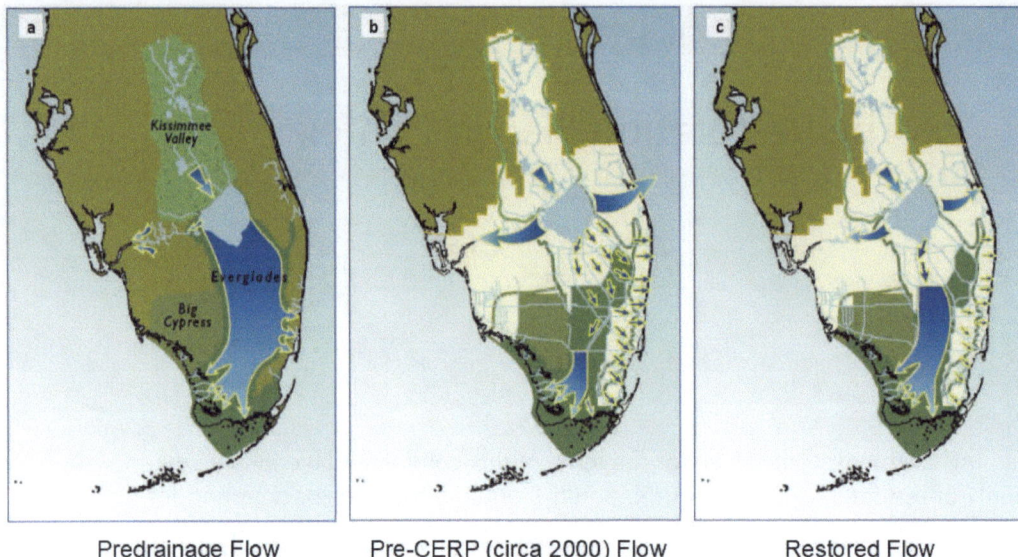

Predrainage Flow   Pre-CERP (circa 2000) Flow   Restored Flow

FIGURE A-1 Water flow in the Everglades under (a) predrainage conditions, (b) pre-CERP (circa 2000) conditions, and (c) conditions envisioned upon completion of the CERP.

NOTES: Restored flow includes less discharge east and west of Lake Okeechobee to the northern estuaries and more flow south into WCA-3 and Everglades National Park compared to pre-CERP conditions. Additionally, the restoration envisioned sheet-flow restoration in WCA-3 rather than conveyance via canals. White areas outline the northern extent of the South Florida ecosystem (see Box 1-1 and Figure 1-2).

SOURCE: Graphics provided by USACE, Jacksonville District.

---

activities was the drastic reduction of natural water storage within the system and an increased susceptibility to drought and desiccation in the southern reaches of the Everglades (NRC, 2005).

After further flooding in 1947 and increasing demands for improved agricultural production and flood management for the expanding population centers on the southeast Florida coast, the U.S. Congress authorized the Central and Southern Florida (C&SF) Project. This project provided flood management and urban and agricultural water supply by straightening 103 miles of the meandering Kissimmee River, expanding the Herbert Hoover Dike, constructing a levee along the eastern boundary of the Everglades to prevent flows into the southeastern urban areas, establishing the 700,000-acre Everglades Agricultural Area (EAA) south of Lake Okeechobee, and creating a series of Water Conservation Areas (WCAs) in the remaining space between the lake and Everglades National Park

(Light and Dineen, 1994). The eastern levee isolated about 100,000 acres of the Everglades ecosystem, making it available for development (Lord, 1993). In total, urban and agricultural development have reduced the Everglades to about one-half its predrainage area (see Figure 1-1b; Davis and Ogden, 1994) and have contaminated its waters with chemicals such as phosphorus, nitrogen, sulfur, mercury, and pesticides. Associated drainage and flood management structures, including the Central and Southern Florida Project, have diverted large quantities of water directly east and west to the northern estuaries, thereby reducing the dominantly southward freshwater flows and natural water storage that defined the ecosystem (see Figure A-1b).

The profound hydrologic alterations were accompanied by many changes in the biotic communities in the ecosystem, including reductions and changes in the composition, distribution, and abundance of the populations of wading birds. Today, the federal government has listed 78 plant and animal species in South Florida as threatened or endangered, with many more included on state lists. Some distinctive Everglades habitats, such as custard apple forests and peripheral wet prairie, have disappeared altogether, while other habitats are severely reduced in area (Davis and Ogden, 1994; Marshall et al., 2004). Approximately 1 million acres are contaminated with mercury from atmospheric deposition (McPherson and Halley, 1996; Orem et al., 2011).

Phosphorus from agricultural runoff has impacted water quality in large portions of the Everglades and has been particularly problematic in Lake Okeechobee (Flaig and Reddy, 1995). The Caloosahatchee and St. Lucie estuaries, including parts of the Indian River Lagoon, have been greatly altered by high and extremely variable freshwater discharges that bring nutrients and contaminants and disrupt salinity regimes (Doering, 1996; Doering and Chamberlain, 1999).

At least as early as the 1920s, private citizens were calling attention to the degradation of the Florida Everglades (Blake, 1980). However, by the time Marjory Stoneman Douglas's classic book *The Everglades: River of Grass* was published in 1947 (the same year that Everglades National Park was dedicated), the South Florida ecosystem had already been altered extensively. Beginning in the 1970s, prompted by concerns about deteriorating conditions in Everglades National Park and other parts of the South Florida ecosystem, the public, as well as the federal and state governments, directed increased attention to the adverse ecological effects of the flood management and irrigation projects (Kiker et al., 2001; Perry, 2004). By the late 1980s it was clear that various minor corrective measures undertaken to remedy the situation were insufficient. As a result, a powerful political consensus developed among federal agencies, Native American Tribes, state agencies and commissions, county governments, and conservation organizations that a large restoration effort was needed in the Everglades (Kiker et al., 2001). This recognition culminated in the Comprehensive Everglades Restoration Plan (CERP), authorized by Congress in 2000, which builds on other ongoing restoration activities of the state and federal governments to create what was at the time the most ambitious restoration effort in the nation's history.

RESTORATION GOALS FOR THE EVERGLADES

Several goals have been articulated for the restoration of the South Florida ecosystem, reflecting the various restoration programs. The South Florida Ecosystem Restoration Task Force (hereafter, simply the Task Force), an intergovernmental body established to facilitate coordination in the restoration effort, has three broad strategic goals: (1) "get the water right"; (2) "restore, preserve, and protect natural habitats and species"; and (3) "foster compatibility of the built and natural systems" (SFERTF, 2000). These goals encompass, but are not limited to, the CERP. The Task Force works to coordinate and build consensus among the many non-CERP restoration initiatives that support these broad goals.

The goal of the CERP, as stated in the Water Resources Development Act of 2000 (WRDA 2000), is "restoration, preservation, and protection of the South Florida Ecosystem while providing for other water-related needs of the region, including water supply and flood protection." The Programmatic Regulations (33 CFR §385.3) that guide implementation of the CERP further clarify this goal by defining restoration as "the recovery and protection of the South Florida ecosystem so that it once again achieves and sustains those essential hydrologic and biological characteristics that defined the undisturbed South Florida ecosystem." These defining characteristics include a large areal extent of interconnected wetlands, extremely low concentrations of nutrients in freshwater wetlands, sheet flow, healthy and productive estuaries, resilient plant communities, and an abundance of native wetland animals (DOI and USACE, 2005). Although development has permanently reduced the areal extent of the Everglades ecosystem, the CERP hopes to recover many of the Everglades' original characteristics and natural ecosystem processes in the remnant Everglades. At the same time, the CERP is charged to maintain levels of flood protection (as of 2000) and was designed to provide for other water-related needs, including urban and agricultural water supply (DOI and USACE, 2005).

Although the CERP contributes to each of the Task Force's three goals, it focuses primarily on restoring the hydrologic features of the undeveloped wetlands remaining in the South Florida ecosystem, on the assumption that improvements in ecological conditions will follow. Originally, "getting the water right" had four components—quality, quantity, timing, and distribution. However, the hydrologic properties of flow, encompassing the concepts of direction, velocity, and discharge, have been recognized as an important component of getting the water right that had previously been overlooked (NRC, 2003b; SCT, 2003). Numerous studies have supported the general approach to getting the water right (Davis and Ogden, 1994; NRC, 2005; SSG, 1993), although it is widely recognized that recovery of the native habitats and species in South Florida may require restoration efforts in addition to getting the water right, such as controlling nonnative species and reversing the decline in the spatial extent and compartmentalization of the natural landscape (SFERTF, 2000; SSG, 1993).

The goal of ecosystem restoration can seldom be the exact re-creation of some historical or preexisting state because physical conditions, driving forces, and boundary conditions usually have changed and are not fully recoverable (e.g., see discussions of climate change and CERP goals in NASEM [2016] and NRC [2014]. Rather, restoration is better viewed as the process of assisting the recovery of a degraded ecosystem to the point where it contains sufficient biotic and abiotic resources to continue its functions without further assistance in the form of energy or other resources from humans (NRC, 1996; Society for Ecological Restoration International Science & Policy Working Group, 2004). The term *ecosystem rehabilitation* may be more appropriate when the objective is to improve conditions in a part of the South Florida ecosystem to at least some minimally acceptable level that allows the restoration of the larger ecosystem to advance. However, flood management remains a critical aspect of the CERP design because improving hydrology and sheet flow in extensive wetland areas has the potential, through seepage, to flood adjacent urban and agricultural areas. Artificial storage will be required to replace the lost natural storage in the system (NRC, 2005), and groundwater management also requires attention to boundaries between developed and natural areas. For these and other reasons, even when the CERP is complete, it will require large inputs of energy and human effort to operate and maintain pumps, stormwater treatment areas (STAs), canals and levees, and reservoirs, and to continue to manage nonnative species. Thus, for the foreseeable future, the CERP does not envision ecosystem restoration or rehabilitation that returns the ecosystem to a state where it can "manage itself."

The broad CERP goals should be interpreted in the context of the complex Everglades ecosystem in order to guide restoration efforts. Early restoration was motivated by ambitious, albeit generalized, expectations for the ecosystem. For example, the CERP conceptual plan, also called the Yellow Book (USACE and SFWMD, 1999), stated, "At all levels in the aquatic food chains, the numbers of such animals as crayfish, minnows, sunfish, frogs, alligators, herons, ibis, and otters, will markedly increase." Currently the systemwide goals for the restoration upon which policymakers agree (USACE et al., 2007) are largely qualitative, indicating a desired direction of change for a number of indicators, without a quantitative objective, providing no clear expectation of how the success of restoration efforts should collectively be assessed. Systemwide ecological indicators with quantitative targets have been established by restoration scientists for assessing restoration success (Brandt et al., 2018; Doren et al., 2009), but these targets have not been endorsed for use in restoration planning. Individual CERP projects have project-specific goals, which are also typically qualitative, and NASEM (2018, 2021) noted the need for quantitative restoration objectives for each project, with accompanying expectations of how and when they will be achieved through management actions, to better support assessment of progress and adaptive management. Continued investment in Everglades restoration proceeds based on improving the current undesirable state of the system rather than toward a specific set of quantitative characteristics desired for the future South Florida ecosystem.

An additional factor challenging the ability of the restoration efforts to meet the "essential hydrologic and biological characteristics that defined the undisturbed South Florida ecosystem" is ongoing climate change, including changes in precipitation patterns, sea-level rise, and ocean warming. Not only have irreversible changes occurred since the 19th century, but also, since the development of the CERP, mean sea levels at Key Largo have risen approximately 11 cm[1] and future projections call for further increases of as much as 2 m in South Florida during the 21st century (NOAA, 2017).

Implicit in the understanding of ecosystem restoration is the recognition that natural systems are self-designing and dynamic; therefore, it is not possible to know in advance exactly what can or will be achieved. Thus, ecosystem restoration proceeds in the face of scientific uncertainty and must consider a range of possible future conditions. NASEM (2016) discusses the challenges to restoration goals arising from major changes that have occurred since the inception of the CERP in 1999, and NASEM (2018) recommended that agencies anticipate and design for the Everglades of the future rather than focusing restoration only on the past Everglades.

What Restoration Requires

Restoring the South Florida ecosystem to a desired ecological landscape requires reestablishment of critical processes that sustain its functioning. Although getting the water right is the oft-stated and immediate goal, the restoration ultimately aims to restore the distinctive characteristics of the historical ecosystem to the remnant Everglades (DOI and USACE, 2005). Getting the water right is a means to that end, not the end itself. The hydrologic and ecological characteristics of the historical Everglades serve as general restoration goals for a functional (albeit reduced in size) Everglades ecosystem. The first Committee

---

[1] See https://tidesandcurrents.noaa.gov/sltrends/sltrends_station.shtml?id=8724580.

on Independent Scientific Review of Everglades Restoration Progress (CISRERP) identified five critical components of Everglades restoration (NRC, 2007):

1. enough water storage capacity combined with operations that allow for appropriate volumes of water to support healthy estuaries and the return of sheet flow through the Everglades ecosystem while meeting other demands for water;
2. mechanisms for delivering and distributing the water to the natural system in a way that resembles historical flow patterns, affecting volume, depth, velocity, direction, distribution, and timing of flows;
3. barriers to eastward seepage of water so that higher water levels can be maintained in parts of the Everglades ecosystem without compromising the current levels of flood protection of developed areas as required by the CERP;
4. methods for securing water quality conditions compatible with restoration goals for a natural system that was inherently extremely nutrient poor, particularly with respect to phosphorus; and
5. retention, improvement, and expansion of the full range of habitats by preventing further losses of critical wetland and estuarine habitats, and by protecting lands that could usefully be part of the restored ecosystem.

NRC (2007) concluded that if these five critical components of restoration are achieved and the difficult problem of invasive species is managed, then the basic physical, chemical, and biological processes that created the historical Everglades can once again work to create and sustain a functional mosaic of biotic communities that resemble what was distinctive about the historical Everglades, albeit of a reduced scale.

The history of the Everglades and ongoing global climate change and sea-level rise will make replication of the predrainage system impossible. Because of the historical changes that have occurred through engineered structures, urban development, introduced species, and other factors, the paths taken by the ecosystem and its components in response to restoration efforts will not retrace the paths taken to reach current conditions. End results will also often differ from the historical system because climate change and sea-level rise, permanently established invasive species, and other factors have moved the ecosystem away from its historical state (Hiers et al., 2012) and will continue to change the restored system in the future. The specific nature and extent of the functional mosaic thus depends on not only the degree to which the five critical components can be achieved but also future precipitation patterns, rising sea levels, marine incursion into estuaries and coastal wetlands, and continued investment in water and ecological management.

Even if the restored system does not exactly replicate the historical system, or reach all the biological, chemical, and physical targets, the reestablishment of natural processes and dynamics should result in a viable and valuable Everglades ecosystem under current conditions. The central principle of ecosystem management is to provide for the natural processes that historically shaped an ecosystem, because ecosystems are characterized by the processes that regulate them. How the reestablished processes interact with future changes within and external to the system will determine the future character of the ecosystem, its species, and communities.

RESTORATION ACTIVITIES

Several restoration programs, including the largest of the initiatives, the CERP, are now under way. The CERP often builds upon non-CERP activities (also called "foundation projects"), many of which are essential to the effectiveness of the CERP. The following section provides a brief overview of the CERP and some of the major non-CERP activities.

Comprehensive Everglades Restoration Plan

WRDA 2000 authorized the CERP as the framework for modifying the Central and Southern Florida Project. Considered a blueprint for the restoration of the South Florida ecosystem, the CERP is led by two organizations with considerable expertise managing the water resources of South Florida— the USACE, which built most of the canals and levees throughout the region, and the South Florida Water Management District (SFWMD), the state agency with primary responsibility for operating and maintaining this complicated water collection and distribution system.

The CERP conceptual plan (or Yellow Book; USACE and SFWMD, 1999) proposes major alterations to the Central and Southern Florida Project in an effort to reverse decades of ecosystem decline. The Yellow Book includes 68 project components to be constructed at an estimated cost of approximately $23.2 billion (in 2019 dollars, including program coordination and monitoring costs; USACE and DOI, 2020; Figure A-2). Major components of the restoration plan focus on restoring the quantity, quality, timing, and distribution of water for the South Florida ecosystem. The Yellow Book outlines the major CERP components, including the following:

- **Conventional surface-water storage reservoirs.** The Yellow Book includes plans for approximately 1.5 million acre-feet (AF) of storage, located north of Lake Okeechobee, in the St. Lucie and Caloosahatchee basins, in the EAA, and in Palm Beach, Broward, and Miami-Dade counties.

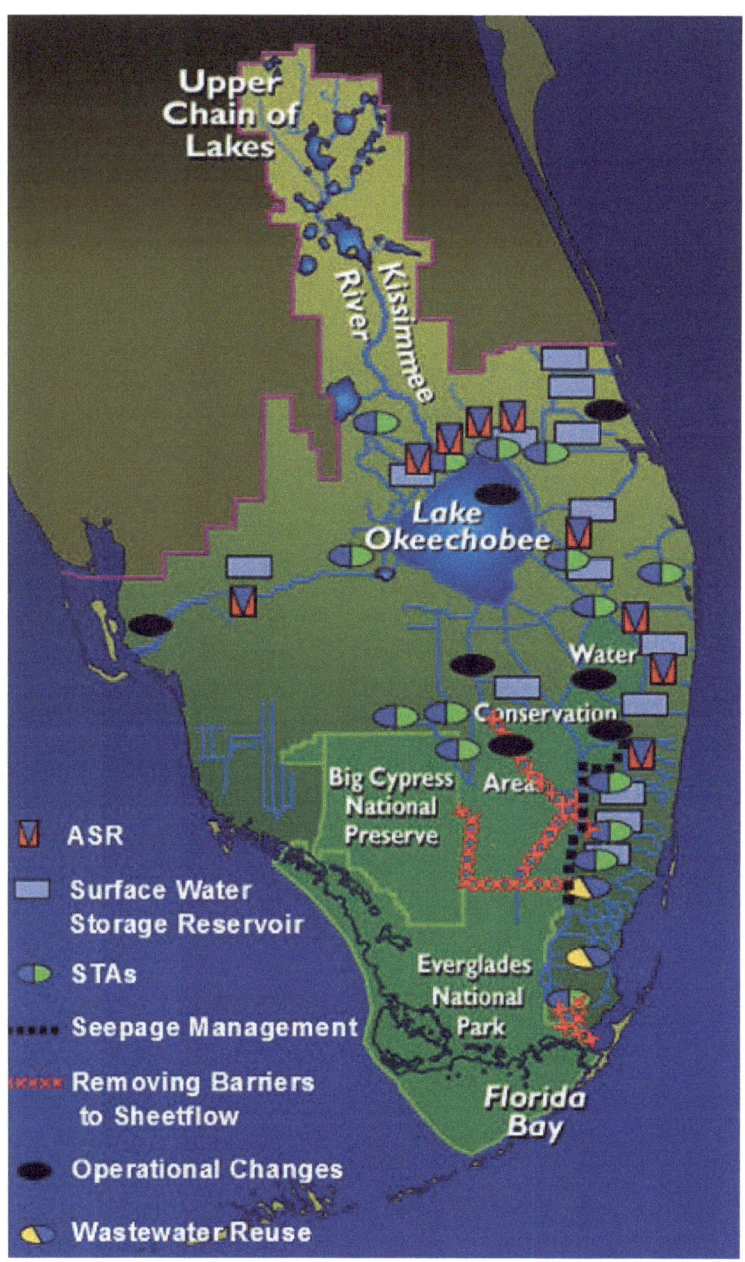

FIGURE A-2 Major project components of the CERP as outlined in 1999.

SOURCE: Courtesy of L. Mahoney, USACE.

- **Aquifer storage and recovery (ASR).** The Yellow Book proposes to provide substantial water storage through ASR, a highly engineered approach that would use a large number of wells built around Lake Okeechobee, in Palm Beach County, and in the Caloosahatchee Basin to store water approximately 1,000 feet below ground.
- **In-ground reservoirs.** The Yellow Book proposes additional water storage in quarries created by rock mining.
- **STAs.** The CERP contains plans for additional constructed wetlands that will treat agricultural and urban runoff water before it enters natural wetlands.[2]
- **Seepage management.** The Yellow Book outlines seepage management projects to prevent unwanted loss of water from the remnant Everglades through levees and groundwater flow. The approaches include adding impermeable barriers to the levees, installing pumps near levees to redirect lost water back into the Everglades, and holding water levels higher in undeveloped areas between the Everglades and the developed lands to the east.
- **Removing barriers to sheet flow.** The CERP includes plans for removing 240 miles of levees and canals to reestablish shallow sheet flow of water through the Everglades ecosystem.
- **Rainfall-driven water management.** The Yellow Book includes operational changes in the water delivery schedules to the WCAs and Everglades National Park to mimic more natural patterns of water delivery and flow through the system.
- **Water reuse and conservation.** To address shortfalls in water supply, the Yellow Book proposes two advanced wastewater treatment plants so that the reclaimed water could be discharged to wetlands along Biscayne Bay or used to recharge the Biscayne aquifer.

The largest portion of the budget is devoted to storage projects and to acquiring the lands needed for them. Implementation progress and documented natural system responses are discussed in Chapter 2.

---

[2] Although some STAs are included among CERP projects, the USACE has clarified its policy on federal cost sharing for water quality features. A memo from the Assistant Secretary of the Army (Civil Works) (USACE, 2007) states, "Before there can be a Federal interest to cost share a WQ [water quality] improvement feature, the State must be in compliance with WQ standards for the current use of the water to be affected and the work proposed must be deemed essential to the Everglades restoration effort." The memo goes on to state, "[T]he Yellow Book specifically envisioned that the State would be responsible for meeting water quality standards." However, the Secretary of the Army can recommend to Congress that project features deemed "essential to Everglades restoration" be cost shared. In such cases, the state is responsible for 100 percent of the costs to treat water to state standards for its current use, and federal cost sharing is determined based on the additional treatment needed to meet the requirements of Everglades restoration (K. Taplin, USACE, personal communication, 2018).

The modifications to the Central and Southern Florida Project embodied in the CERP were originally expected to take more than three decades to complete (and will likely now take much longer), and to be effective they require a clear strategy for managing and coordinating restoration efforts. The Everglades Programmatic Regulations (33 CFR §385) state that decisions on CERP implementation are made by the USACE and the SFWMD (or any other local project sponsors), in consultation with the U.S. Department of the Interior, the Environmental Protection Agency (EPA), the Department of Commerce, the Miccosukee Tribe of Indians of Florida, the Seminole Tribe of Florida, the Florida Department of Environmental Protection, and other federal, state, and local agencies (33 CFR §385).

WRDA 2000 endorses the use of an adaptive management framework for the restoration process, and the Programmatic Regulations (33 CFR §385.31[a]) formally establish an adaptive management program that will

> assess responses of the South Florida ecosystem to implementation of the Plan; . . . [and] seek continuous improvement of the Plan based upon new information resulting from changed or unforeseen circumstances, new scientific and technical information, new or updated modeling; information developed through the assessment principles contained in the Plan; and future authorized changes to the Plan. . . .

An interagency body called Restoration, Coordination, and Verification (RECOVER; discussed further in Chapter 5) was established early in the development of the CERP to ensure that sound science is used in the restoration. The RECOVER leadership group oversees the monitoring and assessment program that will evaluate the progress of the CERP toward restoring the natural system and will assess the need for changes to the plan through the adaptive management process.

Non-CERP Restoration Activities

When Congress authorized the CERP in WRDA 2000, the SFWMD, the USACE, the National Park Service, and the U.S. Fish and Wildlife Service were already implementing several activities intended to restore key aspects of the Everglades ecosystem, which are critical to the overall restoration progress. In fact, the CERP's effectiveness was predicated upon the completion of many of these projects, which include Modified Water Deliveries to Everglades National Park, C-111 South Dade, and state water quality treatment projects developed under the Everglades Construction Project (see Figure A-3). Additional restoration efforts have also been launched outside of the CERP since that time, including state efforts to improve the quality of water flowing into the remnant Everglades under the Restoration Strategies program (see Chapter 3). Recent progress on key non-CERP projects with critical linkages to the CERP are described in Chapters 2 and 3.

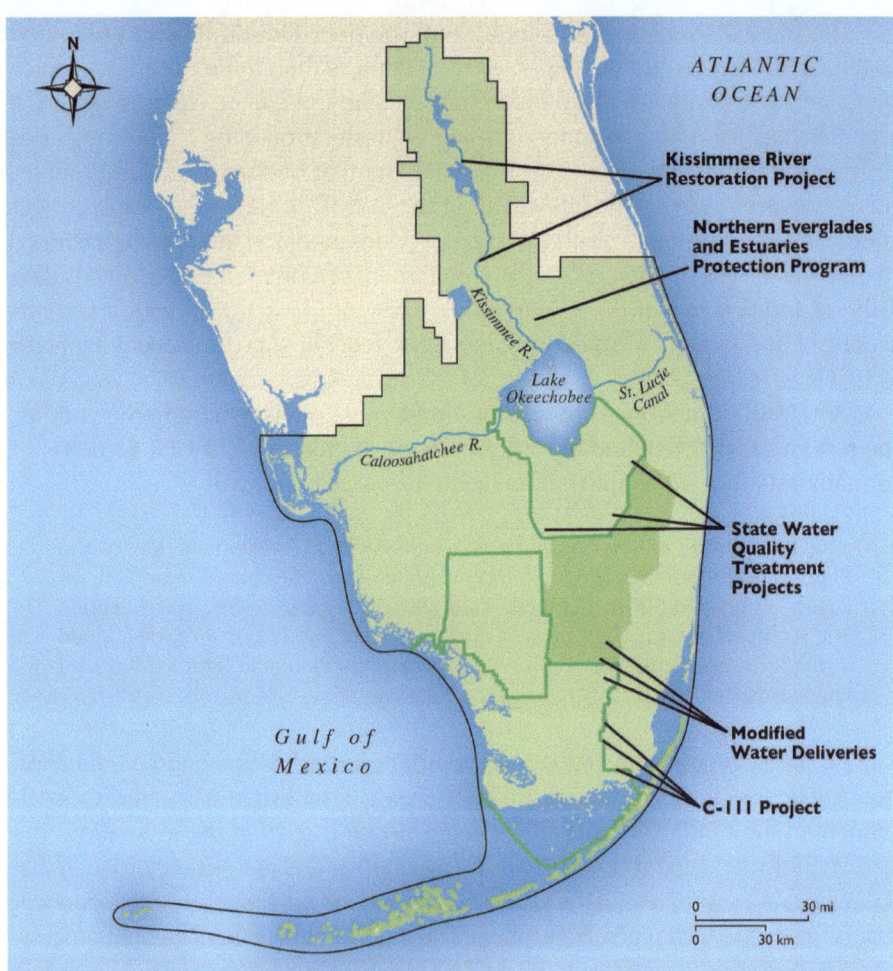

FIGURE A-3 Locations of major non-CERP initiatives.

SOURCE: Map by International Mapping.

Major Developments and Changing Context Since 2000

Several major program-level developments have occurred since the CERP was launched that have affected the pace and focus of CERP efforts. In 2004, Florida launched Acceler8, a plan to hasten the pace of project implementation that was bogged down by the slow federal planning process (for further discussion of Acceler8, see NRC, 2007). Acceler8 originally included 11 CERP project components and 1 non-CERP project, and although the state was unable to complete all the original tasks, the program led to increased state investment and expedited project construction timelines for several CERP projects.

Operation of Lake Okeechobee has been modified twice since the CERP was developed in ways that have reduced total storage. In April 2000, the Water Supply and Environment regulation schedule was implemented to reduce high-water impacts on the lake's littoral zone and harmful high discharges to the St. Lucie and Caloosahatchee estuaries. The regulation schedule was changed again in 2008 to reduce the risk of failure of the Herbert Hoover Dike until the USACE could make critical repairs. The 2008 changes to the regulation schedule resulted in a loss of 564,000 AF of potential storage from the regional system (see NASEM, 2016). With the 2023 completion of the Herbert Hoover Dike rehabilitation efforts and several CERP projects, a revised Lake Okeechobee regulation schedule was developed and implemented in 2024 (see Chapter 2).

In the years since the CERP was launched, the State of Florida has increasingly encouraged the use of alternative water supplies—including wastewater, stormwater, and excess surface water—to meet future water demands (e.g., FDEP, 2015). In 2006, the SFWMD passed the Lower East Coast Regional Water Availability Rule, which caps groundwater withdrawals at 2006 levels, requiring urban areas to meet increased demand through a combination of conservation and alternative water supplies. In 2007, the Florida legislature mandated that ocean wastewater discharges in South Florida be eliminated and 60 percent of those discharges be reused by 2025 (Fla. Stat. §403.086[9]), representing approximately 180 million gallons per day (MGD) of new water supply for the Lower East Coast. As of 2018, South Florida utilities achieved 9.8 MGD in potable water use offsets through additional water reuse under this mandate, although Miami Dade County, the largest discharger, stated that "it cannot technically, environmentally, and economically meet the statutorily required reuse" (FDEP, 2020). It remains unclear whether or how these initiatives and mandates will affect the expectations for agricultural and urban water supply from the CERP, particularly because the capture of excess surface water is a key element of the CERP.

In 2010, the EPA issued its court-ordered Amended Determination, which directed the State of Florida to correct deficiencies in meeting the narrative and numeric nutrient criteria in the Everglades Protection Area (EPA, 2010). In 2012, the State of Florida launched its Restoration Strategies Regional Water Quality Plan, which was approved by the EPA and the Court as an alternative means to address the Amended Determination. The State of Florida is currently constructing approximately 6,500 acres of new STAs and three flow equalization basins (116,000 AF; see Chapter 3). These water quality treatment improvements are designed so that water leaving the STAs will meet a new water quality–based effluent limit (WQBEL) to comply with the 10 parts per billion (ppb) total phosphorus water quality criterion for the Everglades Protection Area (see Chapter 3).[3]

Changing Understanding of Restoration Challenges

Much new knowledge has been gained since the launch of the CERP that enhances understanding of restoration challenges and opportunities and informs future restoration planning and management. Of the many advances in knowledge since 1999, climate change and sea-level rise are among the most significant. As outlined in NASEM (2016), changes in precipitation and evapotranspiration are expected to have substantial impacts on CERP outcomes. Downscaled precipitation projections remain uncertain and range from modest increases to sizable decreases for South Florida, and research continues locally and nationally to improve these projections. Sea-level rise is already affecting the distribution of Everglades habitats and causing coastal flooding in some low-lying urban areas. CERP planners are now evaluating all future restoration benefits in the context of low, medium, and high sea-level rise projections, although recent CISRERP reports (NASEM, 2016, 2018; NRC, 2014) have recommended greater consideration of climate change and sea-level rise in CERP project and program planning.

Since the CERP was developed, the significance of invasive species management for the success of restoration has also been recognized by the Task Force and its member agencies.[4] Nonnative species constitute a substantial proportion of the current biota of the Everglades. The approximately 250 nonnative plant species comprise about 16 percent of the regional flora (see NRC, 2014). South Florida has a subtropical climate with habitats that are similar to those from which many of the invaders originate, with relatively few native species in many taxa to compete with introduced ones. Some species, especially of introduced vascular plants and reptiles, have had dramatic effects on the structure and functioning of Everglades ecosystems, and therefore necessitate aggressive management and early detection of new high-risk invaders to ensure that ongoing CERP efforts to get the water right allow native species to prosper instead of simply enhancing conditions for invasive species.

---

[3] The WQBEL is a numeric discharge limit used to regulate permitted discharges from the STAs so as to not exceed a long-term geometric mean of 10 ppb within the Everglades Protection Area. This numeric value is now translated into a flow-weighted mean (FWM) total phosphorus (TP) concentration and applied to each STA discharge point, which now must meet the following: the STAs are in compliance with WQBEL when the TP concentration of STA discharge point does not exceed an (1) annual FWM of 13 ppb in more than 3 out of 5 years and (2) annual FWM of 19 ppb in any water year (EPA, 2010; FDEP, 2017).

[4] See http://www.evergladesrestoration.gov/content/ies.

SUMMARY

The Everglades ecosystem is one of the world's ecological treasures, but for more than a century the installation of an extensive water management infrastructure has changed the geography of South Florida and has facilitated extensive agricultural and urban development. These changes have had profound ancillary effects on regional hydrology, vegetation, and wildlife populations. The CERP, a joint effort led by the state and federal governments and launched in 2000, seeks to reverse the general decline of the ecosystem. Since 2000, the legal context for the CERP and other major Everglades restoration efforts has evolved and the scientific understanding of Everglades restoration and its current and future stressors has expanded, and the programs continue to adapt. Implementation progress is discussed in detail in Chapter 2.

Appendix B

The National Academies of Sciences, Engineering, and Medicine Everglades Reports

This report represents the 18th report by the National Academies of Sciences, Engineering, and Medicine on Everglades restoration. This appendix recaps key findings of the previous reports.

Progress Toward Restoring the Everglades: The Ninth Biennial Review, 2022 (2023)

The 2022 report noted that record funding levels had accelerated Comprehensive Everglades Restoration Plan (CERP) implementation to a remarkable pace, and the Combined Operational Plan had fostered larger-scale restoration of the natural system. The report stated that ecosystem responses were evident over large areas of the central and western Everglades after implementation of recent restoration initiatives. This progress in implementation has increased the importance of analyzing and synthesizing natural system responses. The report also highlighted two primary challenges that deserve additional attention moving forward—water quality and climate change. Specifically, the committee highlighted the need for rigorous scientific support for water quality improvement in stormwater treatment areas (STAs). Particularly concerning phosphorus mobility and removal processes, the committee recommended advances in STA management, as well as further analyses of nutrient dynamics along STA flow-ways and cell-by-cell monitoring. To help account for effects of a changing climate, the committee recommended improved modeling for a wider range of plausible climate conditions accounting for changes in coastal wetland and estuarine salinity changes. The 2022 Biennial Review further recommended the development of a multi-agency Everglades restoration science plan to ensure that the needed tools, research, analysis, and synthesis are available to support critical restoration management decisions.

Progress Toward Restoring the Everglades: The Eighth Biennial Review, 2020
(2021)

The 2020 report highlighted signs of restoration progress that were evident from multiple CERP projects, buoyed by increased funding that expedited the pace of project construction. However, the committee noted that assessments of restoration progress were stymied by a lack of monitoring, analysis, and communication of restoration benefits. For the 2020 report, the committee reviewed the recently developed Combined Operational Plan and examined issues facing the northern and southern estuaries, including priorities for science to support restoration decision making.

With several projects nearing completion, the 2020 report noted that the CERP was pivoting from a focus primarily on project planning and construction toward an expanding emphasis on operational decisions, evaluation of restoration success, adaptive management, and learning. This transition requires a strong organizational foundation for science, systematic monitoring and assessment, effective communication, and new strategies to support decision making. From this analysis, key principles emerged that are relevant across different projects and regional contexts. First, effective monitoring and ongoing data analysis are critical to support assessments of restoration progress, learning, and adaptive management. Synthesis, improved integration of modeling and monitoring, and enhanced applications of modeling tools can be used to turn available information into better understanding to evaluate trade-offs and strengthen decision making. Finally, strong science leadership and appropriate staffing are key elements of an organizational infrastructure to maximize learning and to support more nimble decision making. Investments in the science and decision-making infrastructure for the CERP would improve the value of information developed through monitoring, modeling, and synthesis and would lead to more effective restoration outcomes.

Progress Toward Restoring the Everglades: The Seventh Biennial Review, 2018
(2018)

In the 2018 report, the committee noted that a vision for CERP storage, at least in the northern portion of the system, was now becoming clear, although the future storage to be provided by Lake Okeechobee remains unresolved. The committee concluded that documentation and analysis of incremental restoration benefits from project implementation to date have been inadequate, primarily because of limitations in project-level monitoring and assessment efforts. Improvements to the monitoring and assessment program, at both project and systemwide scales, were recommended to increase the usefulness of monitoring data for CERP decision makers. The report also recommended a mid-course assessment that analyzes projected CERP outcomes in the context of future stressors. Rather than continuing its primary focus on restoring predrainage conditions and basing decisions on the ability to achieve those conditions under contemporary climate (1965–2005), the report recommended that the CERP emphasize restoration focused on the future of the South Florida ecosystem and build upon the accumulating knowledge base to support successful implementation of this program. This effort requires an integrated assessment of the performance of planned CERP projects under future climate and sea level–rise scenarios and other stressors. With seven large projects authorized and awaiting appropriations for construction and three additional projects nearing the end of their planning processes, the report stated that the time is right for a mid-course assessment. This information could then inform robust decisions about future planning, funding, sequencing, and adaptive management. Implementing a restoration program that is resilient to future conditions also requires a science program that can bring the latest information and tools into CERP planning and implementation.

Progress Toward Restoring the Everglades: The Sixth Biennial Review, 2016 (2016)

The 2016 biennial report found that, 16 years into the CERP, completed components of the plan are beginning to show ecosystem benefits, but the committee had several concerns regarding progress. There has been insufficient attention to refining long-term systemwide goals and objectives and the need to adapt the CERP to radically changing system and planning constraints. It now is known that the natural system was historically much wetter than previously assumed, bringing into question some of the hydrologic goals embedded in the CERP. Sea-level rise will reduce the footprint of the system, temperature and evaporative water losses will increase, rainfall may become more variable, and more storage will likely be needed to accommodate future increases or decreases in the quantity and intensity of runoff.

Review of the Everglades Aquifer Storage and Recovery Regional Study (2015)

The Florida Everglades is a large and diverse aquatic ecosystem that has been greatly altered over the past century by an extensive water control infrastructure designed to increase agricultural and urban economic productivity. The CERP, launched in 2000, is a joint effort led by the state and federal government to reverse the decline of the ecosystem. Increasing water storage is a critical component of the restoration, and the CERP included projects that would drill more than 330 aquifer storage and recovery (ASR) wells to store up to 1.65 billion gallons per day in porous and permeable units in the aquifer system during wet periods for recovery during seasonal or longer-term dry periods.

To address uncertainties regarding regional effects of large-scale ASR implementation in the Everglades, the U.S. Army Corps of Engineers (USACE) and the South Florida Water Management District (SFWMD) conducted an 11-year ASR Regional Study, with focus on the hydrogeology of the Floridan aquifer system, water quality changes during aquifer storage, possible ecological risks posed by recovered water, and the regional capacity for ASR implementation. At the request of the USACE, this report reviewed the ASR Regional Study Technical Data Report and assessed progress in reducing uncertainties related to full-scale CERP ASR implementation. This report considered the validity of the data collection and interpretation methods, integration of studies, evaluation of scaling from pilot- to regional-scale application of ASR, and the adequacy and reliability of the study as a basis for future applications of ASR.

Progress Toward Restoring the Everglades: The Fifth Biennial Review, 2014 (2014)

This report is the fifth biennial evaluation of progress being made in the CERP. Despite exceptional project planning accomplishments, over the past 2 years progress toward restoring the Everglades has been slowed by frustrating financial and procedural constraints. The Central Everglades Planning Project is an impressive strategy to accelerate Everglades restoration and avert further degradation by increasing water flow to the ecosystem. However, timely authorization, funding, and creative policy and implementation strategies will be essential to realize important near-term restoration benefits. At the same time, climate change and the invasion of nonnative plant and animal species further challenge the Everglades ecosystem. The impacts of changing climate—especially sea-level rise—add urgency to restoration efforts to make the Everglades more resilient to changing conditions.

Progress Toward Restoring the Everglades: The Fourth Biennial Review, 2012 (2012)

The 2012 report found that, 12 years into the CERP, little progress has been made in restoring the core of the remaining Everglades ecosystem; instead, most project construction so far has occurred along its periphery. To reverse ongoing ecosystem declines, it will be necessary to expedite restoration projects that target the central Everglades and to improve both the quality and quantity of the water in the ecosystem. The Central Everglades Planning Project offers an innovative approach to this challenge, although additional analyses are needed at the interface of water quality and water quantity to maximize restoration benefits within existing legal constraints.

Progress Toward Restoring the Everglades: The Third Biennial Review, 2010 (2010)

The 2010 report found that while natural system restoration progress from the CERP remains slow, in the past 2 years, there have been noteworthy improvements in the pace of implementation and in the relationship between the federal and state partners. Continued public support and political commitment to long-term funding will be needed for the restoration plan to be completed. The science program continues to address important issues, but more transparent mechanisms for integrating science into decision making are needed. Despite such progress, several important challenges related to water quality and water quantity have become increasingly clear, highlighting the difficulty of achieving restoration goals simultaneously for all ecosystem components. Achieving these goals will be enormously costly and will take decades at least. Rigorous scientific analyses of potential conflicts among the hydrologic requirements of Everglades landscape features and species, and the trade-offs between water quality and quantity, considering timescales of reversibility, are needed to inform future prioritization and funding decisions. Understanding and communicating these trade-offs to stakeholders are critical.

Progress Toward Restoring the Everglades: The Second Biennial Review, 2008 (2008)

The report concluded that budgeting, planning, and procedural matters are hindering a federal and state effort to restore the Florida Everglades ecosystem, which is making only scant progress toward achieving its goals. Good science has been developed to support restoration efforts, but future progress is likely to be limited by the availability of funding and current authorization mechanisms. Despite the accomplishments that lay the foundation for CERP construction, no CERP projects have been completed to date. To begin reversing decades of decline, managers should address complex planning issues and move forward with projects that have the most potential to restore the natural ecosystem.

Progress Toward Restoring the Everglades: The First Biennial Review, 2006 (2007)

This report is the first in a congressionally mandated series of biennial evaluations of the progress being made by the CERP. The report found that progress has been made in developing the scientific basis and management structures needed to support a massive effort to restore the Florida Everglades ecosystem.

However, some important projects have been delayed due to several factors, including budgetary restrictions and a project planning process that can be stalled by unresolved scientific uncertainties. The report outlined an alternative approach that can help the initiative move forward even as it resolves remaining scientific uncertainties. The report called for a boost in the rate of federal spending if the restoration of Everglades National Park and other projects are to be completed on schedule.

Re-Engineering Water Storage in the Everglades: Risks and Opportunities (2005)

Human settlements and flood control structures have significantly reduced the Everglades, which once encompassed more than 3 million acres of slow-moving water enriched by a diverse biota. The CERP was formulated in 1999 with the goal of restoring the original hydrologic conditions of the remaining Everglades. A major feature of this plan is providing enough storage capacity to meet human and ecological needs. This report reviewed and evaluated not only storage options included in the plan but also other options not considered in the plan. Along with providing hydrologic and ecological analyses of the size, location, and functioning of water storage components, the report also discussed and made recommendations on related critical factors, such as timing of land acquisition, intermediate states of restoration, and trade-offs among competing goals and ecosystem objectives.

The CERP imposes some constraints on sequencing of its components. The report concluded that two criteria are most important in deciding how to sequence components of such a restoration project: (1) protecting against additional habitat loss by acquiring or protecting critical lands in and around the Everglades and (2) providing ecological benefits as early as possible.

There is a considerable range in the degree to which various proposed storage components involve complex design and construction measures, rely on active controls and frequent equipment maintenance, and require fossil fuels or other energy sources for operation. The report recommended that, to the extent possible, the CERP should develop storage components that have fewer of those requirements and are thus less vulnerable to failure and more likely to be sustainable in the long term.

Furthermore, as new information becomes available and as the effectiveness and feasibility of various restoration components become clearer, some of the earlier adaptation and compromises might need to be revisited. The report recommended that methods be developed to allow for assessment of trade-offs over broad spatial and long temporal scales, especially for the entire ecosystem, and gives an example of what an overall performance indicator for the Everglades system might look like.

Adaptive Monitoring and Assessment for the Comprehensive Everglades Restoration Plan (2003)

A key premise of the CERP is that restoring the historical hydrologic regime in the remaining wetlands will reverse declines in many native species and biological communities. Given the uncertainties that will attend future responses of Everglades ecosystems to restored water regimes, a research, monitoring, and adaptive management program is planned. This report assessed the extent to which the restoration effort's "monitoring and assessment plan" included the following elements crucial to any adaptive management scheme: (1) clear restoration goals and targets, (2) a sound baseline description and conceptualization of the system, (3) an effective process for learning from management actions, and (4) feedback mechanisms for improving management based on the learning process.

The report concluded that monitoring needs must be prioritized, because many goals and targets that have been agreed to may not be achievable or internally consistent. Priorities could be established based on the degree of flexibility or reversibility of a component and its potential impact on future management decisions. Such a prioritization should be used for scheduling and sequencing of projects, for example. Monitoring that meets multiple objectives (e.g., adaptive management, regulatory compliance, and a "report card") should be given priority.

Ecosystem-level, systemwide indicators should be developed, such as land cover and land use measures, an index of biotic integrity, and diversity measures. Regionwide monitoring of human and environmental drivers of the ecosystem, especially population growth, land use change, water demand, and sea-level rise, is recommended. Monitoring, modeling, and research should be well integrated, especially with respect to defining the restoration reference state and using "active" adaptive management.

Does Water Flow Influence Everglades Landscape Patterns? (2003)

A commonly stated goal of the CERP is to "get the water right." This has largely meant restoring the timing and duration of water levels and the water quality in the Everglades. Water flow (speed, discharge, direction) has been considered mainly in the coastal and estuarine system but not elsewhere. Should the restoration plan be setting targets for flows in other parts of the Everglades as well?

There are legitimate reasons why flow velocities and discharges have thus far not received greater emphasis in the plan. These include a relative lack of field information and poor resolution of numerical models for flows. There are, however, compelling reasons to believe that flow has important influences in the central Everglades ecosystem. The most important reason is the existence of major, ecologically important landforms—parallel ridges, sloughs, and "tree islands"—that are aligned with present and inferred past flow directions. There are difficulties in interpreting this evidence, however, as it is essentially circumstantial and not quantitative. Alternative mechanisms by which flow may influence this landscape can to some extent be evaluated from short-term research on underlying bedrock topography, detailed surface topographic mapping, and accumulation rates of suspended organic matter. Nonetheless, more extensive and long-term research will also be necessary, beginning with the development of alternative conceptual models of the formation and maintenance of the landscape to guide a research program. Research on maintenance rather than evolution of the landscape should have higher priority because of its direct impact on restoration. Monitoring should be designed for the full range of flow conditions, including extreme events.

Overall, flows approximating historical discharges, velocities, timing, and distribution should be considered in restoration design, but quantitative flow-related performance measures are not appropriate until there is a better scientific understanding of the underlying science. At present, neither a minimum nor a maximum flow to preserve the landscape can be established.

Science and the Greater Everglades Ecosystem Restoration: An Assessment of the Critical Ecosystems Study Initiative (2003)

The Everglades represents a unique ecological treasure, and a diverse group of organizations is currently working to reverse the effects of nearly a century of wetland drainage and impoundment. The path to restoration will not be easy, but sound scientific information will increase the reliability of the restoration, help enable solutions for unanticipated problems, and potentially reduce long-term costs. The investment in scientific research relevant to restoration, however, decreased substantially within some agencies, including one major U.S. Department of the Interior (DOI) science program, the Critical Ecosystem Studies Initiative (CESI). In response to concerns regarding declining levels of funding for scientific research and the adequacy of science-based support for restoration decision making, the U.S. Congress instructed the DOI to commission the National Academies to review the scientific component of the CESI and provide recommendations for program management, strategic planning, and information dissemination.

Although improvements should be made, this report noted that the CESI has contributed useful science in support of the DOI's resource stewardship interests and restoration responsibilities in South Florida. It recommends that the fundamental objectives of the CESI research program remain intact, with continued commitment to ecosystem research. Several improvements in CESI management are suggested, including broadening the distribution of requests for proposals and improving review standards for proposals and research products. The report asserted that funding for CESI science has been inconsistent and as of 2002 was less than that needed to support the DOI's interests in and responsibilities for restoration. The development of a mechanism for comprehensive restoration-wide science coordination and synthesis is recommended to enable improved integration of scientific findings into restoration planning.

Florida Bay Research Programs and Their Relation to the Comprehensive Everglades Restoration Plan (2002)

This report of the Committee on Restoration of the Greater Everglades Ecosystem evaluated Florida Bay studies and restoration activities that potentially affect the success of the CERP. Florida Bay is a large, shallow marine system immediately south of the Everglades, bounded by the Florida Keys and the Gulf of Mexico. Some of the water draining from the Everglades flows directly into northeast Florida Bay. Other freshwater drainage reaches the bay indirectly from the northwest.

For several decades until the late 1980s, clear water and dense seagrass meadows characterized most of Florida Bay. However, beginning around 1987, the seagrass beds began dying in the western and central bay. It is often assumed that increased flows to restore freshwater Everglades habitats will also help restoration of Florida Bay. However, the CERP may actually result in higher salinities in Central Florida Bay than exist presently and thus exacerbate the ecological problems. Furthermore, some percentage of the proposed increase in fresh surface-water flow discharging northwest of the Bay will eventually reach the central Bay, where its dissolved organic nitrogen may lead to algal blooms. Complicating the analysis of such issues is the lack of an operational bay circulation model.

The report noted the importance of additional research in the following areas: estimates of groundwater discharge to the bay; full characterization and quantification of surface runoff in major basins; transport and total loads of nitrogen and phosphorus from freshwater sources, especially in their organic forms; effects on nutrient fluxes of decreasing freshwater flows into the northeastern Bay, and of increasing flows northwest of the Bay; and the development of an operational Florida Bay circulation model to support a bay water quality model and facilitate analysis of CERP effects on the Bay.

Regional Issues in Aquifer Storage and Recovery for Everglades Restoration: A Review of the ASR Regional Study Project Management Plan of the Comprehensive Everglades Restoration Plan (2002)

The report reviewed a comprehensive research plan on Everglades restoration drafted by federal and Florida officials that assesses a central feature of the restoration: a proposal to drill more than 300 wells funneling up to 1.7 billion gallons of water a day into underground aquifers, where it would be stored and then pumped back to the surface to replenish the Everglades during dry periods. The report stated that the research plan goes a long way to providing information needed to settle remaining technical questions and clearly responds to suggestions offered by scientists in Florida and in a previous report by the National Research Council.

Aquifer Storage and Recovery in the Comprehensive Everglades Restoration Plan: A Critique of the Pilot Projects and Related Plans for ASR in the Lake Okeechobee and Western Hillsboro Areas (2001)

ASR is a major component in the CERP, which was developed by the USACE and the SFWMD. The plan would use the upper Floridan aquifer to store large quantities of surface water and shallow groundwater during wet periods for recovery during droughts.

ASR may limit evaporation losses and permit recovery of large volumes of water during multi-year droughts. However, the proposed scale is unprecedented and little subsurface information has been compiled. Key unknowns include impacts on existing aquifer uses, suitability of source waters for recharge, and environmental and/or human health impacts due to water quality changes during subsurface storage.

To address these issues, the USACE and the SFWMD proposed aquifer storage recharge pilot projects in two key areas. The charge to the Committee on Restoration of the Greater Everglades Ecosystem was to examine a draft of its plans from a perspective of adaptive management. The report concluded that regional hydrogeologic assessment should include development of a regional-scale groundwater flow model, extensive well drilling and water quality sampling, and a multiobjective approach to ASR facility siting. It also recommends that water quality studies include laboratory and field bioassays and ecotoxicological studies, studies to characterize organic carbon of the source water and anticipate its effects on subsurface biogeochemical processes, and laboratory studies. Finally, it recommends that pilot projects be part of adaptive assessment.

Appendix C

Biographical Sketches of Committee Members and Staff

James Saiers (*Chair*) is the Clifton R. Musser Professor of Hydrology at the Yale School of Forestry and Environmental Studies. Dr. Saiers studies how human activities and natural processes affect the quality of drinking-water resources and alter freshwater flows within aquifers, wetlands, and river basins. His recent research projects address water quality impacts of fossil-fuel development, carbon and nutrient transport through watersheds, radionuclide migration in groundwater, and nature-based solutions for carbon capture. He served as a member of the Hydraulic Fracturing Research Advisory Panel of the Environmental Protection Agency Science Advisory Board. He earned his B.S. in geology from the Indiana University of Pennsylvania and his M.S. and Ph.D. in environmental sciences from the University of Virginia. Dr. Saiers has served on the Committee on Independent Scientific Review of Everglades Restoration Progress since 2012, and he chaired the Committee to Review the Florida Aquifer Storage and Recovery Regional Study Technical Data Report.

Casey Brown is provost professor in the Department of Civil and Environmental Engineering at the University of Massachusetts Amherst. His primary research interest is the development of analytical methods for improving the use of scientific observations and data in decision making, with a focus on climate and water resources, and he has worked extensively on projects around the world in this regard. He chairs the Water and Society Technical Committee of the American Geophysical Union Hydrology Section and the Water Resources Planning under Climate Change Technical Committee of the American Society of Civil Engineers Environmental and Water Resources Institute Systems Committee. He earned his B.S. in civil engineering from the University of Notre Dame, his M.S. from the University of Massachusetts Amherst, and his Ph.D. in environmental engineering science from Harvard University.

John Callaway is professor emeritus in the Department of Environmental Science at the University of San Francisco (USF). He recently served as the Delta Lead Scientist (2017-2020) for the Delta Science Program and Delta Stewardship Council, and previously served as the associate director of the Pacific Estuarine Research Laboratory at San Diego State University. Dr. Callaway conducts research on wetland restoration, climate change effects on tidal wetlands, and wetland carbon dynamics. In 2013, he was awarded USF's Distinguished Research Award. He is associate editor for *Estuaries and Coasts* and was co-chair for the 2023 meeting of the Coastal and Estuarine Research Federation in Portland, Oregon. He has served on a number of advisory panels on wetland restoration and management issues, including in Louisiana, southern California, and the San Francisco Bay area. Dr. Callaway received an M.A. in biology from San Francisco State University and his Ph.D. in oceanography and coastal sciences from Louisiana State University.

Philip M. Dixon is university professor in the Department of Statistics at Iowa State University. His research centers on developing and evaluating statistical methods to answer biological questions. His research interests include ecological and environmental statistics, mathematical biology, and computational modeling. He previously worked as a biostatistician at the Savannah River Ecology Lab administered by the University of Georgia. He earned his A.B. in biology from the University of California, Berkeley, an M.S. in statistics from Cornell University, and a Ph.D. in ecology and evolutionary biology from Cornell University.

Charles T. Driscoll, Jr. (NAE) is university professor in the Department of Civil and Environmental Engineering at Syracuse University, where he also serves as the director of the Center for Environmental Systems Engineering. His teaching and research interests are in the area of environmental chemistry, biogeochemistry, and environmental quality modeling. A principal research focus has been the response of forest, aquatic, and coastal ecosystems to disturbance, including air pollution, land use change, and elevated inputs of nutrients and mercury. Dr. Driscoll is currently an investigator of the National Science Foundation's Long-Term Ecological Research Network's project at the Hubbard Brook Experimental Forest in New Hampshire. He is a member of the National Academy of Engineering. He is a fellow of the American Academy for the Advancement of Science. Dr. Driscoll received his B.S. in civil engineering from the University of Maine and his M.S. and Ph.D. in environmental engineering from Cornell University.

Marla R. Emery is scientific advisor with the Norwegian Institute for Nature Research, where she works to integrate social and ecological sciences to support environmental policy. Dr. Emery previously was a research geographer with the U.S. Department of Agriculture Forest Service for 25 years, where much of her research was conducted in partnership with Indigenous communities. Her research interests include contemporary uses of wild plants and mushrooms in the United States, the ecosystem services that foraging practices provide, and their implications for land management. Dr. Emery co-chaired the Intergovernmental Science-Policy Platform on Biodiversity and Ecosystem Services Assessment of Sustainable Use of Wild Species (2018–2022), for which she led an international team of scientists to assess and synthesize the state of knowledge on factors affecting the global sustainability of human uses of animals, fungi, and plants. In this effort, she also headed up the work to incorporate Indigenous and local knowledge. She served as technical consultant on nontimber forest products to the fourth Indian Forest Management Assessment Team report to the U.S. Congress (2020–2022), developing the protocol for consultations with Tribal leaders and resource staff throughout the contiguous United States and Alaska. Dr. Emery received a Ph.D. in geography from Rutgers University.

Margaret W. Gitau is professor of agricultural and biological engineering at Purdue University. Her research focuses on water resources with emphasis on water quality, integrated hydrologic and water quality modeling, and data-driven decision making and management. In particular, she works on assessing and predicting long- and short-term watershed responses to perturbations such as land use/land cover and climate change and developing strategies and solutions to enable similar analysis in data scarce areas. Dr. Gitau has served on the Advisory Board of University of Florida's Department of Agricultural and Biological Engineering and serves on the editorial board of the *Journal of Soil and Water Conservation*. She holds a Ph.D. in agricultural and biological engineering from the Pennsylvania State University.

Matthew C. Harwell is a supervisory ecologist with the U.S. Environmental Protection Agency (EPA) Office of Research and Development (ORD), where he has worked for 13 years. He currently manages the ORD coastal environmental research laboratory in Newport, Oregon. Prior to his work at the EPA, Dr. Harwell spent 10 years with the U.S. Fish and Wildlife Service working on Greater Everglades Ecosystem restoration. Cumulatively, Dr. Harwell has spent over 25 years working on restoration science in multiple systems including Chesapeake Bay, Lake Okeechobee, South Florida and the Greater Everglades, and the Pacific Northwest. His areas of specialization include ecosystem restoration, ecosystem services, ecosystem assessment, integration and communication of science for decision makers, and adaptive management. Dr. Harwell has been on the planning team for the National Conference on Ecosystem Restoration since 2004 and was a foundational board member of the Large Scale Ecosystem Restoration Section of the Society of Ecological Restoration. Dr. Harwell has a B.S. in biology from University of South Florida and a Ph.D. in marine sciences from Virginia Institute of Marine Sciences.

William A. Hopkins III is professor in the Department of Fish and Wildlife Conservation at Virginia Tech. He is also the associate executive director of the Fralin Life Sciences Institute and the founding director of the Global Change Center at Virginia Tech. Prior to joining the faculty at Virginia Tech, Dr. Hopkins was faculty at the University of Georgia's Savannah River Ecology Laboratory. His research focuses on how anthropogenic disturbances such as climate change, pollution, and habitat loss affect wildlife. He is an award-winning educator, researcher, and leader, to include the highest awards offered to faculty at both Virginia Tech and in the Commonwealth of Virginia. He received a B.S. in biology from Mercer University, an M.S. in zoology from Auburn University, and a Ph.D. from the University of South Carolina. He has previously served on National Academies' committees addressing issues related to freshwater resources, mining, and research data quality in federal agencies.

Tracy Quirk is an associate professor and wetland ecologist in the Department of Oceanography and Coastal Sciences at Louisiana State University (LSU). Prior to joining LSU in 2014, she was an assistant professor in the Biodiversity, Earth, and Environmental Sciences Department at Drexel University in Pennsylvania. Dr. Quirk's research focuses on wetland plant ecology and soil biogeochemistry including the effects of environmental stresses, human impacts, and wetland restoration. She has had more than 30 grants funded totaling more than $5 million and has approximately 40 publications in peer-reviewed journals. She has served as an expert panelist and on advisory boards for several regional and national committees She is interested in the Everglades' ecology, management, and restoration, and she teaches a 2-week field course in the Everglades to LSU undergraduate and graduate students every other year. Dr. Quirk received her Ph.D. in marine biosciences from the University of Delaware in 2010, her M.S. in ecology and environmental sciences from the University of Louisiana in 2005, and her B.S. in wildlife and fisheries biology from the University of Vermont in 1998.

K. Ramesh Reddy is graduate research professor and director at the School of Natural Resources and Environment at the University of Florida. His research areas include biogeochemistry, soil and water quality, ecological indicators, restoration of wetlands, and aquatic systems. Dr. Reddy investigates biogeochemical cycling of macro-nutrients in natural ecosystems, including wetlands, shallow lakes, estuaries, and constructed wetlands, as related to soil and water quality, carbon sequestration, and greenhouse gas emissions. He has served as a member of the U.S. National Committee for Soil Sciences and on the U.S. Environmental Protection Agency's Science Advisory Board Panel. He serves as co-chair for the biennial Greater Everglades Ecosystem Restoration science conference. Dr. Reddy earned his Ph.D. in agronomy and soil science from Louisiana State University in 1976.

Helen M. Regan is professor of biology at the University of California, Riverside. Dr. Regan's research areas span quantitative conservation ecology and probabilistic risk assessment. She uses integrated modeling frameworks to link empirical data, population models, species distribution models, climate data, urban growth models, and fire data to investigate the impacts of climate change, habitat loss and fragmentation, and altered fire regime on plants. Dr. Regan's prior work has focused on the characterization and treatment of uncertainty in the assessment of risks in ecology and conservation biology. She currently serves on the Standards and Petitions Committee for the International Union for Conservation of Nature (IUCN) Red List Categories and Criteria and co-chairs the IUCN Species Survival Commission Climate Change Specialist Group. Dr. Regan received her B.S. from LaTrobe University and her Ph.D. from the University of New England, both in Australia.

Alan D. Steinman is the Allen and Helen Hunting Research Professor of Water Resources at the Annis Water Resources Institute (AWRI), Grand Valley State University. Previously, he was director of AWRI for 22 years and prior to that director of the Lake Okeechobee Restoration Program at the South Florida Water Management District. Dr. Steinman's research interests include aquatic ecosystem restoration, harmful algal blooms, phosphorus cycling, and water resources policy. He is a fellow of the Society of Freshwater Science. Dr. Steinman was awarded a postdoctoral fellowship from Oak Ridge National Laboratory and earned a Ph.D. in botany/aquatic ecology from Oregon State University, an M.S. in botany from the University of Rhode Island, and a B.S. in botany from the University of Vermont.

Jeffrey R. Walters is the Harold Bailey Professor of Biology at Virginia Tech, a position he has held since 1994. His professional experience includes assistant, associate, and full professorships at North Carolina State University from 1980 until 1994. His research interests are in the behavioral ecology, population biology, and conservation of birds, and his recent work has focused on cooperative breeding, dispersal behavior, and endangered species issues. Dr. Walters has done extensive research on endangered red-cockaded woodpeckers in

North Carolina and Florida, and he chaired an American Ornithologists' Union Conservation Committee Review that looked at the biology, status, and management of the Cape Sable seaside sparrow, an endangered bird endemic to the Everglades. Dr. Walters served on two panels of the Sustainable Ecosystems Institute that addressed issues with endangered birds in the Everglades restoration in addition to previously serving as a member of the National Academies' Committee on Restoration of the Greater Everglades Ecosystem. He holds a B.A. from West Virginia University and a Ph.D. from the University of Chicago.

David L. Wegner is a senior scientist at Woolpert Engineering. He is retired from a senior staff position on water, energy, and transportation committees in the U.S. House of Representatives, where he worked on legislation that directly affected administration policy and federal agency actions related to the U.S. Army Corps of Engineers, the U.S. Department of the Interior (DOI), the Environmental Protection Agency (EPA), the Tennessee Valley Authority, and the U.S. Department of Energy. Prior to serving in Washington, DC, he worked for more than 20 years for DOI managing water and science programs in the Colorado River basin and the Grand Canyon, where he was instrumental in formulating the adaptive management approach for other river systems impacted by dams and river operations. He serves on the EPA Environmental Finance Advisory Board and sits on the advisory boards for the Alliance for Global Water Adaptation and the International Association of Hydro-Environment Research. Mr. Wegner received his M.S. in engineering/fluvial geomorphology from Colorado State University.

STAFF

Stephanie E. Johnson (*Study Director*) is a senior program officer with the Water Science and Technology Board. Since joining the National Research Council in 2002, she has worked on a wide range of water-related studies, on topics such as desalination, wastewater reuse, contaminant source remediation, coal and uranium mining, coastal risk reduction, and ecosystem restoration. She has served as study director for many studies, including the Panel to Review the Critical Ecosystem Studies Initiative and all nine Committees on Independent Scientific Review of Everglades Restoration Progress. Dr. Johnson received her B.A. from Vanderbilt University in chemistry and geology and her M.S. and Ph.D. in environmental sciences from the University of Virginia.

Noel Walters is an associate program officer in the Division on Earth and Life Studies for the Water Science and Technology Board and the Board on Earth Sciences and Resources. She previously worked in the National Academies' Gulf Research Program developing grant programs for the Gulf Health and Resilience

Unit and the Gulf Environmental Protection and Stewardship Unit. Prior to joining the National Academies, Walters worked as a data quality analyst for the Energy Information Administration reviewing monthly and annual electric power industry reports. She earned her B.A. from McDaniel College in sociology and her M.A. in sociology from the George Washington University, with concentrations in social inequality and urban sociology.

Emily Bermudez is a senior program assistant with the Water Science and Technology Board and Board on Earth Sciences and Resources. Prior to joining the National Academies in August 2022, she conducted geomorphology research at Oberlin College using sediment fingerprinting to test if erosion rates in Cuba changed due to the transition to organic agriculture from sugarcane monoculture. She received her B.A. in geology from Oberlin College.

Samuel Kraft is a senior program assistant with the Water Science and Technology Board and Board on Earth Sciences and Resources. Prior to joining the National Academies in April 2024, he worked as a ground penetrating radar analyst for a company that specializes in concrete services. Kraft received his B.A. in geology from Millersville University of Pennsylvania.